Finite Element Analysis

Finite Element Analysis

G. Lakshmi Narasaiah

Prof. & Head Aeronautical Engineering Dept.
MLR Institute of Technology
Dundigal, Hyderabad 500 043

Formerly Senior Manager,
Corp. R & D, BHEL Vikasnagar, Hyderabad 500 093

CRC Press
Taylor & Francis Group
Boca Raton London New York

CRC Press is an imprint of the
Taylor & Francis Group, an **informa** business

First published 2009 by BS Publications

Published 2018 by CRC Press
Taylor & Francis Group
6000 Broken Sound Parkway NW, Suite 300
Boca Raton, FL 33487-2742

First issued in paperback 2018

© by Author
CRC Press is an imprint of Taylor & Francis Group, an Informa business

No claim to original U.S. Government works

ISBN 13: 978-1-138-11809-6 (pbk)
ISBN 13: 978-1-4200-9510-4 (hbk)

Visit the Taylor & Francis Web site at
http://www.taylorandfrancis.com

and the CRC Press Web site at
http://www.crcpress.com

Distributed in India, Pakistan, Nepal, Myanmar (Burma), Bhutan, Bangladesh and Sri Lanka by **BS Publications**

Distributed in the rest of the world by
CRC Press LLC, Taylor and Francis Group,

Dedicated to my

Beloved father

Gogineni Raja Gopala Rao
(1911-1985)

Prof. P.N. Murthy

Advisor/TCS

Prof (Retd.), Aeronautical Engg

I.I.T., Kanpur.

Foreword

In the field of structural analysis, the advent of strain energy and finite element methods marked a watershed. This, with the introduction of computers, made it possible to handle many complex and large structures with confidence. Any structure which is responding to a system of loads is assumed to respond by moving into a deformed shape with minimum strain energy. Variational expression of this is used in the analytical process to decipher the stress and deformation fields in a loaded structure. The assumption in this is that an infinitesimal disturbance, at the extremum point, to the deflected shape will not alter the strain energy. Similarly, if one can construct a valid potential function, one can use variational methods to determine the stress, strain and displacement fields to the required approximation.

An important assumption that is made with respect to a structure is that a real structure can be represented by a reasonable model, which can be expected to yield the same behaviour under loading as the actual structure. The other assumption is that a structure can be divided into elements, which, when assembled together, can be treated as representing the real structure or equivalent model. It is now realized that such an assumption is valid only in equilibrium and linear conditions. In the case of non-linear and non-equilibrium situations, the assumption that the behaviour of the sum of parts is the same as that of the whole is not valid.

Prof. Lakshmi Narasaiah's book deals with the finite element analysis of elastic structures with small deformations. His treatment of the subject is elegant for a beginner whose main aim is to get familiar with the subject. The author avoids the urge to delve into the nuances of theoretical elaborations. He remains always within the bounds of his goal of making a student familiar with the basics of the method. For this, he resorts to the familiar mechanism of presenting plenty of illustrative examples with suitable, relevant explanations. This makes it easy to read and absorb, and gives a good foundation to build upon. He also deals with an FEM package to make the student understand the way in which a package is created and used. This is useful information for one who will be working with many elaborate software packages for structural analysis. He can work with them with the confidence of basic knowledge.

I wish the effort of Prof. Lakshmi Narasaiah will benefit a large number of students and other enthusiasts involved in structural analysis.

I thank him for giving me the opportunity to read his book and introduce it to his audience.

(P.N. Murthy)

Preface

Many books are available on this subject of Finite Element Method, popularly known as FEM. In many of these books, the subject is treated with a rigorous mathematical orientation, using variational calculus. Initially, during '70s and '80s, this subject was offered for postgraduate students and research scholars in select institutes like IITs, IISc etc. The mathematical approach followed in most of these books was essential for a good understanding of this subject so that the students can write tailor-made programs for their specific applications as well as pursue their research work.

With the availability of general purpose commercial software such as ANSYS, NISA, ADINA, PAFEC, NASTRAN,,.. during late '80s and '90s supported on faster and cheaper computer systems, FEM has become an essential subject for aeronautical, civil and mechanical engineering graduates. The need has undergone a major paradigm shift from a detailed mathematical background so as to write computer programs for their specific requirement to a general awareness of the method for effective use of the available software in the design and analysis of complicated products to make them competitive in the market.

FEM is, therefore, introduced in many universities during this decade for undergraduate engineering students in Aeronautical, Civil and Mechanical branches. Since they are neither exposed to variational calculus nor exposed to theory of elasticity, a strong need for a simple book on FEM ideally suited for these students was felt. This book is an effort in this direction.

This book helps a young engineer in effectively using available general-purpose software and includes aspects of design, which are usually covered in 'Mechanics of Solids', for enabling proper interpretation of the results obtained through these software.

Mere calculation of displacements and stresses at various points of a component using one such software is not the end in itself, since it does not indicate safety of a product in operation. *The codes are based on stresses across some critical sections rather than on stresses at individual points of a component.* Hence, a brief description of the mandatory codes applicable for design and operation of pressure vessels has been included in Chapter-10, which is essential for validation of any design.

In practice, design of a component is neither limited to static loads nor limited to steady state operation. These loads are treated in transient heat conduction, forced vibrations, fatigue, creep, etc. All these topics cannot be included in the subject of FEA. A brief coverage is provided in Chapter-10, to provide a broader perspective for the design engineer about other types of analyses for designing a safe product.

Most of the software in Computational Fluid Dynamics (CFD) were based on Finite Difference Method (FDM) for a long time. FEM is slowly replacing FDM in this area due to its flexibility in modeling complicated boundaries. Chapter-11 deals briefly with CFD and includes a comparison of different methods.

Application of the method, using general-purpose software ANSYS, for analysis of some simple problems with standard solutions for verification of results is given in Chapter-12.

Due to the abundant availability of general-purpose software with interactive graphics options, listing of a sample program is not considered relevant for the present-day students as well as practicing engineers.

The book is a compilation of my notes while teaching this subject in engineering colleges during the last 6 years and modified by incorporating the needs of practicing engineers, based on my experience in the analysis of many critical components of thermal, hydro, nuclear and fuel cell power plants, during my 25 years of service in BHEL at Tiruchy & Corporate R&D, Hyderabad.

In spite of my best efforts, some numerical mistakes and typing mistakes might have crept in. I will be grateful for your comments and suggestions for improving the presentation and content in the subsequent editions.

G. Lakshmi Narasaiah

14.11.2007

Ph : 040-66881678

e-mail : gogineni_ln@yahoo.co.in

Acknowledgements

The author wishes to acknowledge many persons who have contributed to the presentation of this subject in this form. Due to space constraints, mention is made of very few people here. The author is greatly indebted to the following :

Prof. P.N.Murthy who laid the foundation for this subject at IIT, Kanpur.

Sri N.S.Kurup, who gave practical significance and physical understanding of results of various analysis at BHEL, Tiruchy.

Discussions with colleagues and students in some engineering colleges, where I taught the subject, have highlighted the need for a simplified book of this nature. The author expresses his sincere thanks to all those who have indirectly contributed to finalizing the scope and examples included in this book.

Last but not the least is the support and encouragement from my wife Sandhya Rani and sons Chaitanya and Rajesh, which helped me in bringing out this book in the present form.

-Author

Contents

Foreword .... **(vii)**

Preface .... **(ix)**

Acknowledgements .... **(xi)**

Chapter 1

Introduction   1-34

1.1 Design and Analysis of a Component 1

1.2 Approximate Method vs. Exact Method 4

1.3 Weighted Residual Methods 5

1.4 Variational Method or Rayleigh – Ritz Method 12

1.5 Principle of Minimum Potential Energy 22

1.6 Origin of FEM 26

1.7 Principle of FEM 27

1.8 Classification of FEM 31

1.9 Types of Analyses 32

1.10 Summary 33

Objective Questions 33

Chapter 2

Matrix Operations   35-60

2.1 Types of Matrices 35

2.2 Matrix Algebra 37

2.3   Determinant 39

2.4   Inversion of a Matrix 41

2.5   Methods of Solution of Simultaneous Equations 41

    2.5.1   By Inversion of the Coefficient Matrix 42

    2.5.2   Direct Methods 44

    2.5.3   Iterative Methods 53

2.6   Eigen Values and Eigen Vectors 55

2.7   Matrix Inversion Through Characteristic Equation 59

2.8   Summary 60

Chapter 3

Theory of Elasticity   61-84

3.1   Degrees of Freedom 61

3.2   Rigid Body Motion 62

3.3   Discrete Structures 62

3.4   Continuum Structures 62

3.5   Material Properties 63

3.6   Linear Analysis 63

3.7   Non-linear Analysis 63

3.8   Stiffness and Flexibility 64

3.9   Principle of Minimum Potential Energy 65

3.10 Stress and Strain at a Point 65

3.11 Principal Stresses 68

3.12 Mohr's Circle for Representation of 2-D Stresses 68

3.13 VonMises Stress 71

3.14 Theory of Elasticity 72

3.15 Summary 82

Chapter 4

Discrete (1-D) Elements   85-125

4.1   Degrees of Freedom of Different Elements 85

4.2   Calculation of Stiffness Matrix by Direct Method 86

4.3   Calculation of Stiffness Matrix by Variational Principle 88

4.4   Transformation Matrix 92

4.5   Assembling Element Stiffness Matrices 94

4.6   Boundary Conditions 96

4.7   Beam Element Stiffness Matrix by Variational Approach 98

4.8   General Beam Element 100

4.9   Pipe Element 103

4.10   Summary 104

Objective Questions 105

Solved Problems 108

Chapter 5

Continuum (2-D & 3-D) Elements   127-158

5.1   2-D Elements Subjected to In-plane Loads 127

5.2   Simplex, Complex and Multiplex Elements 128

5.3   Stiffness Matrix of a CST Element 129

       5.3.1   Stiffness Matrix of a Right Angled Triangle 131

5.4   Convergence Conditions 133

5.5   Geometric Isotropy 136

5.6   Aspect Ratio 138

5.7   Inter-Element Compatibility 139

5.8   2-D Elements Subjected to Bending Loads 141

5.9   3-D Elements 143

5.10 Axi-symmetric Elements 144

5.11 Summary 146

Objective Questions 147

Solved Problems 152

Chapter 6

Higher Order and Iso-Parametric Elements   159-199

6.1 Higher Order Elements 159

6.2 Isoparametric Elements 169

6.3 Stiffness Matrices of Some Iso-parametric Elements 170

6.4 Jacobian 182

6.5 Strain-displacement Relations 184

6.6 Summary 188

Objective Questions 189

Solved Problems 190

Chapter 7

Factors Influencing Solution   201-236

7.1 Distributed Loads 201

7.2 Statically Equivalent Loads vs. Consistent Loads 202

7.3 Consistent Loads for a Few Common Cases 206

7.4 Assembling Element Stiffness Matrices 209

7.5 Automatic Mesh Generation 214

7.6 Optimum Mesh Model 215

7.7 Gaussian Points & Numerical Integration 216

7.8 Modelling Techniques 220

7.9 Boundary Conditions for Continuum Analysis 224

7.10 Transition Element 228

7.11   Substructuring or Super Element Approach 230

7.12   Deformed and Undeformed Plots 231

7.13   Summary 232

Objective Questions **233**

Chapter 8

Dynamic Analysis (undamped free vibrations)   237-253

8.1   Normalising Eigenvectors 239

8.2   Modelling for Dynamic Analysis 240

8.3   Mass Matrix 240

8.4   Summary 251

Objective Questions **252**

Chapter 9

Steady State Heat Conduction   255-276

9.1   Governing Equations 255

9.2   1-D Heat Conduction 257

        9.2.1   Heat Conduction Through a Wall 258

        9.2.2   Heat Transfer Through a Fin 268

9.3   2-D heat Conduction in a Plate 273

9.4   Summary 275

Objective Questions **276**

Chapter 10

Design Validation and Other Types of Analysis   277-291

10.1   Compliance with Design Codes 277

10.2   Transient Heat Condition 281

10.3   Buckling of Columns 282

10.4 Fatigue Analysis 283

10.5 Creep Analysis 285

10.6 Damped Free Vibration 287

10.7 Forced Vibration 288

10.8 Torsion of a Non-circular Rod 289

Chapter 11

Computational Fluid Dynamics   293-307

11.1 Introduction 293

11.2 Governing Equation 294

11.3 Finite Difference Method (FDM) 296

11.4 Elliptic Equations (or boundary value problems) 297

11.5 Finite Volume Method (FVM) 305

11.6 FDM vs. FEM 307

Chapter 12

Practical Analysis Using a Software   309-329

12.1 Using a General Purpose Software 309

12.2 Some Examples with ANSYS 311

Objective Questions 326

Answers 331

References for Additional Reading 333

Index 335

CHAPTER 1

INTRODUCTION

1.1   DESIGN AND ANALYSIS OF A COMPONENT

Mechanical design is the design of a component for optimum size, shape, etc., *against failure* under the application of operational loads. A good design should also minimise the cost of material and cost of production. Failures that are commonly associated with mechanical components are broadly classified as:

(a) Failure by breaking of brittle materials and fatigue failure (when subjected to repetitive loads) of ductile materials.

(b) Failure by yielding of ductile materials, subjected to non-repetitive loads.

(c) Failure by elastic deformation.

The last two modes cause change of shape or size of the component rendering it useless and, therefore, refer to functional or *operational failure*. Most of the design problems refer to one of these two types of failures. Designing, thus, involves estimation of stresses and deformations of the components at different critical points of a component for the specified loads and boundary conditions, so as to satisfy operational constraints.

Design is associated with the calculation of dimensions of a component to withstand the applied loads and perform the desired function. *Analysis* is associated with the estimation of displacements or stresses in a component of assumed dimensions so that adequacy of assumed dimensions is validated. *Optimum design* is obtained by many iterations of modifying dimensions of the component based on the calculated values of displacements and/or stresses *vis-à-vis* permitted values and re-analysis.

An analytic method is applied to a model problem rather than to an *actual physical problem*. Even many laboratory experiments use models. A *geometric model* for analysis can be devised after the physical nature of the problem has been understood. A model excludes superfluous details such as bolts, nuts, rivets, but includes all essential features, so that analysis of the model is not unnecessarily complicated and yet provides results that describe the actual problem with sufficient accuracy. A geometric model becomes a ***mathematical model*** when its behaviour is described or approximated by incorporating restrictions such as homogeneity, isotropy, constancy of material properties and mathematical simplifications applicable for small magnitudes of strains and rotations.

Several methods, such as method of joints for trusses, simple theory of bending, simple theory of torsion, analyses of cylinders and spheres for axisymmetric pressure load etc., are available for designing/analysing simple components of a structure. These methods try to obtain exact solutions of second order partial differential equations and are based on several assumptions on sizes of the components, loads, end conditions, material properties, likely deformation pattern etc. Also, these methods are not amenable for generalisation and effective utilisation of the computer for repetitive jobs.

Strength of materials approach deals with a single beam member for different loads and end conditions (free, simply supported and fixed). In a space frame involving many such beam members, each member is analysed independently by an assumed distribution of loads and end conditions.

For example, in a 3-member structure (portal frame) shown in Fig. 1.1, the (horizontal) beam is analysed for deflection and bending stress by strength of materials approach considering its both ends simply supported. The load and moment reactions obtained at the ends are then used to calculate the deflections and stresses in the two columns separately.

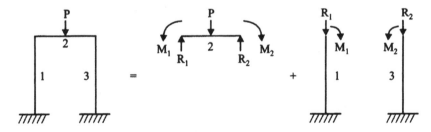

FIGURE 1.1 Analysis of a simple frame by strength of materials approach

Simple supports for the beam imply that the columns do not influence slope of the beam at its free ends (valid when bending stiffness of columns = 0 or the column is highly flexible). Fixed supports for the beam imply that the slope of the beam at its ends is zero (valid when bending stiffness of columns = ∞ or the column is extremely rigid). But, the ends of the horizontal beam are neither simply supported nor fixed. The degree of fixity or influence of columns on the slope of the beam at its free ends is based on a finite, non-zero stiffness value.

Thus, the maximum deflection of the beam depends upon the relative stiffness of the beam and the columns at the two ends of the beam.

For example, in a beam of length 'L', modulus of elasticity 'E', moment of inertia 'I' subjected to a uniformly distributed load of 'p' (Refer Fig. 1.2).

Deflection, $\delta = \dfrac{5\,pL^4}{384\,EI}$ with simple supports at its two ends (case (a))

$\qquad\quad = \dfrac{pL^4}{384\,EI}$ with fixed supports at its two ends (case (b))

Case (a) : Simple supports   Case (b) : Fixed supports

FIGURE 1.2 Deflection of a beam with different end conditions

If, in a particular case,
L = 6 m, E = 2 × 10^{11} N/m^2, Moment of inertia for beam I_B = 0.48 × 10^{-4} m^4

Moment of inertia for columns I_C = 0.48 × 10^{-4} m^4 and distributed load p = 2 kN/m,

$\qquad$ δ_{max} = 3.515 mm with simple supports at its two ends

and $\quad$ δ_{max} = 0.703 mm with fixed supports at its two ends

whereas, deflection of the same beam, when analysed along with columns by FEM,

$\qquad$ δ_{max} = 1.8520 mm, when $I_B = I_C$ (Moments of inertia for beam & columns)

$\qquad\qquad$ = 1.0584 mm, when 5 $I_B = I_C$

and $\qquad$ = 2.8906 mm, when $I_B = 5\,I_C$

All the three deflection values clearly indicate presence of columns with finite and non-zero stiffness and, hence, the deflection values are in between those of beam with free ends and beam with fixed ends.

Thus, designing a single beam member of a frame leads to under-designing if fixed end conditions are assumed while it leads to over-designing if simple supports are assumed at its ends. Simply supported end conditions are, therefore, normally used for a conservative design in the conventional approach. Use of strength of materials approach for designing a component is, therefore, associated with higher factor of safety. The individual member method was acceptable for civil structures, where weight of the designed component is not a serious constraint. A more accurate analysis of discrete structures with few members is carried out by the potential energy approach. *Optimum beam design is achieved by analysing the entire structure which naturally considers finite stiffness of the columns, based on their dimensions and material, at its ends.* This approach is followed in the Finite Element Method (FEM).

1.2 APPROXIMATE METHOD VS. EXACT METHOD

An analytical solution is a mathematical expression that gives the values of the desired unknown quantity at any location of a body and hence is valid for an infinite number of points in the component. However, it is not possible to obtain analytical mathematical solutions for many engineering problems.

For problems involving complex material properties and boundary conditions, numerical methods provide approximate but acceptable solutions (with reasonable accuracy) for the unknown quantities – only at discrete or finite number of points in the component. Approximation is carried out in two stages :

(a) In the formulation of the mathematical model, w.r.t. the physical behaviour of the component. Example : Approximation of joint with multiple rivets at the junction of any two members of a truss as a pin joint, assumption that the joint between a column and a beam behaves like a simple support for the beam,.... The results are reasonably accurate far away from the joint.

(b) In obtaining numerical solution to the simplified mathematical model. The methods usually involve approximation of a functional (such as Potential energy) in terms of unknown functions (such as displacements) at finite number of points. There are two broad categories:

   (i) **Weighted residual methods** such as Galerkin method, Collocation method, Least squares method, etc.

   (ii) **Variational method** (Rayleigh-Ritz method, FEM). FEM is an improvement of Rayleigh–Ritz method by choosing a variational function valid over a small element and *not* on the entire component, which will be discussed in detail later. These methods also use the principle of minimum potential energy.

   (iii) *Principle of minimum potential energy :* Among all possible kinematically admissible displacement fields (satisfying compatibility and boundary conditions) of a conservative system, the one corresponding to stable equilibrium state has minimum potential energy. For a component in static equilibrium, this principle helps in the evaluation of unknown displacements of deformable solids (continuum structures).

Some of these methods are explained here briefly to understand the historical growth of analysis techniques.

1.3 WEIGHTED RESIDUAL METHODS

Most structural problems end up with differential equations. Closed form solutions are not feasible in many of these problems. Different approaches are suggested to obtain approximate solutions. One such category is the weighted residual technique. Here, an approximate solution, in the form $y = \Sigma N_i.C_i$ for $i = 1$ to n where C_i are the unknown coefficients or weights (constants) and N_i are functions of the independent variable satisfying the given kinematic boundary conditions, is used in the differential equation. Difference between the two sides of the equation with known terms, on one side (usually functions of the applied loads), and unknown terms, on the other side (functions of constants C_i), is called the residual, R. This residual value may vary from point to point in the component, depending on the particular approximate solution. Different methods are proposed based on how the residual is used in obtaining the best (approximate) solution. Three such popular methods are presented here.

(a) Galerkin Method

It is one of the weighted residual techniques. In this method, solution is obtained by equating the integral of the product of function N_i and residual R over the entire component to zero, for each N_i. Thus, the 'n' constants in the approximate solution are evaluated from the 'n' conditions $\int N_i.R.dx = 0$ for $i = 1$ to n. The resulting solution may match with the exact solution at some points of the component and may differ at other points. The number of terms N_i used for approximating the solution is arbitrary and depends on the accuracy desired. This method is illustrated through the following examples of beams in bending.

Example 1.1

Calculate the maximum deflection in a simply supported beam, subjected to concentrated load 'P' at the center of the beam. (Refer Fig. 1.3)

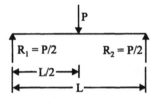

FIGURE 1.3

Solution

$y = 0$ at $x = 0$ and $x = L$ are the kinematic boundary conditions of the beam. So, the functions N_i are chosen from $(x - a)^p.(x - b)^q$, with different positive integer values for p and q; and $a = 0$ and $b = L$.

(i) *Model-1 (1-term approximation)* : The deflection is assumed as

$$y(x) = N.c$$

with the function, $N = x(x - L)$,

which satisfies the end conditions $y = 0$ at $x = 0$ and $y = 0$ at $x = L$.
The load-deflection relation for the beam is given by

$$EI\left(\frac{d^2y}{dx^2}\right) = M$$

where $M = (P/2).x$ for $0 \leq x \leq L/2$
and $M = (P/2).x - P.[x - (L/2)] = (P/2).(L - x)$ for $L/2 \leq x \leq L$
Thus, taking $y = x.(x - L).c$,

$$\frac{d^2y}{dx^2} = 2c$$

and the residual of the equation, $R = EI\left(\frac{d^2y}{dx^2}\right) - M = EI . 2c - M$

Then, the unknown constant 'c' in the function 'N' is obtained from

$$\int_0^{\frac{L}{2}} N.R.dx + \int_{\frac{L}{2}}^{L} N.R.dx = 0$$

(two integrals needed, since expression for M changes at $x = \dfrac{L}{2}$)

$$\int_0^{\frac{L}{2}} [x.(x-L)].\left[EI.2c - \left(\frac{P}{2}\right).x\right]dx + \int_{\frac{L}{2}}^{L} [x.(x-L)].\left[EI.2c - \left(\frac{P}{2}\right).(L-x)\right]dx = 0$$

$$\Rightarrow c = \frac{5\,PL}{64\,EI}$$

Therefore, $y = x(x - L).\dfrac{5\,PL}{64\,EI}$

At $x = \dfrac{L}{2}, \quad y = y_{max} = -\dfrac{5\,PL^3}{256\,EI}$ or $\dfrac{-PL^3}{51.2\,EI}$

This approximate solution is close to the exact solution of $\dfrac{-PL^3}{48\,EI}$

obtained by double integration of $EI\left(\dfrac{d^2y}{dx^2}\right) = M = \left(\dfrac{P}{2}\right).x$, with

appropriate end conditions.

(ii) *Model-2 (2-term approximation):* The deflection is assumed as

$$y(x) = N_1.c_1 + N_2.c_2$$

with the functions $N_1 = x(x - L)$ and $N_2 = x.(x - L)^2$

which satisfy the given end conditions.

Thus, taking $y = x.(x - L).c_1 + x.(x - L)^2.c_2$,

$$\left(\dfrac{d^2y}{dx^2}\right) = 2c_1 + 2.(3x - 2L).c_2$$

and the residual of the equation,

$$R = EI.\left(\dfrac{d^2y}{dx^2}\right) - M = EI.[2c_1 + 2.(3x - 2L).c_2] - M$$

where $M = (P/2).x$ for $0 \le x \le L/2$

and $M = (P/2).x - P.[x - (L/2)] = (P/2).(L - x)$ for $L/2 \le x \le L$

Then, the unknown constants 'c_1' and 'c_2' in the functions 'N_i' are obtained from

$$\int_0^L N_1.R.dx = \int_0^{\frac{L}{2}} [x.(x - L)]\left\{ EI.[2c_1 + 2.(3x - 2L)c_2] - \left(\dfrac{P}{2}\right).x \right\} dx$$

$$+ \int_{\frac{L}{2}}^L [x.(x - L)]\left\{ EI.[2c_1 + 2.(3x - 2L)c_2] - \left(\dfrac{P}{2}\right).(L - x) \right\} dx = 0$$

and $$\int_0^L N_2.R.dx = \int_0^{\frac{L}{2}} [x.(x - L)^2].\left\{ EI.[2c_1 + 2.(3x - 2L)c_2] - \left(\dfrac{P}{2}\right).x \right\} dx$$

$$+ \int_{\frac{L}{2}}^L [x.(x - L)^2]\left\{ EI.[2c_1 + 2.(3x - 2L)c_2] - \left(\dfrac{P}{2}\right).(L - x) \right\} dx = 0$$

Simplifying these equations, we get

$$2c_1 - c_2.L = \dfrac{5PL}{16EI} \quad \text{and} \quad 5c_1 - 4c_2.L = \dfrac{75PL}{192EI}$$

Solving these two simultaneous equations, we get

$$c_1 = \frac{55PL}{192EI} \quad \text{and} \quad c_2 = \frac{25P}{96EI}$$

Thus, we get

$$y = x(x - L) . \frac{55PL}{192EI} + x(x - L)^2 . \frac{25P}{96EI}$$

and   at $x = L/2$, $y = y_{max} = \dfrac{PL^3(-55 + 25)}{4 \times 192EI}$ or $-\dfrac{PL^3}{25.6EI}$

Note : The bending moment M is a function of x. The exact solution of y should be a minimum of 3rd order function so that $\dfrac{d^2y}{dx^2} = \dfrac{M}{EI}$ is a function of x.

(b) Collocation Method

In this method, also called as the ***point collocation method***, the residual is equated to zero at 'n' select points of the component other than those at which the displacement value is specified, where 'n' is the number of unknown coefficients in the assumed displacement field, i.e., $R(\{c\}, x_i) = 0$ for i = 1, ..n. It is also possible to apply collocation method on some select surfaces or volumes. In that case, the method is called ***sub-domain collocation method.***

i.e., $\int R(\{c\}, x).dS_j = 0$ for $j = 1, ..m$

or $\int R(\{c\}, x). dV_k = 0$ for $k = 1, ..m$

These methods also result in 'n' algebraic simultaneous equation in 'n' unknown coefficients, which can be easily evaluated.

The simpler of the two for manual calculation, point collocation method, is explained better through the following example.

Example 1.2

Calculate the maximum deflection in a simply supported beam, subjected to concentrated load 'P' at the center of the beam. (Refer Fig. 1.4)

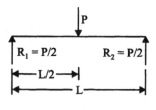

FIGURE 1.4

Solution

$y = 0$ at $x = 0$ and $x = L$ are the kinematic boundary conditions of the beam. So, the functions N_i are chosen from $(x - a)^p.(x - b)^q$, with different positive integer values for p and q; and a = 0 and b = L.

(i) *Model-1 (1-term approximation):* The deflection is assumed as
$$y(x) = N.c$$

with the function $N = x(x - L)$,

which satisfies the end conditions $y = 0$ at $x = 0$ and $y = 0$ at $x = L$

The load-deflection relation for the beam is given by

$$EI\left(\frac{d^2y}{dx^2}\right) = M$$

where $\qquad M = (P/2).x \qquad\qquad\qquad$ for $0 \le x \le L/2$

and $\qquad M = (P/2).x - P.[x - (L/2)] = (P/2).(L - x) \quad$ for $L/2 \le x \le L$

Thus, taking $y = x.(x - L).c$, $\quad \dfrac{d^2y}{dx^2} = 2c$

and the residual of the equation, $R = EI\left(\dfrac{d^2y}{dx^2}\right) - M = EI\left(\dfrac{d^2y}{dx^2}\right) - \left(\dfrac{P}{2}\right).x$

Then, the unknown constant 'c' in the function 'N' is obtained by choosing the value of residual at some point, say $x = L/2$, as zero.

i.e., $\qquad R(c,x) = EI.2c - \left(\dfrac{P}{2}\right).x = 0$ at $x = \dfrac{L}{2} \quad \Rightarrow \quad c = \dfrac{PL}{8EI}$

Therefore, $\quad y = x(x - L).\dfrac{PL}{8EI}$

At $\qquad x = \dfrac{L}{2}, \quad y = y_{max} = -\dfrac{PL^3}{32EI}$

(ii) *Model-2 (2-term approximation)* : The deflection is assumed as
$$y(x) = N_1.c_1 + N_2.c_2$$

with the functions $N_1 = x(x - L)$ and $N_2 = x.(x - L)^2$,

which satisfy the given end conditions.

Thus, taking $y = x.(x - L).c_1 + x.(x - L)^2.c_2$,

$$\frac{d^2y}{dx^2} = 2c_1 + 2.(3x - 2L)c_2$$

and the residual of the equation,

$$R = EI\left(\frac{d^2y}{dx^2}\right) - M = EI.[2c_1 + 2.(3x - 2L).c_2] - M$$

where $M = (P/2).x$ for $0 \le x \le L/2$

and $M = (P/2).x - P.[x - (L/2)] = (P/2).(L - x)$ for $L/2 \le x \le L$

Then, the unknown constants 'c_1' and 'c_2' in the functions 'N_i' are obtained from

$$R(\{c\},x) = EI.[2c_1 + 2.(3x - 2L).c_2] - (P/2).x = 0 \qquad \text{at } x = L/4$$

and $R(\{c\},x) = EI.[2c_1 + 2.(3x - 2L).c_2] - (P/2).(L - x)\} = 0$ at $x = 3L/4$

or $4c_1 - 5L.c_2 = \dfrac{PL}{4EI}$ and $4c_1 + L.c_2 = \dfrac{PL}{4EI}$

$\Rightarrow c_1 = \dfrac{PL}{16EI}$ and $c_2 = 0$

$\Rightarrow y_{max} = -\dfrac{PL^3}{64EI}$ at $x = \dfrac{L}{2}$

Choosing some other collocation points, say $x = \dfrac{L}{3}$ and $x = \dfrac{2L}{3}$,

$c_1 - L.c_2 = \dfrac{PL}{12EI}$ and $c_1 + 0 = \dfrac{PL}{12EI}$

$\Rightarrow c_1 = \dfrac{PL}{12EI}$ and $c_2 = 0$

At $x = \dfrac{L}{2}$, $y_{max} = \dfrac{-PL^3}{48EI}$,   which matches exactly with closed form solution

(c) **Least Squares Method**

In this method, integral of the residual over the entire component is minimized. i.e., $\dfrac{\partial I}{\partial c_i} = 0$ for $i = 1, ..n$,

where $I = \int [R(\{a\}, x)]^2.dx$

This method also results in 'n' algebraic simultaneous equation in 'n' unknown coefficients, which can be easily evaluated.

Example 1.3

Calculate the maximum deflection in a simply supported beam, subjected to concentrated load 'P' at the center of the beam. (Refer Fig. 1.5)

Solution

Again, y = 0 at x = 0 and y = 0 at x = L are the kinematic boundary conditions of the beam. So, the functions N_i are chosen from $(x - a)^p.(x - b)^q$, with different positive integer values for p and q.

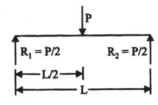

FIGURE 1.5

1-term approximation : The deflection is again assumed as   y(x) = N.C,

with the function N = x(x – L),

which satisfies the end conditions   y = 0 at x = 0   and   y = 0 at x = L

The load-deflection relation for the beam is given by

$$EI \left(\frac{d^2y}{dx^2} \right) = M$$

where   M = (P/2).x   for $0 \le x \le L/2$

and   M = (P/2).x – P.[x – (L/2)] = (P/2).(L – x)   for $L/2 \le x \le L$

Thus, taking $y = x.(x – L).c$, $\dfrac{d^2y}{dx^2} = 2c$

and the residual of the equation, $R = EI \left(\dfrac{d^2y}{dx^2} \right) – M = EI \,.\, 2c – M$

Then, $I = \int [R(\{c\}, x)]^2 .dx$ and the constant 'c' in the function y(x) is obtained from

$$\frac{\partial I}{\partial a_i} = \frac{\partial}{\partial c_i} \int_0^{\frac{L}{2}} \left[[R(\{c\}, x)]^2 - \left(\frac{P}{2} \right).x \right] dx + \frac{\partial}{\partial c_i} \int_{\frac{L}{2}}^{L} \left[[R(\{c\}, x)]^2 - \left(\frac{P}{2} \right).(L - x) \right] dx$$

$$\Rightarrow c = \frac{PL}{8EI} \quad \text{and } y_{max} = \frac{-PL^3}{32EI} \text{ at } x = \frac{L}{2}$$

1.4 VARIATIONAL METHOD OR RAYLEIGH - RITZ METHOD

This method involves choosing a displacement field over the entire component, usually in the form of a polynomial function, and evaluating unknown coefficients of the polynomial for minimum potential energy. It gives an approximate solution. Practical application of this method is explained here through three different examples, involving

  (a) uniform bar with concentrated load,

  (b) bar of varying cross section with concentrated load,   and

  (c) uniform bar with distributed load (self-weight).

Example 1.4

Calculate the displacement at node 2 of a fixed beam shown in Fig. 1.6, subjected to an axial load 'P' at node 2.

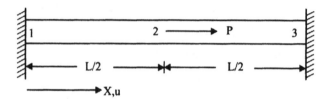

FIGURE 1.6

Solution

Method - 1

The total potential energy for the linear elastic one-dimensional rod with built-in ends, when body forces are neglected, is

$$\pi = \frac{1}{2} \int EA \left(\frac{du}{dx} \right)^2 dx - P\,u_1$$

Let us assume $u = a_1 + a_2x + a_3x^2$ as the polynomial function for the displacement field.

Kinematically admissible displacement field must satisfy the natural boundary conditions

$u = 0$ at $x = 0$ which implies $a_1 = 0$

and $u = 0$ at $x = L$ which implies $a_2 = -a_3 L$

At $x = \dfrac{L}{2}$, $u_1 = a_2 \left(\dfrac{L}{2} \right) + a_3 \left(\dfrac{L}{2} \right)^2 = -a_3 \dfrac{L^2}{4}$

Therefore,

$$\pi = \frac{1}{2}\int_0^L EA\left(\frac{du}{dx}\right)^2 dx - P\,u_1$$

$$= \frac{1}{2}\int_0^L EA(a_2 + 2a_3 x)^2 dx - P\left(-a_3 \frac{L^2}{4}\right)$$

$$= \frac{1}{2} EAa_3^2 \int_0^L (2x - L)^2 dx + P\,a_3 \frac{L^2}{4}$$

$$= EAa_3^2 \frac{L^3}{6} + P\,a_3 \frac{L^2}{4}$$

For stable equilibrium, $\quad \dfrac{\partial \pi}{\partial a_3} = 0 \quad$ gives $\quad a_3 = \dfrac{-3P}{4EAL}$

Displacement at node 2, $\quad u_2 = -\dfrac{a_3 L^2}{4} = \dfrac{3PL}{16AE}$

$$= \left(\frac{3}{4}\right)\left(\frac{PL}{4AE}\right)$$

It differs from the exact solution by a factor of $\dfrac{3}{4}$. Exact solution is obtained when a piece-wise polynomial interpolation is used in the assumption of displacement field, u.

Stress in the bar, $\sigma = E\left(\dfrac{du}{dx}\right) = E(a_2 + a_3 x) = E(x - L)a_3$

$$= \frac{-3\,PE(x - L)}{4\,EAL} = \frac{3\,P(L - x)}{4\,AL}$$

$$= +\left(\frac{3}{4}\right)\left(\frac{P}{A}\right) \text{ at } x = 0 \quad \text{and} \quad -\left(\frac{3}{4}\right)\left(\frac{P}{A}\right) \text{ and } x = L$$

or $\qquad\qquad \pm\left(\dfrac{3}{2}\right)\left(\dfrac{P}{2A}\right)$

Due to the assumption of a quadratic displacement field over the system, stress is found to vary along the length of the bar. However, stress is expected to be constant (tensile from 1 to 2 and compressive from 2 to 3). Hence, the solution is not exact.

Method - 2

In order to compare the accuracy of the solution obtained by Rayleigh-Ritz method, the beam is analysed considering it to be a system of two springs in series as shown in Fig. 1.7 and using the stiffness of the axially loaded bar in the potential energy function.

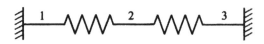

FIGURE 1.7

The stiffness of each spring is obtained from

$$K = \frac{P}{u} = \frac{(\sigma.A)}{\left[\varepsilon\left(\frac{L}{2}\right)\right]} = \frac{2AE}{L}$$

Total potential energy of the system is given by

$$\pi = \left(\frac{1}{2}k_1u_2^2 + \frac{1}{2}k_2u_2^2\right) + \left(-Pu_2\right) = K.u_2^2 - Pu_2$$

For equilibrium of this 1-DOF system,

$$\frac{\partial \pi}{\partial u_2} = 2K.u_2 - P = 0$$

or $$u_2 = \frac{P}{2K} = \frac{PL}{4AE}$$

Stress in the beam is given by,

$$\sigma = E\varepsilon = E.\left[\frac{u_2}{\left(\frac{L}{2}\right)}\right] = \frac{2Eu_2}{L} = \frac{P}{2A}$$

The displacement at 2 by Rayleigh-Ritz method differs from the exact solution by a factor of $\frac{3}{4}$, while the maximum stress in the beam differs by a factor of $\frac{3}{2}$. The stresses obtained by this approximate method are thus on the conservative side. Exact solution is obtained when a piece-wise polynomial interpolation is used in the assumption of displacement field, u. The results are plotted in Fig. 1.8.

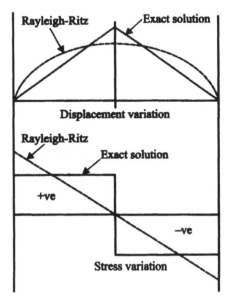

FIGURE 1.8

Method - 3

If the assumed displacement field is confined to a single element or segment of the component, it is possible to choose a more accurate and convenient polynomial. This is done in finite element method (FEM). Since total potential energy of each element is positive, minimum potential energy theory for the entire component implies minimum potential energy for each element. Stiffness matrix for each element is obtained by using this principle and these matrices for all the elements are assembled together and solved for the unknown displacements after applying boundary conditions. A more detailed presentation of FEM is provided in chapter 4.

Applying this procedure in the present example, let the displacement field in each element of the 2-element component be represented by $u = a_0 + a_1.x$. With this assumed displacement field, stiffness matrix of each axial loaded element of length (L/2) is obtained as

$$[K] = \left(\frac{2AE}{L}\right)\begin{bmatrix} 1 & -1 \\ -1 & 1 \end{bmatrix} \quad \text{and} \quad \{P\} = [K]\{u\}$$

The assembled stiffness matrix for the component with two elements is then obtained by placing the coefficients of the stiffness matrix in the appropriate locations as

$$\left(\frac{2AE}{L}\right)\begin{bmatrix} 1 & -1 & 0 \\ -1 & 1+1 & -1 \\ 0 & -1 & 1 \end{bmatrix}\begin{Bmatrix} u_1 \\ u_2 \\ u_3 \end{Bmatrix} = \begin{Bmatrix} P_1 \\ P_2 \\ P_3 \end{Bmatrix}$$

Applying boundary conditions $u_1 = 0$ and $u_3 = 0$, we get

$$\left(\frac{2AE}{L}\right) 2\,u_2 = P_2 = P \implies u_2 = \frac{PL}{4AE}$$

The potential energy approach and Rayleigh-Ritz method are now of only academic interest. FEM is a better generalisation of these methods and extends beyond discrete structures.

Examples of Rayleigh-Ritz method, with variable stress in the members

These examples are referred again in higher order 1-D truss elements, since they involve stress or strain varying along the length of the bar.

Example 1.5

Calculate displacement at node 2 of a tapered bar, shown in Fig 1.9, with area of cross-section A_1 at node 1 and A_2 at node 2 subjected to an axial tensile load 'P'.

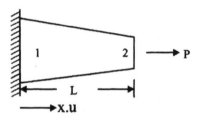

FIGURE 1.9

Solution

Different approximations are made for the displacement field and comparison made, in order to understand the significance of the most reasonable assumption.

(a) Since the bar is identified by 2 points, let us choose a first order polynomial (with 2 unknown coefficients) to represent the displacement field. Variation of A along the length of the bar adds additional computation. Let $A(x) = A_1 + (A_2 - A_1).x/L$

Let $\quad u = a_1 + a_2.x \qquad$ At $x = 0$, $u = a_1 = 0$

Then, $\quad u = a_2.x \quad ; \quad \dfrac{du}{dx} = a_2 \quad$ and $\quad u_2 = a_2.L$

Therefore,

$$\pi = \frac{1}{2} \int_0^L EA(x)\left(\frac{du}{dx}\right)^2 dx - P\,u_2$$

$$= \frac{1}{2} \int_0^L E\left[A_1 + (A_2 - A_1)\frac{x}{L}\right]a_2^2\, dx - P.a_2.L$$

$$= \frac{1}{2} E\left[A_1 L + (A_2 - A_1)\frac{L}{2}\right]a_2^2 - P.a_2.L$$

$$= \frac{1}{2} E\left[(A_1 + A_2)\frac{L}{2}\right]a_2^2 - P.a_2.L$$

For stable equilibrium, $\quad \dfrac{\partial \pi}{\partial a_2} = E\left[(A_1 + A_2)\dfrac{L}{2}\right]a_2 - P.L = 0$

from which a_2 can be evaluated as, $\quad a_2 = \dfrac{P}{\left[\dfrac{E(A_1 + A_2)}{2}\right]}$

Then, $\quad u_2 = a_2.L = \dfrac{P.L}{\left[\dfrac{E.(A_1 + A_2)}{2}\right]}$

and $\quad \sigma_1 = \sigma_2 = E\left(\dfrac{du}{dx}\right) = E\,a_2 = \dfrac{P}{\left[\dfrac{(A_1 + A_2)}{2}\right]}$

For the specific data of $A_1 = 40$ mm^2, $A_2 = 20$ mm^2 and $L = 200$ mm, we obtain,

$$u_2 = \frac{6.667\,P}{E} \quad \text{and} \quad \sigma_1 = \sigma_2 = 0.0333\,P$$

(b) Choosing displacement field by a first order polynomial gave constant strain (first derivative) and hence constant stress. Since a tapered bar is expected to have a variable stress, it is implied that the displacement field should be expressed by a minimum of 2nd order polynomial. Therefore, the solution is repeated with

$$u = a_1 + a_2.x + a_3.x^2 \qquad \text{At} \quad x = 0,\ u = a_1 = 0$$

Then, $u = a_2.x + a_3.x^2$; $\dfrac{du}{dx} = a_2 + 2a_3.x$ and $u_2 = a_2.L + a_3.L^2$

Therefore, $\pi = \dfrac{1}{2}\displaystyle\int_0^L EA(x).\left(\dfrac{du}{dx}\right)^2 dx - P u_2$

$$= \dfrac{1}{2}\int_0^L E\left[A_1 + (A_2 - A_1).\dfrac{x}{L}\right](a_2 + 2a_3 x)^2 dx - P\left(a_2.L + a_3.L^2\right)$$

For stable equilibrium, $\dfrac{\partial \pi}{\partial a_2} = 0$

$$\Rightarrow\ 3\, a_2\, (3A_1 - A_2) + 2a_3\, L\, (5A_1 - 2A_2) = \dfrac{6P}{E}$$

and $\dfrac{\partial \pi}{\partial a_3} = 0 \Rightarrow\ a_2\, (5A_1 - 2A_2) + a_3\, L\, (7A_1 - 3A_2) = \dfrac{3P}{E}$

For the specific data of $A_1 = 40$ mm^2, $A_2 = 20$ mm^2 and $L = 200$ mm, we obtain,

$$u_2 = a_2.L + a_3.L^2 = 6.652\,\dfrac{P}{E}$$

$$\sigma_1 = a_2 E \qquad\qquad = 0.0339\ P$$

and $\sigma_2 = E(a_2 + 2a_3 L) = 0.03518\ P$

(c) This problem can also be solved by assuming a 2nd order displacement function, satisfying a linearly varying stress along the length but with 2 unknown coefficients as

$$u = a_1 + a_2.x^2 \qquad \text{At } x = 0,\, u = a_1 = 0$$

Then, $u = a_2.x^2$; $\dfrac{du}{dx} = 2a_2.x$ and $u_2 = a_2.L^2$

Therefore,

$$\pi = \dfrac{1}{2}\int_0^L EA\left(\dfrac{du}{dx}\right)^2 dx - Pu_2$$

$$= \dfrac{1}{2}\int_0^L E\left[A_1 + (A_2 - A_1).\dfrac{x}{L}\right](2a_2\, x)^2 dx - P\left(a_2.L^2\right)$$

$$= \dfrac{1}{2}\, E\left[\dfrac{[A_1 L^3]}{3} + \dfrac{(A_2 - A_1)L^3}{4}\right](4a_2^2) - P\left(a_2.L^2\right)$$

$$= a_2^2\, L\, E\,\dfrac{(A_1 + 3A_2)}{6} - P\, a_2.L^2$$

For stable equilibrium, $\dfrac{\partial \pi}{\partial a_2} = \dfrac{a_2 L E (A_1 + 3A_2)}{3 - P L^2} = 0$ which gives an

expression for a_2 as

$$a_2 = \frac{P\,L}{\left[\dfrac{E\,(A_1 + 3A_2)}{3}\right]}$$

For the same set of data for A_1, A_2 and L, we get

$$u_2 = a_2.L^2 = \frac{6\,P}{E}$$

$$\sigma_1 = E.\,\epsilon_1 = E.\left(\frac{du}{dx}\right)_{x=0}$$

$$= E.(2a_2 x)_{x=0} = 0$$

$$\sigma_2 = E.(2a_2 x)_{x=L} = 2Ea_2 L = 0.06\,P$$

Sr.No.	Displacement polynomial	Displacement, u_2	Stress at 1, σ_1	Stress at 2, σ_2
1	$u = a_1 + a_2.x$	6.667 P/E	0.0333 P	0.0333 P
2	$u = a_1 + a_2.x + a_3.x^2$	6.652 P/E	0.0339 P	0.03518 P
3	$u = a_1 + a_2.x^2$	6.0 P/E	0.0	0.06 P
4	$\sigma_i = P/A_i$ (Exact solution)		0.025 P	0.05 P

These three assumed displacement fields gave different approximate solutions. These are plotted graphically here, for a better understanding of the differences. Exact solution depends on how closely the assumed displacement field matches with the actual displacement field.

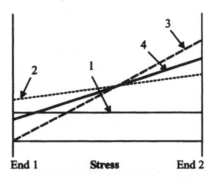

The most appropriate displacement field should necessarily include constant term, linear term and then other higher order terms.

Example 1.6

Calculate the displacement at node 2 of a vertical bar, shown in Fig. 1.10, due to its self-weight. Let the weight be w N/m of length.

Solution

Since the load is distributed, varying linearly from zero at the free end to maximum at the fixed end, it implies that the stress also varies linearly from the free end to fixed end. As shown in the last example, therefore, a quadratic displacement is the most appropriate. However, work potential needs to be calculated through integration of product of varying load and corresponding displacement, along the length.

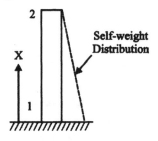

FIGURE 1.10

(a) Let $u = a_1 + a_2.x + a_3.x^2$ At $x = 0$, $u_1 = a_1 = 0$

$$\frac{du}{dx} = a_2 + 2a_3.x$$

Since applied load is zero at the free end,

Strain at $x = L$, $\left(\dfrac{du}{dx}\right)_2 = a_2 + 2a_3.L = 0 \;\Rightarrow\; a_2 = -2a_3L$

Then, $u = a_3.(x^2 - 2Lx)$; and $\dfrac{du}{dx} = 2a_3.(x - L)$

Let $P = -w(L - x)$ acting along –ve x-direction

Therefore,

$$\pi = \frac{1}{2}\int_0^L EA\left(\frac{du}{dx}\right)^2 dx - \int_0^L P\,du$$

$$= \frac{1}{2}\int_0^L EA.[2a_3.(x-L)]^2 dx - \int_0^L [-w(L-x)]2a_3.(x-L)\,dx$$

$$= \frac{2EA.a_3^2.L^3}{3} - \frac{2w.a_3.L^3}{3}$$

For stable equilibrium, $\dfrac{\partial \pi}{\partial a_3} = 0 \Rightarrow a_3 = \dfrac{w}{2EA}$

At $x = L$, $u_2 = -a_3.L^2 = \dfrac{-w.L^2}{2EA}$

Stress, $\sigma = E.\dfrac{du}{dx} = E.[2a_3.(x-L)] = \left(\dfrac{w}{A}\right).(x-L)$

At $x = 0$, $\sigma_1 = -\left(\dfrac{w.L}{A}\right)$ compressive

And at $x = L$, $\sigma_2 = 0$

Example 1.7

Calculate the displacement at node 2 of a vertical bar supported at both ends, shown in Fig. 1.11, due to its self-weight. Let the weight be w N/m of length.

Solution

As explained in the last example, a quadratic displacement is the most appropriate to represent linearly varying stress along the bar.

(a) Let $u = a_1 + a_2.x + a_3.x^2$ At $x = 0$, $u_1 = a_1 = 0$

At $x = L$, $u_2 = a_2.L + a_3.L^2 = 0$ $\Rightarrow a_2 = -a_3L$

Then, $u = a_3.(x^2 - Lx)$

and $\dfrac{du}{dx} = a_3.(2x - L)$

Let $P = -w(L-x)$ acting along –ve x-direction

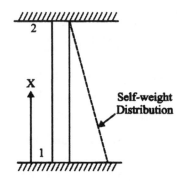

FIGURE 1.11

Therefore,

$$\pi = \frac{1}{2}\int_0^L EA\left(\frac{du}{dx}\right)^2 dx - \int_0^L P\, du$$

$$= \frac{1}{2}\int_0^L EA.[a_3.(2x-L)]^2\, dx + \int_0^L w(L-x)a_3.(2x-L)dx$$

$$= \frac{1}{2}EA.a_3^2.\frac{L^3}{3} - 2w.a_3.\frac{L^3}{3}$$

For stable equilibrium,

$$\frac{\partial \pi}{\partial a_3} = 0 \Rightarrow a_3 = \frac{2w}{EA}$$

At $\quad x = \dfrac{L}{2}, \quad u = -a_3.\dfrac{L^2}{4} = -w.\dfrac{L^2}{2EA}$

Stress, $\quad \sigma = E.\dfrac{du}{dx} = E.[a_3.(2x-L)] = \left(\dfrac{2w.}{A}\right).(2x-L)$

At $\quad x = 0, \quad \sigma_1 = -(2w.L/A) \quad$ compressive

and at $\quad x = L, \quad \sigma_2 = (2w.L/A) \quad$ tensile

1.5 PRINCIPLE OF MINIMUM POTENTIAL ENERGY

The total potential energy of an elastic body (π) is defined as the sum of total strain energy (U) and work potential (W).

i.e., $\qquad \pi = U + W,$

where $\qquad U = \left(\dfrac{1}{2}\right)\int_V \sigma\, \varepsilon\, dV$

and $\qquad W = -\int_V u^T F\, dV - \int_S u^T\, T\, dS - \sum u_i\, P_i$

Here, 'F' is the distributed body force, 'T' is the distributed surface force and 'P$_i$' are the concentrated loads applied at points i = 1, ..n. One or more of them may be acting on the component at any instant.

For a bar with axial load, if stress σ and strain ε are assumed uniform throughout the bar,

$$U = \left(\frac{1}{2}\right)\sigma\, \varepsilon\, v = \left(\frac{1}{2}\right)\sigma\, \varepsilon\, A\, L = \left(\frac{1}{2}\right)(\sigma\, A)(\varepsilon\, L) = \left(\frac{1}{2}\right)F\delta = \left(\frac{1}{2}\right)k\, \delta^2 \quad .....(1.1)$$

The work potential, $W = -\int q^T f \, dV - \int q^T T \, ds - \sum u_i^T P_i$ .....(1.2)

for the body force, surface traction and point loads, respectively.

Application of this method is demonstrated through the following simple examples. Since FEM is an extension of this method, more examples are included in this category.

Example 1.8

Calculate the nodal displacements in a system of four springs shown in Fig. 1.12

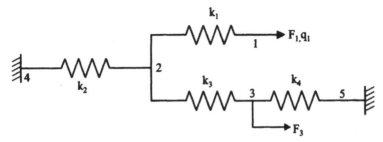

FIGURE 1.12 Example of a 5-noded spring system

Solution

The total potential energy is given by

$$\pi = \left(\frac{1}{2} k_1 \delta_1^2 + \frac{1}{2} k_2 \delta_2^2 + \frac{1}{2} k_3 \delta_3^2 + \frac{1}{2} k_4 \delta_4^2 \right) + \left(-F_1 q_1 - F_3 q_3 \right)$$

where, q_1, q_2, q_3 are the three unknown nodal displacements.
At the fixed points
$$q_4 = q_5 = 0$$
Extensions of the four springs are given by
$$\delta_1 = q_1 - q_2 \quad ; \qquad \delta_2 = q_2$$
$$\delta_3 = q_3 - q_2 \quad ; \qquad \delta_4 = -q_3$$
For equilibrium of this 3-DOF system,
$$\frac{\partial \pi}{\partial q_i} = 0 \text{ for } i = 1, 2, 3$$

or

$$\frac{\partial \pi}{\partial q_1} = k_1 (q_1 - q_2) - F_1 = 0$$

$$\frac{\partial \pi}{\partial q_2} = -k_1(q_1 - q_2) + k_2 q_2 - k_3 (q_3 - q_2) = 0$$

$$\frac{\partial \pi}{\partial q_3} = k_3(q_3 - q_2) + k_4 q_3 - F_3 = 0$$

These three equilibrium equations can be rewritten and expressed in matrix form as

$$\begin{bmatrix} k_1 & -k_1 & 0 \\ -k_1 & k_1 + k_2 + k_3 & -k_3 \\ 0 & -k_3 & k_3 + k_4 \end{bmatrix} \begin{Bmatrix} q_1 \\ q_2 \\ q_3 \end{Bmatrix} = \begin{Bmatrix} F_1 \\ 0 \\ F_2 \end{Bmatrix}$$

Considering free body diagrams of each node separately, represented by the following figures,

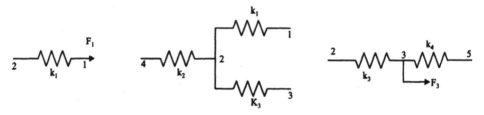

the equilibrium equations are $k_1 \delta_1 = F_1$

$$k_2 \delta_2 - k_1 \delta_1 - k_3 \delta_3 = 0$$

$$k_3 \delta_3 - k_4 \delta_4 = F_3$$

These equations, expressed in terms of nodal displacements q, are similar to the equations obtained earlier by the potential energy approach.

Example 1.9

Determine the displacements of nodes of the spring system (Fig. 1.13).

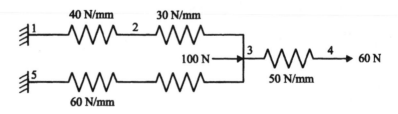

FIGURE 1.13 Example of a 4-noded spring system

Solution

Total potential energy of the system is given by

$$\pi = \left(\frac{1}{2}k_1\delta_1^2 + \frac{1}{2}k_2\delta_2^2 + \frac{1}{2}k_3\delta_3^2 + \frac{1}{2}k_4\delta_4^2\right) + \left(-F_3q_3 - F_4q_4\right)$$

where q_2, q_3, q_4 are the three unknown nodal displacements.

At the fixed points

$$q_1 = q_5 = 0$$

Extensions of the four springs are given by,

$$\delta_1 = q_2 - q_1 \; ; \; \delta_2 = q_3 - q_2 \; ; \; \delta_3 = q_4 - q_3 \; ; \; \delta_4 = q_3 - q_5$$

For equilibrium of this 3-DOF system,

$$\frac{\partial \pi}{\partial q_i} = 0 \quad \text{for } i = 2,3,4$$

or $\dfrac{\partial \pi}{\partial q_2} = k_1 q_2 - k_2(q_3 - q_2) = 0$ .....(a)

$$\frac{\partial \pi}{\partial q_3} = k_2(q_3 - q_2) - k_3(q_4 - q_3) + k_4 q_3 - F_3 = 0 \qquad \text{.....(b)}$$

$$\frac{\partial \pi}{\partial q_4} = k_3(q_4 - q_3) - F_4 = 0 \qquad \text{.....(c)}$$

These equilibrium equations can be expressed in matrix form as

$$\begin{bmatrix} k_1 + k_2 & -k_2 & 0 \\ -k_2 & k_2 + k_3 + k_4 & -k_3 \\ 0 & -k_3 & k_3 \end{bmatrix} \begin{Bmatrix} q_2 \\ q_3 \\ q_4 \end{Bmatrix} = \begin{Bmatrix} 0 \\ F_3 \\ F_4 \end{Bmatrix}$$

or $\begin{bmatrix} 40+30 & -30 & 0 \\ -30 & 30+50+60 & -50 \\ 0 & -50 & 50 \end{bmatrix} \begin{Bmatrix} q_2 \\ q_3 \\ q_4 \end{Bmatrix} = \begin{Bmatrix} 0 \\ 100 \\ 60 \end{Bmatrix}$

Substituting

$$q_4 = \frac{F_4}{k_3} + q_3 = \left(\frac{60}{50}\right) + q_3 = 1.2 + q_3 \qquad \text{from eq. (c)}$$

and $q_2 = \dfrac{k_2 q_3}{k_1 + k_2} = \dfrac{30 q_3}{(30 + 40)} = \dfrac{3 q_3}{7}$ from eq. (a)

in eq. (b), we get

$$k_2 \left[q_3 - \frac{3q_3}{7} \right] - k_3 \left[1.2 + q_3 \right] + k_4 . q_3 - F_3 = 0$$

which gives, $q_3 = 2.0741$ mm

and then, $q_2 = 0.8889$ mm ; $q_4 = 3.2741$ mm

WHY FEM ?

The Rayleigh-Ritz method and potential energy approach are now of only academic interest. For a big problem, it is difficult to deal with a polynomial having as many coefficients as the number of DOF. FEM is a better generalization of these methods and extends beyond the discrete structures. Rayleigh-Ritz method of choosing a polynomial for displacement field and evaluating the coefficients for minimum potential energy is used in FEM, at the individual element level to obtain element stiffness matrix (representing load-displacement relations) and assembled to analyse the structure.

1.6   ORIGIN OF FEM

The subject was developed during 2nd half of 20th century by the contribution of many researchers. It is not possible to give chronological summary of their contributions here. Starting with application of force matrix method for swept wings by S. Levy in 1947, significant contributions by J.H.Argyris, H.L.Langhaar,   R.Courant,   M.J.Turner,   R.W.Clough,   R.J.Melosh, J.S.Przemieniecki, O.C.Zienkiewicz, J.L.Tocher, H.C.Martin, T.H.H.Pian, R.H.Gallaghar, J.T.Oden, C.A.Felippa, E.L.Wilson, K.J.Bathe, R.D.Cook etc... lead to the development of the method, various elements, numerical solution techniques, software development and new application areas.

Individual member method of analysis, being over-conservative, provides a design with bigger and heavier members than actually necessary. This method was followed in civil structures where weight is not a major constraint. Analysis of the complete structure was necessitated by the need for a better estimation of stresses in the design of airplanes with minimum factor of safety (and, hence, minimum weight), during World War-II. Finite element method, popular as FEM, was developed initially as Matrix method of structural analysis for discrete structures like trusses and frames.

FEM is also extended later for continuum structures to get better estimation of stresses and deflections even in components of variable cross-section as well as with non-homogeneous and non-isotropic materials, allowing for optimum design of complicated components. While matrix method was limited to a few discrete structures whose load-displacement relationships are derived from basic strength of materials approach, FEM was a generalisation of the method on the basis of variational principles and energy theorems and is applicable to all types of structures – discrete as well as continuum. It is based on conventional theory of elasticity (equilibrium of forces and Compatibility of displacements) and variational principles.

In FEM, the entire structure is analysed without using assumptions about the degree of fixity at the joints of members and hence better estimation of stresses in the members was possible. This method generates a large set of simultaneous equations, representing load-displacement relationships. Matrix notation is ideally suited for computerising various relations in this method. Development of numerical methods and availability of computers, therefore, helped growth of matrix method. Sound knowledge of strength of materials, theory of elasticity and matrix algebra are essential pre-requisites for understanding this subject.

1.7   PRINCIPLE OF FEM

In FEM, actual component is replaced by a simplified model, identified by a finite number of *elements* connected at common points called *nodes*, with an assumed behaviour or response of each element to the set of applied loads, and evaluating the unknown field variable (displacement, temperature) at these finite number of points.

Example 1.10

The first use of this physical concept of representing a given domain as a collection of discrete parts is recorded in the evaluation of π from superscribed and inscribed polygons (Refer Fig. 1.14) for measuring circumference of a circle, thus approaching correct value from a higher value or a lower value (*Upper bounds/Lower bounds*) and improving accuracy as the number of sides of polygon increased (*convergence*). Value of π was obtained as 3.16 or $10^{1/2}$ by 1500 BC and as 3.1415926 by 480 AD, using this approach.

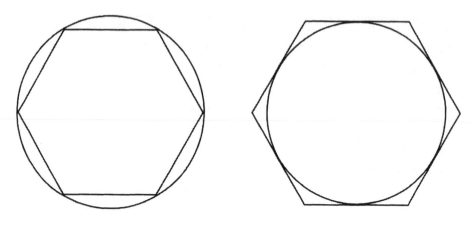

Case (a) Inscribed polygon   Case (b) Superscribed polygon

FIGURE 1.14 Approximation of a circle by an inscribed and a superscribed polygon

Perimeter of a circle of diameter 10 cm = πD = 31.4 cm

Case-A : The circle of radius 'r' is now approximated by an inscribed regular polygon of side 's'. Then, using simple trigonometric concepts, the length of side 's' of any regular n-sided polygon can now be obtained as s = 2 r sin (360/2n). Actual measurements of sides of regular or irregular polygon inscribed in the circle were carried out in those days, in the absence of trigonometric formulae, to find out the perimeter.

     With a 4-sided regular polygon,   perimeter = 4 s = 28.284
     With a 8-sided regular polygon,   perimeter = 8 s = 30.615
     With a 16-sided regular polygon,   perimeter = 16 s = 31.215

approaching correct value from a lower value, as the number of sides of the inscribed polygon theoretically increases to infinity.

Case-B : The same circle is now approximated by a superscribed polygon of side 's', given by

         s = 2 r tan (360/2n)

Then,   With a 4-sided regular polygon,   perimeter = 4 s = 40
        With a 8-sided regular polygon,   perimeter = 8 s = 33.137
        With a 16-sided regular polygon,   perimeter = 16 s = 31.826

approaching correct value from a higher value, as the number of sides of the circumscribed polygon theoretically increases to infinity.

A better estimate of the value of π (ratio of circumference to diameter) was found by taking average perimeter of inscribed and superscribed polygons, approaching correct value as the number of sides increases.

Thus with a 4-sided regular polygon, perimeter = (40+28.284) / 2 = 34.142

With a 8-sided regular polygon, perimeter = (33.137+30.615) / 2 = 31.876

With a 16-sided regular polygon, perimeter = (31.826+31.215) / 2 = 31.520

Example 1.11

In order to understand the principle of FEM, let us consider one more example, for which closed form solutions are available in every book of 'Strength of materials'. A common application for mechanical and civil engineers is the calculation of tip deflection of a cantilever beam AB of length 'L' and subjected to uniformly distributed load 'p'. For this simple case, closed form solution is obtained by integrating twice the differential equation.

$$EI\frac{d^2y}{dx^2} = M$$

and applying boundary conditions

$$y = 0 \text{ and } \frac{dy}{dx} = 0 \quad \text{at } x = 0 \text{ (fixed end, A)},$$

we get, at $x = L$, $y_{max} = \dfrac{p\,L^4}{8\,EI}$

This distributed load can be approximated as concentrated loads $(P_1, P_2,...P_N)$ acting on 'N' number of small elements, which together form the total cantilever beam. Each of these concentrated loads is the total value of the distributed load over the length of each element $(P_1 = P_2 = ... = P_N = p\,L\,/\,N)$, acting at its mid-point, as shown in Fig 1.15. Assuming that the tip deflection (at B) is small, the combined effect of all such loads can be obtained by linear superposition of the effects of each one of them acting independently. We will again make use of closed form solutions for the tip deflection values of a cantilever beam subjected to concentrated loads at some intermediate points.

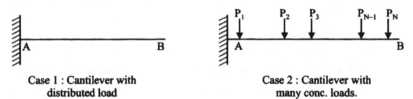

Case 1 : Cantilever with
distributed load

Case 2 : Cantilever with
many conc. loads.

FIGURE 1.15 Cantilever beam with distributed load approximated by many concentrated loads

Tip deflection of the cantilever when subjected to concentrated load P_J at a distance L_J from the fixed end is given by

$$y_B = (y)_J + \left(\frac{dy}{dx}\right)_J .(L - L_J)$$

Closed form solutions for $(y)_J$ and $(dy/dx)_J$ can be obtained by integrating the beam deflection equation with appropriate boundary conditions, as

$$y_J = P_J \frac{L_J^4}{3\,EI}$$

$$\left(\frac{dy}{dx}\right)_J = P_J \frac{L_J^3}{2\,EI}$$

Deflection at B, y_B, due to the combined effect of all the concentrated loads along the length of the cantilever can now be obtained, by linear superposition, as

$$y_B = [(y)_1 + (dy/dx)_1 (L - L_1)] + [(y)_2 + (dy/dx)_2 (L - L_2)] +\ldots +$$
$$[(y)_N + (dy/dx)_N (L - L_N)]$$

The results obtained with different number of elements are given in the table below, for the cantilever of length 200 cm, distributed load of 50 N/cm and $EI = 10^9 \text{ N cm}^2$.

S. No.	No. of elements	Tip displacement, y_B (cm)
1	3	9.815
2	4	9.896
3	5	9.933
4	6	9.954
5	8	9.974
6	10	9.983
7	15	9.993
8	20	9.996

The exact value obtained for the cantilever with uniformly distributed load, from the closed form solution, is $y_B = 10.0$ cm. It can be seen, even in this simple case, that the tip deflection value approaches true solution from a lower value as the number of elements increases. In other words, the tip deflection value even with a small number of elements gives an approximate solution.

This method in this form is not useful for engineering analysis as the approximate solution is lower than the exact value and, in the absence of error estimate, the solution is not practically useful.

FEM approach, based on minimum potential energy theorem, *converges to the correct solution from a higher value* as the number of elements in the model increases. While the number of elements used in a model is selected by the engineer, based on the required accuracy of solution as well as the availability of computer with sufficient memory, FEM has become popular as it ensures usefulness of the results obtained (on a more conservative side) even with lesser number of elements.

Finite Element Analysis (FEA) based on FEM is a simulation, not reality, applied to the mathematical model. Even very accurate FEA may not be good enough, if the mathematical model is inappropriate or inadequate. A mathematical model is an idealisation in which geometry, material properties, loads and/or boundary conditions are simplified based on the analyst's understanding of what features are important or unimportant in obtaining the results required. The error in solution can result from three different sources.

Modelling error – associated with the approximations made to the real problem.

Discretisation error – associated with type, size and shape of finite elements used to represent the mathematical model; can be reduced by modifying mesh.

Numerical error – based on the algorithm used and the finite precision of numbers used to represent data in the computer; most softwares use double precision for reducing numerical error.

It is entirely possible for an unprepared software user to misunderstand the problem, prepare the wrong mathematical model, discretise it inappropriately, fail to check computed output and yet accept nonsensical results. FEA is a solution technique that removes many limitations of classical solution techniques; but does not bypass the underlying theory or the need to devise a satisfactory model. Thus, *the accuracy of FEA depends on the knowledge of the analyst in modelling the problem correctly*.

1.8   CLASSIFICATION OF FEM

The basic problem in any engineering design is to evaluate displacements, stresses and strains in any given structure under different loads and boundary conditions. Several approaches of Finite Element Analysis have been developed to meet the needs of specific applications. The common methods are :

Displacement method – Here the structure is subjected to applied loads and/or specified displacements. The primary unknowns are displacements, obtained by inversion of the stiffness matrix, and the derived unknowns are stresses and strains. Stiffness matrix for any element can be obtained by variational principle, based on minimum potential energy of any stable structure and, hence, this is the most commonly used method.

Force method – Here the structure is subjected to applied loads and/or specified displacements. The primary unknowns are member forces, obtained by inversion of the flexibility matrix, and the derived unknowns are stresses and strains. Calculation of flexibility matrix is possible only for discrete structural elements (such as trusses, beams and piping) and hence, this method is limited in the early analyses of discrete structures and in piping analysis

Mixed method – Here the structure is subjected to applied loads and/or specified displacements. The method deals with large stiffness coefficients as well as very small flexibility coefficients in the same matrix. Analysis by this method leads to numerical errors and is not possible except in some very special cases.

Hybrid method – Here the structure is subjected to applied loads and stress boundary conditions. This deals with special cases, such as airplane door frame which should be designed for stress-free boundary, so that the door can be opened during flight, in cases of emergencies.

Displacement method is the most common method and is suitable for solving most of the engineering problems. The discussion in the remaining chapters is confined to displacement method.

1.9 TYPES OF ANALYSES

Mechanical engineers deal with two basic types of analyses for discrete and continuum structures, excluding other application areas like fluid flow, electromagnetics. FEM helps in modelling the component once and perform both the types of analysis using the same model.

(a) *Thermal analysis* – Deals with steady state or transient heat transfer by conduction and convection, both being linear operations while radiation is a non-linear operation, and estimation of temperature distribution in the component. This result can form one load condition for the structural analysis.

(b) *Structural analysis* – Deals with estimation of stresses and displacements in discrete as well as continuum structures under various types of loads such as gravity, wind, pressure and temperature. Dynamic loads may also be considered.

1.10   SUMMARY

- Finite Element Method, popularly known as FEM, involves analysis of the entire structure, instead of separately considering individual elements with simplified or assumed end conditions. It thus helps in a more accurate estimate of the stresses in the members, facilitating optimum design.

- FEM involves idealizing the given component into a finite number of small elements, connected at nodes. FEM is an extension of Rayleigh-Ritz method, eliminating the difficulty of dealing with a large polynomial representing a suitable displacement field valid over the entire structure. Over each finite element, the physical process is approximated by functions of desired type and algebraic equations, which relate physical quantities at these nodes and are developed using variational approach. Assembling these element relationships in the proper way is assumed to approximately represent relationships of physical quantities of the entire structure.

- FEM is based on minimum potential energy theorem. It approaches true solution from a higher value, as the number of elements increases. Thus, it gives a conservative solution even with a small number of elements, representing a crude idealisation.

OBJECTIVE QUESTIONS

1. The solution by FEM is
   (a) always exact   (b) mostly approximate
   (c) sometimes exact   (d) never exact

2. Discrete analysis covers
   (a) all 2-D trusses & frames
   (b) all 3-D trusses & frames
   (c) all 2-D and 3-D trusses & frames
   (d) no trusses; only frames

3. FEM is a generalization of
    (a) Rayleigh-Ritz method   (b) Weighted residual method
    (c) Finite difference method   (d) Finite volume method

4. Variational principle is the basis for
    (a) Displacement method   (b) Weighted residual method
    (c) Finite difference method   (d) Finite volume method

5. Displacement method is based on minimum
    (a) potential energy
    (b) strain energy
    (c) complementary strain energy
    (d) work done

6. Hybrid method is best suited for problems with prescribed
    (a) displacements   (b) forces   (c) stresses   (d) temperature

7. Primary variable in FEM structural analysis is
    (a) displacement   (b) force   (c) stress   (d) strain

8. Stress boundary conditions can be prescribed in
    (a) displacement method   (b) hybrid method
    (c) force method   (d) mixed method

9. Prescribed loads can form input data in
    (a) displacement method   (b) hybrid method
    (c) force method   (d) mixed method

10. Stiffness matrix approach is used in
    (a) displacement method   (b) stress method
    (c) force method   (d) mixed method

11. Displacement method of FEM for structural analysis gives
    (a) stiffness matrix   (b) flexibility matrix
    (c) conductance matrix   (d) mixed matrix

12. Flexibility matrix approach is used in
    (a) displacement method   (b) stress method
    (c) force method   (d) mixed method

CHAPTER 2

MATRIX OPERATIONS

FEM deals with a large number of linear algebraic equations, which can be conveniently expressed in matrix form. Matrices are also amenable to computer programming. Knowledge of matrix algebra is essential to understand and solve problems in FEM. A brief review of matrix algebra is given below for a good understanding of the remaining text.

A matrix is a group of m × n numbers (scalars or vectors) arranged in m number of rows and n number of columns. It is denoted by [A] or $[A]_{mxn}$ indicating matrix of 'm' rows and 'n' columns or matrix of *order* m × n. Each element of the matrix is identified by a_{ij}, located in the ith row and j^{th} column.

$$[A] = \begin{bmatrix} a_{11} & a_{12} & a_{13} & a_{14} & \cdots & a_{1n} \\ a_{21} & a_{22} & a_{23} & a_{24} & \cdots & a_{2n} \\ a_{31} & a_{32} & a_{33} & a_{34} & \cdots & a_{3n} \\ - & - & - & - & & - \\ a_{m1} & a_{m2} & a_{m3} & a_{m4} & \cdots & a_{mn} \end{bmatrix} \qquad .....(2.1)$$

2.1 TYPES OF MATRICES

Based on the number of rows and columns as well as on the nature of elements, some matrices are distinctly identified as follows. The symbol '∀' indicates 'for all'.

(a) *Square matrix* is a matrix of same number of rows and columns and is usually indicated as $[A]_{n \times n}$ or square matrix of order n i.e., matrix with m = n.

The elements from left top to right bottom .i.e., $a_{11} \cdots a_{nn}$ form the *leading diagonal.*

(b) *Row matrix* is a matrix of only one row and is usually indicated as $\{A\}^T$ i.e., matrix with m = 1.

(c) *Column matrix (or vector)* is a matrix of only one column and is usually indicated as $\{A\}$ i.e., matrix with n = 1.

(d) *Banded matrix* is a square matrix whose off-diagonal elements beyond half bandwidth from the diagonal element are all zeroes i.e., $a_{ij} = 0 \; \forall \, j > i + h$ and $\forall \, j < i - h$,

where   h + 1 is half bandwidth   and   2h + 1 is the bandwidth

For example, a 5 × 5 matrix with half bandwidth of 2 is given by

$$[A] = \begin{bmatrix} a_{11} & a_{12} & a_{13} & 0 & 0 \\ a_{21} & a_{22} & a_{23} & a_{24} & 0 \\ a_{31} & a_{32} & a_{33} & a_{34} & a_{35} \\ 0 & a_{42} & a_{43} & a_{44} & a_{45} \\ 0 & 0 & a_{53} & a_{54} & a_{55} \end{bmatrix} \qquad \text{.....(2.2)}$$

(e) *Diagonal matrix* is a banded matrix of bandwidth equal to 1 i.e., $a_{ij} = 0 \quad \forall \, i \neq j$ or   Banded matrix with h = 0   .....(2.3)

For example, a diagonal matrix of order 5 is given by

$$[A] = \begin{bmatrix} a_{11} & 0 & 0 & 0 & 0 \\ 0 & a_{22} & 0 & 0 & 0 \\ 0 & 0 & a_{33} & 0 & 0 \\ 0 & 0 & 0 & a_{44} & 0 \\ 0 & 0 & 0 & 0 & a_{55} \end{bmatrix} \qquad \text{.....(2.4)}$$

(f) *Identity matrix* is a diagonal matrix with each diagonal element equal to one and is usually indicated by [I] for any square matrix of order n.

i.e., $a_{ij} = 1 \quad \forall \, i = j$ and $a_{ij} = 0 \quad \forall \, i \neq j$ .....(2.5)

(g) *Transpose of matrix*, written as $[A]^T$, is the matrix with each element a_{ij} of matrix [A] written in j^{th} row and i^{th} column or interchanging its rows and columns.

i.e., a_{ij} of matrix $[A] = a_{ji}$ of matrix $[A]^T$

It can be seen that $\left([A]^T \right)^T = [A]$ .....(2.6)

(h) *Symmetric matrix*, defined only for square matrices has elements symmetric about its leading diagonal i.e., $a_{ij} = a_{ji}$

It can be seen that $[A]^T = [A]$ .....(2.7)

(i) *Skew-symmetric matrix*, also defined for square matrices only, has elements equal in magnitude but opposite in sign, about the leading diagonal,

i.e., $\qquad\qquad a_{ij} = -\,a_{ji}$ and elements on leading diagonal $a_{ii} = 0$

It can be seen that $\quad [A]^T = -\,[A] \qquad\qquad\qquad\qquad$.....(2.8)

2.2 MATRIX ALGEBRA

Common mathematical operations on matrices are

(a) *Addition of matrices A and B* (defined only when A and B are of same order) is addition of corresponding elements of both the matrices. If C is the resulting matrix,

$$[C] = [A] + [B] \quad \text{or} \quad c_{ij} = a_{ij} + b_{ij} \qquad\qquad.....(2.9)$$

(b) *Multiplication of a matrix A by a scalar* quantity 's' is multiplication of each element of matrix [A] by the scalar s

i.e., $\qquad s[A] = [s\ a_{ij}] \qquad\qquad\qquad\qquad\qquad$.....(2.10)

Same holds good for division of a matrix A by a scalar 's', since division by 's' is equivalent to multiplication by $\left(\dfrac{1}{s}\right)$ i.e., $\dfrac{[A]}{s} = \left[\dfrac{a_{ij}}{s}\right]$

(c) *Matrix multiplication* is defined only when number of columns in the first matrix A equals number of rows in the second matrix B and is, usually, not commutative. Element c_{ij} is the sum of the products of each element of ith row of matrix A with the corresponding element of the jth column of matrix B.

i.e., $\qquad [A]_{m \times n}\,[B]_{n \times p} = [C]_{m \times p}$

where $\qquad a_{ij}\,b_{jk} = c_{ik} \ \forall\ i = 1, m; j = 1, n; k = 1, p \qquad$.....(2.11)

It can be seen that for any matrix [A] and identity matrix [I], both of order $n \times n$,

$$[A][I] = [I][A] = [A] \qquad\qquad\qquad\qquad.....(2.12)$$

(d) *Transpose of product of matrices*

$$([A][B][C])^T = [C]^T\,[B]^T\,[A]^T \qquad\qquad.....(2.13)$$

(e) *Differentiation of a matrix*, each element of which is a function of x, is differentiation of each element w.r.t. x

i.e., $\qquad \dfrac{d[B(x)]}{dx} = \left[\dfrac{db_{ij}(x)}{dx}\right] \qquad\qquad\qquad$.....(2.14)

(f) *Integration of a matrix*, each element of which is a function of x, is integration of each element w.r.t. x,

i.e., $$\int [B] dx = \left[\int b_{ij} \, dx \right]$$ .....(2.15)

Matrix algebra differs from algebra of real numbers in some respects

For example, matrix multiplication is, in general, not commutative even if both the matrices are of the same order and square,

i.e., $[A][B] \neq [B][A]$

Matrix algebra also differs from vector algebra in some respects.

Vector product is defined as a dot product (giving a scalar result) or a cross product (giving a vector, perpendicular to both the vectors, as the result) whereas matrix multiplication is defined in only one way.

Trace of a matrix, defined for any square matrix of order m, is a scalar quantity and is equal to the sum of elements of its leading diagonal.

i.e., $$Tr[A] = a_{11} + a_{22} + ... + a_{mm} = \sum_{i=1}^{m} a_{ii}$$ .....(2.16)

Minor M_{ij} of a square matrix [A] is the determinant (defined in Section 2.3) of sub matrix of [A] obtained by deleting elements of ith row and j^{th} column from [A].

Cofactor C_{ij} of a square matrix $[A] = (-1)^{i+j} M_{ij}$

Matrix [C] with elements C_{ij} is called *Cofactor matrix*

Adjoint of matrix [A] is defined as $[C]^T$ and is written as Adj [A] or $[A_{ij}]$

Matrix approach is followed in FEM for convenience of representation. Conventionally, stress components, strain components, displacement components etc., are called stress vector, strain vector, displacement vector etc.; but they actually represent column matrices. Each component of these column matrices is a vector quantity representing the corresponding quantity in a particular direction. Hence, matrix algebra, and not vector algebra, is applicable in their case.

Quadratic form – If [A] is an (n × n) matrix and {x} is a (n × 1) vector, then the scalar product $\{x\}^T [A] \{x\}$ is called a quadratic form (having terms like x_1^2, x_2^2,... and $x_1 x_2$, $x_1 x_3$,...).

Example 2.1

Express $q = x_1 - 6x_2 + 3x_1^2 + 5x_1 x_2$ in the matrix form $\frac{1}{2} x^T Q x + C^T x$

Solution

From the data, it can be concluded that vector x has two elements x_1 and x_2. Therefore, matrix Q is of order 2 × 2 and C must be a vector of order 2.

$$\frac{1}{2}\mathbf{x}^T Q\mathbf{x} + C^T\mathbf{x} = \frac{1}{2}[x_1\ x_2]\begin{bmatrix} q_{11} & q_{12} \\ q_{21} & q_{22} \end{bmatrix}\begin{Bmatrix} x_1 \\ x_2 \end{Bmatrix} + [c_1\ c_2]\begin{Bmatrix} x_1 \\ x_2 \end{Bmatrix}$$

$$= \frac{1}{2}[x_1(q_{11}\ x_1 + q_{12}\ x_2) + x_2(q_{21}\ x_1 + q_{22}\ x_2)] + (c_1\ x_1 + c_2\ x_2)$$

$$= c_1\ x_1 + c_2\ x_2 + \left(\frac{q_{11}}{2}\right)x_1^2 + \left(\frac{q_{12}}{2} + \frac{q_{21}}{2}\right)x_1 x_2 + \left(\frac{q_{22}}{2}\right)x_2^2$$

Comparing these coefficients with those of the given expression

$$x_1 - 6x_2 + 3x_1^2 + 5x_1 x_2$$

we get $c_1 = 1$; $c_2 = -6$; $q_{11} = 6$; $q_{22} = 0$; $q_{12} + q_{21} = 10$ or $q_{12} = q_{21} = 5$

$$\therefore x_1 - 6x_2 + 3x_1^2 + 5x_1 x_2 = \frac{1}{2}[x_1\ x_2]\begin{bmatrix} 6 & 5 \\ 5 & 0 \end{bmatrix}\begin{Bmatrix} x_1 \\ x_2 \end{Bmatrix} + [1\ -6]\begin{Bmatrix} x_1 \\ x_2 \end{Bmatrix}$$

2.3 DETERMINANT

It is a scalar quantity defined only for square matrices and is written as $|A|$ or $\det[A]_{n\times n}$. It is defined as the sum of the products $(-1)^{i+j} a_{ij} M_{ij}$ where a_{ij} are the elements along any one row or column and M_{ij} are the corresponding minors. For example, if [A] is a square matrix of order 2,

$$\begin{vmatrix} a_{11} & a_{12} \\ a_{21} & a_{22} \end{vmatrix} = (-1)^{1+1} a_{11} . a_{22} + (-1)^{1+2} a_{12} . a_{21} \quad \text{expanding by 1}^{st} \text{ row}$$

or $(-1)^{1+1} a_{11}\ a_{22} + (-1)^{2+1} a_{21} . a_{12}$ expanding by 1st column

$$= a_{11}\ a_{22} - a_{12} . a_{21}$$

If [A] is a square matrix of order 3,

$$\begin{vmatrix} a_{11} & a_{12} & a_{13} \\ a_{21} & a_{22} & a_{23} \\ a_{31} & a_{32} & a_{33} \end{vmatrix} = (-1)^{1+1} a_{11} A_{11} + (-1)^{1+2} a_{12} A_{12} + (-1)^{1+3} a_{13} A_{13}$$

expanding by 1st row

$$= (-1)^{1+1} a_{11} \begin{vmatrix} a_{22} & a_{23} \\ a_{32} & a_{33} \end{vmatrix} + (-1)^{1+2} a_{12} \begin{vmatrix} a_{21} & a_{23} \\ a_{31} & a_{33} \end{vmatrix} + (-1)^{1+3} a_{13} \begin{vmatrix} a_{21} & a_{22} \\ a_{31} & a_{32} \end{vmatrix}$$

$$= a_{11}(a_{22}.a_{33} - a_{23}.a_{32}) - a_{12}(a_{21}.a_{33} - a_{23}.a_{31}) + a_{13}(a_{21}.a_{32} - a_{22}.a_{31})$$

It can also be expressed as

$$\begin{vmatrix} a_{11} & a_{12} & a_{13} \\ a_{21} & a_{22} & a_{23} \\ a_{31} & a_{32} & a_{33} \end{vmatrix} = (-1)^{1+1} a_{11} A_{11} + (-1)^{2+1} a_{21} A_{21} + (-1)^{3+1} a_{31} A_{31} \text{ expanding}$$

by 1st column

$$= (-1)^{1+1} a_{11} \begin{vmatrix} a_{22} & a_{23} \\ a_{32} & a_{33} \end{vmatrix} + (-1)^{2+1} a_{21} \begin{vmatrix} a_{12} & a_{13} \\ a_{32} & a_{33} \end{vmatrix} + (-1)^{3+1} a_{31} \begin{vmatrix} a_{12} & a_{13} \\ a_{22} & a_{23} \end{vmatrix}$$

$$= a_{11}(a_{22}.a_{33} - a_{23}.a_{32}) - a_{21}(a_{12}.a_{33} - a_{13}.a_{32}) + a_{31}(a_{12}.a_{23} - a_{13}.a_{22})$$

and so on.

It can be seen that,

(a)   Det ([A] [B][C]) = Det [A] . Det [B] . Det [C]   .....(2.17)

(b)   If [L] is a lower triangular matrix with $l_{ii}=1$ and $l_{ij} = 0 \ \forall \ i > j$,

   Det [L] = $l_{11}C_{11} + l_{12}C_{12} + l_{13}C_{13} = l_{11}C_{11} = l_{11} .(l_{22}. l_{33} - 0. l_{32}) = l_{11}. l_{22}. l_{33}$
   = 1

(c)   In the same way, Det [I] = 1

(d)   If [A] = [L][D] [L]^T,

   where [L] is a lower triangular matrix with $l_{ii}=1$ and $l_{ij} = 0 \ \forall \ i > j$

   and [D] is a diagonal matrix

   then,

   Det [A] = Det [L] . Det [D] . Det [L]^T = Det [D]   .....(2.18)

(e)   The determinant of a matrix is not affected by row or column modifications.

   A matrix whose determinant is zero is defined as a *singular matrix*

Example 2.2

Given that the area of a triangle with corners at (x_1, y_1), (x_2, y_2) and (x_3, y_3) can be written in the form,

$$\text{Area} = \frac{1}{2} \text{ Det} \begin{bmatrix} 1 & x_1 & y_1 \\ 1 & x_2 & y_2 \\ 1 & x_3 & y_3 \end{bmatrix}$$

determine the area of the triangle with corners at (1,1), (4,2) and (2,4)

Solution :

$$\text{Area} = \frac{1}{2} \begin{vmatrix} 1 & 1 & 1 \\ 1 & 4 & 2 \\ 1 & 2 & 4 \end{vmatrix} = \frac{1}{2} \begin{vmatrix} 1 & 1 & 1 \\ 0 & 3 & 1 \\ 0 & 1 & 3 \end{vmatrix}$$ with row modifications,
$$R_2 = R_2 - R_1$$
and $$R_3 = R_3 - R_1$$

$$= \frac{1}{2} \left(1 \times (3 \times 3 - 1 \times 1) - 0 \times (1 \times 3 - 1 \times 1) + 0 \times (1 \times 1 - 3 \times 1) \right) = 4$$

2.4   INVERSION OF A MATRIX

It is defined only for a square and non-singular matrix and is defined as

$$[A]^{-1} = \text{Adj } [A] \,/\, \text{Det } [A] \;= [C]^T \,/\, |A| \qquad\qquad .....(2.19)$$

It can be seen that product of any square matrix and its inverse is a unity matrix and the unity matrix being symmetric, the product is commutative.

i.e., $$[A]^{-1}[A] = [A][A]^{-1} = [I] \qquad\qquad .....(2.20)$$

A *necessary and sufficient condition for [A]⁻¹ to exist* is that there be no non-zero vector {x} such that   [A] {x} = {0}. The converse is also true. If there exists a non-zero vector {x} such that [A] {x} = {0}, then [A] does not have an inverse.

2.5   METHODS OF SOLUTION OF SIMULTANEOUS EQUATIONS

Finite element method gives rise to a set of n independent equations in n unknowns, expressed in matrix form as [K] {x} = {P}, where [K] is the stiffness matrix, {x} is the displacement vector and {P} is the applied load vector.

This system is called a *homogeneous* system if all the elements of {P} are zero. This system has a *trivial solution* $x_1 = x_2 = x_3 \ldots\ldots = x_n = 0$.

It is called a *non-homogeneous* system if at least one element of {P} is non-zero.

The system is said to be *consistent* if [K] is a square, non-singular matrix.

Most engineering problems generate a consistent system of equations and have a unique solution for the displacements {x}. F

These unknowns {x} can be evaluated by two basic approaches :

(a)   Calculating inverse of the stiffness matrix [K] and multiplying by the load vector   i.e., $\{x\} = [K]^{-1} \{P\}$

or

(b)   Solving directly the system of equations [K] {x} = {P}.

These methods are broadly classified as:

(i)   direct methods which give exact solution while requiring more computer memory, and

(ii) iterative methods which give approximate solutions with minimum computer memory requirement.

Some of the numerical methods, based on these approaches, are presented here.

2.5.1   BY INVERSION OF THE COEFFICIENT MATRIX

(a) *Method of Cofactors* – It involves calculation of n × n cofactors and determinant for a n × n matrix [A]. It is a costly process in terms of number of calculations and is not commonly used with large sized matrices in finite element analysis.

Example 2.3

Find the inverse of a square matrix [A] by the method of cofactors, if

$$[A] = \begin{bmatrix} 2 & 3 & 4 \\ 4 & 3 & 1 \\ 1 & 2 & 4 \end{bmatrix}$$

Solution

$$M_{11} = \begin{vmatrix} 3 & 1 \\ 2 & 4 \end{vmatrix} = 3 \times 4 - 1 \times 2 = 10 \; ;$$

$$M_{12} = \begin{vmatrix} 4 & 1 \\ 1 & 4 \end{vmatrix} = 4 \times 4 - 1 \times 1 = 15 \; ;$$

$$M_{13} = \begin{vmatrix} 4 & 3 \\ 1 & 2 \end{vmatrix} = 4 \times 2 - 3 \times 1 = 5$$

Similarly,

$$M_{21} = \begin{vmatrix} 3 & 4 \\ 2 & 4 \end{vmatrix} = 4 \; ; \quad M_{22} = \begin{vmatrix} 2 & 4 \\ 1 & 4 \end{vmatrix} = 4 \; ; \quad M_{23} = \begin{vmatrix} 2 & 3 \\ 1 & 2 \end{vmatrix} = 1$$

$$M_{31} = \begin{vmatrix} 3 & 4 \\ 3 & 1 \end{vmatrix} = -9 \; ; \quad M_{32} = \begin{vmatrix} 2 & 4 \\ 4 & 1 \end{vmatrix} = -14 \; ; \quad M_{33} = \begin{vmatrix} 2 & 3 \\ 4 & 3 \end{vmatrix} = -6$$

$$|A| = a_{11} M_{11} - a_{12} M_{12} + a_{13} M_{13} = 2 \times 10 - 3 \times 15 + 4 \times 5 = 20 - 45 + 20 = -5$$

$$[A]^{-1} = \frac{Adj[A]}{|A|} = \frac{1}{|A|} \begin{bmatrix} M_{11} & -M_{21} & M_{31} \\ -M_{12} & M_{22} & -M_{32} \\ M_{13} & -M_{23} & M_{33} \end{bmatrix} = -\left(\frac{1}{5}\right) \begin{bmatrix} 10 & -4 & -9 \\ -15 & 4 & 14 \\ 5 & -1 & -6 \end{bmatrix}$$

(b) *Gauss Jordan method* of inverting a square matrix [A] involves adding an identity matrix [I] of the same order as [A] and carrying out a sequence of row operations like

$$R_1^{(1)} = R_1/a_{11} \quad ; R_2^{(1)} = R_2 - a_{21} R_1^{(1)} \quad ; \ldots\ldots R_n^{(1)} = R_n - a_{n1} R_1^{(1)}$$

$$R_2^{(2)} = R_2^{(1)}/a_{22} \; ; R_1^{(2)} = R_1^{(1)} - a_{12} R_2^{(2)} ; \ldots\ldots R_n^{(2)} = R_n^{(1)} - a_{n2} R_2^{(2)}$$

$$R_3^{(3)} = R_3^{(2)}/a_{33} \; ; R_1^{(3)} = R_1^{(2)} - a_{13} R_3^{(3)} ; \ldots\ldots R_n^{(3)} = R_n^{(2)} - a_{n3} R_3^{(3)} \text{ etc..}$$

on this combination such that [A] is transformed to [I]. This will result in the identity matrix [I] initially appended to [A] getting transformed into $[A]^{-1}$

Example 2.4

Find inverse of matrix [A] by Gauss Jordan method, if

$$[A] = \begin{bmatrix} 2 & 3 & 4 \\ 4 & 3 & 1 \\ 1 & 2 & 4 \end{bmatrix}$$

Solution

$$[A:I] = \begin{bmatrix} 2 & 3 & 4 & : & 1 & 0 & 0 \\ 4 & 3 & 1 & : & 0 & 1 & 0 \\ 1 & 2 & 4 & : & 0 & 0 & 1 \end{bmatrix}$$

Row operations $R_1^{(1)} = R_1/a_{11}$; $R_2^{(1)} = R_2 - a_{21} R_1^{(1)}$ and $R_3^{(1)} = R_3 - a_{31} R_1^{(1)}$ will give

$$= \begin{bmatrix} 1 & \dfrac{3}{2} & 2 & : & \dfrac{1}{2} & 0 & 0 \\ 0 & -3 & -7 & : & -2 & 1 & 0 \\ 0 & \dfrac{1}{2} & 2 & : & -\dfrac{1}{2} & 0 & 1 \end{bmatrix}$$

Row operations $R_2^{(2)} = R_2^{(1)}/a_{22}$; $R_1^{(2)} = R_1^{(1)} - a_{12} R_2^{(2)}$ and $R_3^{(2)} = R_3^{(1)} - a_{32} R_2^{(2)}$ will give

$$= \begin{bmatrix} 1 & 0 & -\dfrac{3}{2} & : & -\dfrac{1}{2} & \dfrac{1}{2} & 0 \\ 0 & 1 & \dfrac{7}{3} & : & \dfrac{2}{3} & -\dfrac{1}{3} & 0 \\ 0 & 0 & \dfrac{5}{6} & : & -\dfrac{5}{6} & \dfrac{1}{6} & 1 \end{bmatrix}$$

Row operations $R_3^{(3)} = R_3^{(2)}/a_{33}$; $R_1^{(3)} = R_1^{(2)} - a_{13} R_3^{(3)}$ and $R_2^{(3)} = R_2^{(2)} - a_{23} R_3^{(3)}$ will give

$$= \begin{bmatrix} 1 & 0 & 0 & : & -2 & \dfrac{4}{5} & \dfrac{9}{5} \\ 0 & 1 & 0 & : & 3 & -\dfrac{4}{5} & -\dfrac{14}{5} \\ 0 & 0 & 1 & : & -1 & \dfrac{1}{5} & \dfrac{6}{5} \end{bmatrix} = \left[I : A^{-1} \right]$$

Thus,

$$[A]^{-1} = \begin{bmatrix} -2 & \dfrac{4}{5} & \dfrac{9}{5} \\ 3 & -\dfrac{4}{5} & -\dfrac{14}{5} \\ -1 & \dfrac{1}{5} & \dfrac{6}{5} \end{bmatrix} = -\left(\dfrac{1}{5}\right) \begin{bmatrix} 10 & -4 & -9 \\ -15 & 4 & 14 \\ 5 & -1 & -6 \end{bmatrix}$$

which is same as the inverted matrix of [A] obtained by the method of cofactors.

2.5.2   DIRECT METHODS

Many engineering problems in FEM will result in a set of simultaneous equations represented by [K] {x} = {P} where [K] is the stiffness matrix of the entire structure and {x} is the vector of nodal displacements due to the applied loads {P}. In many situations, a component may have to be analysed for multiple loads, represented by [P]. Corresponding to each column of [P], representing one set of nodal loads, a particular solution vector {x} exists. Matrix inversion approach is much faster in such cases than the direct solution techniques presented below.

(a) *Cramer's Rule* gives direct solution to a set of simultaneous equations
   [A]{x} = {B}.

   In a system of n equations in n unknowns, this method involves in calculating n + 1 determinants of n × n matrices. It is a very costly method for large sized problems and is, therefore, not popular.

$$x_1 = \frac{|A_1|}{|A|} \; ; \; x_2 = \frac{|A_2|}{|A|}; \; \ldots\ldots \; x_n = \frac{|A_n|}{|A|}$$

   where [A$_i$] is obtained by replacing column i of matrix [A] with the elements of {B}.

Example 2.5

Solve the following system of equations

$$x + y + z = 11$$
$$2x - 6y - z = 0$$
$$3x + 4y + 2z = 0$$

Solution

The coefficient matrix is

$$[A] = \begin{bmatrix} 1 & 1 & 1 \\ 2 & -6 & -1 \\ 3 & 4 & 2 \end{bmatrix}$$

and the vector of constants is   {B} = [11 0 0]^T

Then,

$$[A_1] = \begin{bmatrix} 11 & 1 & 1 \\ 0 & -6 & -1 \\ 0 & 4 & 2 \end{bmatrix}; \quad [A_2] = \begin{bmatrix} 1 & 11 & 1 \\ 2 & 0 & -1 \\ 3 & 0 & 2 \end{bmatrix}; \quad [A_3] = \begin{bmatrix} 1 & 1 & 11 \\ 2 & -6 & 0 \\ 3 & 4 & 0 \end{bmatrix}$$

and

$$|A| = 1\,[(-6) \times 2 - (-1) \times 4] - 1\,[(2) \times 2 - (-1) \times 3] + 1\,[2 \times 4 - (-6) \times 3] = 11$$

$|A_1| = 11\ [(-6) \times 2 -(-1) \times 4] -0\ [1 \times 2 - 4 \times 1] + 0\ [1 \times (-1) - (-6) \times 1] = -88$

$|A_2| = -11\ [2 \times 2 - (-1) \times 3] + 0\ [1 \times 2 - 3 \times 1] - 0\ [1 \times (-1) - 2 \times 1] = -77$

$|A_3| = 11\ [2 \times 4 - (-6) \times 3] - 0\ [1 \times 4 - 3 \times 1] + 0\ [1 \times (-6) - 2 \times 1] = 286$

Therefore, $x = \dfrac{|A_1|}{|A|} = -\dfrac{88}{11} = -8$

$y = \dfrac{|A_2|}{|A|} = -\dfrac{77}{11} = -7$

and $z = \dfrac{|A_3|}{|A|} = \dfrac{286}{11} = 26$

(b) *Gauss Jordan method* for solving a set of non-homogeneous equations involves adding the vector of constants {B} to the right of square matrix [A] of the coefficients and carrying out a sequence of row operations equivalent to multiplication with $[A]^{-1}$ like

$R_1^{(1)} = R_1/a_{11}$; $R_2^{(1)} = R_2 - a_{21}\ R_1^{(1)}$; .... $R_n^{(1)} = R_n - a_{n1}\ R_1^{(1)}$

$R_2^{(2)} = R_2^{(1)}/a_{22}$; $R_1^{(2)} = R_1^{(1)} - a_{12}\ R_2^{(2)}$; .... $R_n^{(2)} = R_n^{(1)} - a_{n2}\ R_2^{(2)}$

$R_3^{(3)} = R_3^{(2)}/a_{33}$; $R_1^{(3)} = R_1^{(2)} - a_{13}\ R_3^{(3)}$; ..... $R_n^{(3)} = R_n^{(2)} - a_{n3}\ R_3^{(3)}$ etc..

on this combination such that [A] is transformed to [I]. This will result in the vector of constants {B} getting transformed into the solution vector {x} i.e. $[A:B] \;\rightarrow\; [I:x]$

It involves up to $\dfrac{2n^3}{3}$ multiplications and is a costly method for large matrices

Example 2.6

Find the solution to the following set of non-homogeneous equations by calculating inverse of matrix [A] by Gauss Jordan method.

$x + y + z = 11$

$2x - 6y - z = 0$

$3x + 4y + 2z = 0$

Solution

The set of equations can be expressed in the form [A] {x} = {B}
where,

$$[A] = \begin{bmatrix} 1 & 1 & 1 \\ 2 & -6 & -1 \\ 3 & 4 & 2 \end{bmatrix}; \quad \{x\} = \begin{Bmatrix} x \\ y \\ z \end{Bmatrix} \text{ and } \{B\} = \begin{Bmatrix} 11 \\ 0 \\ 0 \end{Bmatrix}$$

$$\text{Then, } [A:B] = \begin{bmatrix} 1 & 1 & 1 & : & 11 \\ 2 & -6 & -1 & : & 0 \\ 3 & 4 & 2 & : & 0 \end{bmatrix}$$

Row operations $R_1^{(1)} = R_1/a_{11}$; $R_2^{(1)} = R_2 - a_{21} R_1^{(1)}$ and $R_3^{(1)} = R_3 - a_{31} R_1^{(1)}$ will give

$$= \begin{bmatrix} 1 & 1 & 1 & : & 11 \\ 0 & -8 & -3 & : & -22 \\ 0 & 1 & -1 & : & -33 \end{bmatrix}$$

Row operations $R_2^{(2)} = R_2^{(1)}/a_{22}$; $R_1^{(2)} = R_1^{(1)} - a_{12} R_2^{(2)}$ and $R_3^{(2)} = R_3^{(1)} - a_{32} R_2^{(2)}$ will give

$$= \begin{bmatrix} 1 & 0 & \dfrac{5}{8} & : & \dfrac{33}{4} \\ 0 & 1 & \dfrac{3}{8} & : & \dfrac{11}{4} \\ 0 & 0 & -\dfrac{11}{8} & : & -\dfrac{143}{4} \end{bmatrix}$$

Row operations $R_3^{(3)} = R_3^{(2)}/a_{33}$; $R_1^{(3)} = R_1^{(2)} - a_{13} R_3^{(3)}$ and $R_2^{(3)} = R_2^{(2)} - a_{23} R_3^{(3)}$ will give

$$= \begin{bmatrix} 1 & 0 & 0 & : & -8 \\ 0 & 1 & 0 & : & -7 \\ 0 & 0 & 1 & : & 26 \end{bmatrix} = [I:x]$$

Thus,

$$\{x\} = \begin{bmatrix} -8 \\ -7 \\ 26 \end{bmatrix}$$

which is same as the solution vector $\{x\}$ obtained by the method of cofactors.

(c) *Gauss elimination method* is a direct method for solving $[K]\{x\} = \{P\}$ where $[K]$ is a square matrix of order n. The method involves reduction of the coefficient matrix to upper triangular form and back substitution.

Example 2.7

Solve the following system of simultaneous equations by Gauss elimination method

$$x + y + z = 11$$
$$2x - 6y - z = 0$$
$$3x + 4y + 2z = 0$$

Solution

The given equations can be expressed in matrix form as

$$\begin{bmatrix} 1 & 1 & 1 \\ 2 & -6 & -1 \\ 3 & 4 & 2 \end{bmatrix} \begin{Bmatrix} x \\ y \\ z \end{Bmatrix} = \begin{Bmatrix} 11 \\ 0 \\ 0 \end{Bmatrix}$$

The elements of [K] are reduced to upper diagonal form by the following operations, performing on one column at a time.

Column-1

$$R_2 = R_2 - R_1 \left(\frac{k_{21}}{k_{11}} \right)$$

and $R_3 = R_3 - R_1 \left(\frac{k_{31}}{k_{11}} \right)$ give $\begin{bmatrix} 1 & 1 & 1 \\ 0 & -8 & -3 \\ 0 & 1 & -1 \end{bmatrix} \begin{Bmatrix} x \\ y \\ z \end{Bmatrix} = \begin{Bmatrix} 11 \\ -22 \\ -33 \end{Bmatrix}$

Column-2 $R_3 = R_3 - R_2 (k_{31}/k_{22})$ give $\begin{bmatrix} 1 & 1 & 1 \\ 0 & -8 & -3 \\ 0 & 0 & -\dfrac{11}{8} \end{bmatrix} \begin{Bmatrix} x \\ y \\ z \end{Bmatrix} = \begin{Bmatrix} 11 \\ -22 \\ -\dfrac{143}{4} \end{Bmatrix}$

Back substitution

Row-3 $z = \left(-\dfrac{143}{4} \right) \left(-\dfrac{8}{11} \right) = 26$

Row-2 $y = \dfrac{-22 + 3.z}{(-8)} = -7$

Row-1 $x = \dfrac{11 - y - z}{1} = -8$

which is same as the solution vector {x} obtained by the earlier methods.

(d) *LU factorisation method* is based on expressing square matrix [K] as a product of two triangular matrices L and U.

    i.e.,   [K] = [L] [U],

    where   [L] is a lower triangular matrix

    and   [U] is an upper triangular matrix

This method involves $n^3/3 + n^2$ operations and thus is an *economical method* for large size problems. There are three variations of this method.

(i) *Doolittle method* wherein diagonal elements of lower triangular matrix L are equal to 1

Step-1 $\{P\} = [K]\{x\} = [L][U]\{x\} = [L]\{v\}$ .....(2.21)

$$\begin{bmatrix} k_{11} & k_{12} & k_{13}...k_{1n} \\ k_{21} & k_{22} & k_{23}...k_{2n} \\ k_{31} & k_{32} & k_{33}...k_{3n} \\ . & . & ..... \\ k_{n1} & k_{n2} & k_{n3}...k_{nn} \end{bmatrix} = \begin{bmatrix} 1 & 0 & 0...0 \\ l_{21} & 1 & 0...0 \\ l_{31} & l_{32} & 1...0 \\ . & . & ..... \\ l_{n1} & l_{n2} & l_{n3}...1 \end{bmatrix} \begin{bmatrix} u_{11} & u_{12} & u_{13}...u_{1n} \\ 0 & u_{22} & u_{23}...u_{2n} \\ 0 & 0 & u_{33}...u_{3n} \\ . & . & ..... \\ 0 & 0 & 0...u_{nn} \end{bmatrix}$$

The elements of matrices [L] and [U] are obtained using the following equations of matrix multiplication

$$k_{11} = 1 \cdot u_{11} \qquad \Rightarrow u_{11}$$
$$k_{12} = 1 \cdot u_{12} \qquad \Rightarrow u_{12}$$
.... $$k_{1n} = 1 \cdot u_{1n} \qquad \Rightarrow u_{1n}$$
$$k_{21} = l_{21} \cdot u_{11} \qquad \Rightarrow l_{21}$$
$$k_{22} = l_{21} \cdot u_{12} + 1 \cdot u_{22} \qquad \Rightarrow u_{22}$$
$$k_{23} = l_{21} \cdot u_{13} + 1 \cdot u_{23} \qquad \Rightarrow u_{23}$$
..... $$k_{2n} = l_{21} \cdot u_{1n} + 1 \cdot u_{2n} \qquad \Rightarrow u_{2n}$$

and so on

Step-2 : Intermediate solution vector $\{v\}$ is calculated from $\{P\} = [L]\{v\}$

i.e $$\begin{Bmatrix} P_1 \\ P_2 \\ P_3 \\ \vdots \\ P_n \end{Bmatrix} = \begin{bmatrix} 1 & 0 & 0...0 \\ l_{21} & 1 & 0...0 \\ l_{31} & l_{32} & 1...0 \\ . & . & .... . \\ l_{n1} & l_{n2} & l{n_3}...1 \end{bmatrix} \begin{Bmatrix} v_1 \\ v_2 \\ v_3 \\ \vdots \\ v_n \end{Bmatrix}$$

from which, elements of $\{v\}$ can be obtained as

$$v_1 = P_1$$

$$v_2 = P_2 - l_{21} \cdot v_1$$

$$v_3 = P_3 - l_{31} \cdot v_1 - l_{32} \cdot v_2 \quad \text{and so on}$$

Step-3 : Actual solution vector {x} is calculated from $\{v\} = [U]\{x\}$

i.e.,
$$\begin{Bmatrix} v_1 \\ v_2 \\ v_3 \\ \vdots \\ v_n \end{Bmatrix} = \begin{bmatrix} u_{11} & u_{12} & u_{13} & \cdots & u_{1n} \\ 0 & u_{22} & u_{23} & \cdots & u_{2n} \\ 0 & 0 & u_{33} & \cdots & u_{3n} \\ \vdots & \vdots & \vdots & \vdots & \vdots \\ 0 & 0 & 0 & \cdots & u_{nn} \end{bmatrix} \begin{Bmatrix} x_1 \\ x_2 \\ x_3 \\ \vdots \\ x_n \end{Bmatrix}$$

from which, elements of { x } can be obtained as

$$x_n = \frac{v_n}{u_{nn}}$$

$$x_{n-1} = \frac{\left(v_{n-1} - u_{n-1,n} \, x_n\right)}{u_{n-1,\; n-1}}$$

$$x_{n-2} = \frac{\left(v_{n-2} - u_{n-2,\; n-1} \, x_{n-1} - u_{n-2,\; n} \, x_n\right)}{u_{n-2,\; n-2}} \qquad \text{and so on}$$

Example 2.8

Solve the following system of simultaneous equations by L-U factorisation method

$$x + y + z = 11$$
$$2x - 6y - z = 0$$
$$3x + 4y + 2z = 0$$

Solution

The coefficient matrix is

$$[K] = \begin{bmatrix} 1 & 1 & 1 \\ 2 & -6 & -1 \\ 3 & 4 & 2 \end{bmatrix} = [L][U] = \begin{bmatrix} l_{11} & 0 & 0 \\ l_{21} & l_{22} & 0 \\ l_{31} & l_{32} & l_{33} \end{bmatrix} \begin{bmatrix} u_{11} & u_{12} & u_{13} \\ 0 & u_{22} & u_{23} \\ 0 & 0 & u_{33} \end{bmatrix}$$

The elements of [L] and [U] are obtained from

$$\begin{aligned}
&k_{11} = l_{11} \cdot u_{11} \text{ and } l_{11} = 1 &&\Rightarrow && u_{11} = 1 \\
&k_{12} = l_{11} \cdot u_{12} \text{ and } l_{11} = 1 &&\Rightarrow && u_{12} = 1 \\
&k_{13} = l_{11} \cdot u_{13} \text{ and } l_{11} = 1 &&\Rightarrow && u_{13} = 1 \\
&k_{21} = l_{21} \cdot u_{11} &&\Rightarrow && l_{21} = 2 \\
&k_{22} = l_{21} \cdot u_{12} + l_{22} \cdot u_{22} \text{ and } l_{22} = 1 &&\Rightarrow && u_{22} = -8 \\
&k_{23} = l_{21} \cdot u_{13} + l_{22} \cdot u_{23} \text{ and } l_{22} = 1 &&\Rightarrow && u_{23} = -3
\end{aligned}$$

$$k_{31} = l_{31} \cdot u_{11}$$

$$k_{32} = l_{31} \cdot u_{12} + l_{32} \cdot u_{22}$$

$$k_{33} = l_{31} \cdot u_{13} + l_{32} \cdot u_{23} + l_{33} \cdot u_{33} \text{ and } l_{33} = 1$$

$$\Rightarrow \quad l_{31} = 3$$

$$\Rightarrow \quad l_{32} = -1/8$$

$$\Rightarrow \quad u_{33} = -11/8$$

Thus $\{P\} = [K]\{X\} = [L][U][X]$

or

$$\begin{Bmatrix} 11 \\ 0 \\ 0 \end{Bmatrix} = \begin{bmatrix} 1 & 0 & 0 \\ 2 & 1 & 0 \\ 3 & -\dfrac{1}{8} & 0 \end{bmatrix} \begin{bmatrix} 1 & 1 & 1 \\ 0 & -8 & -3 \\ 0 & 0 & -\dfrac{11}{8} \end{bmatrix} \begin{Bmatrix} x \\ y \\ z \end{Bmatrix}$$

Let $\{P\} = [L]\{v\}$

Then, elements of vector $\{v\}$ are obtained from

$$v_1 = P_1 \qquad\qquad = 11$$

$$v_2 = P_2 - l_{21} \cdot v_1 \qquad = -22$$

$$v_3 = P_3 - l_{31} \cdot v_1 - l_{32} \cdot v_2 = 143/4$$

Actual solution vector $\{x\}$ is calculated from $\{v\} = [U]\{x\}$

$$z = \frac{v_3}{u_{33}} \qquad\qquad = 26$$

$$y = \frac{(v_2 - u_{23}\, z)}{u_{22}} \qquad = -7$$

$$x = \frac{(v_1 - u_{12}\, y - u_{13}\, z)}{u_{11}} \qquad = -8$$

(ii) *Crout's method* is another factorisation method wherein diagonal elements of upper triangular matrix U are equal to 1. The remaining procedure is identical to Doolittle method.

(iii) *Cholesky method* is also a direct method for solving $[K]\{x\} = \{P\}$, where *[K] is symmetric and positive definite.*

    i.e. for any vector $\{x\}$, $\{x\}^T [A] \{x\} > 0$

Here, $\{P\} = [K]\{x\} = [L][L]^T\{x\}$ where [L] is a lower triangular matrix This method is also a factorisation method except that upper triangular matrix [U] is replaced by the transpose of lower triangular matrix [L]. Hence, the diagonal elements of [L] will not be equal to 1. Since one matrix [L] only need to be stored, it *requires less memory* and is especially useful in large-size problems.

Step-1

$$
\begin{bmatrix} k_{11} & k_{12} & k_{13}\ldots k_{1n} \\ k_{21} & k_{22} & k_{23}\ldots k_{2n} \\ k_{31} & k_{32} & k_{33}\ldots k_{3n} \\ \vdots & \vdots & \vdots\ldots \\ k_{n1} & k_{n2} & k_{n3}\ldots k_{nn} \end{bmatrix} = \begin{bmatrix} l_{11} & 0 & 0\ldots 0 \\ l_{21} & l_{22} & 0\ldots 0 \\ l_{31} & l_{32} & l_{33}\ldots 0 \\ \vdots & \vdots & \vdots \\ l_{n1} & l_{n2} & l_{n3}\ldots l_{nm} \end{bmatrix} \begin{bmatrix} l_{11} & l_{21} & l_{31}\ldots l_{n1} \\ 0 & l_{22} & l_{32}\ldots l_{n2} \\ 0 & 0 & l_{33}\ldots l_{n3} \\ \vdots & \vdots & \vdots \\ 0 & 0 & 0\ldots l_{nn} \end{bmatrix}
$$

The elements of matrix [L] are obtained using the following equations of matrix multiplication

$$k_{11} = l_{11} \cdot l_{11} \qquad\qquad \Rightarrow l_{11}$$

$$k_{12} = l_{11} \cdot l_{21} \qquad\qquad \Rightarrow l_{21}$$

.... $$k_{1n} = l_{11} \cdot l_{n1} \qquad\qquad \Rightarrow l_{n1}$$

$k_{21} = l_{21} \cdot l_{11}$ is same as $k_{12} = l_{11} \cdot l_{21}$ since [K] is symmetric and is redundant

$$k_{22} = l_{21} \cdot l_{21} + l_{22} \cdot l_{22} \qquad \Rightarrow l_{22}$$

$$k_{23} = l_{21} \cdot l_{31} + l_{22} \cdot l_{32} \qquad \Rightarrow l_{32}$$

..... $$k_{2n} = l_{21} \cdot l_{1n} + l_{22} \cdot l_{n2} \qquad \Rightarrow l_{n2}$$

and so on

Step-2 : Intermediate solution vector {v} is calculated from $\{P\} = [L]\{v\}$

i.e.,

$$
\begin{Bmatrix} P_1 \\ P_2 \\ P_3 \\ \vdots \\ P_n \end{Bmatrix} = \begin{bmatrix} l_{11} & 0 & 0\ldots 0 \\ l_{21} & l_{22} & 0\ldots 0 \\ l_{31} & l_{32} & l_{33}\ldots 0 \\ \vdots & \vdots & \vdots \\ l_{n1} & l_{n2} & l_{n3}\ldots l_{nn} \end{bmatrix} \begin{Bmatrix} v_1 \\ v_2 \\ v_3 \\ \vdots \\ v_n \end{Bmatrix}
$$

from which, elements of {v} can be obtained as

$$v_1 = \frac{P_1}{l_{11}}$$

$$v_2 = \frac{(P_2 - l_{21} \cdot v_1)}{l_{22}}$$

$$v_3 = \frac{(P_3 - l_{31} \cdot v_1 - l_{32} \cdot v_2)}{l_{33}} \quad \text{and so on}$$

Step-3 : Actual solution vector $\{x\}$ is calculated from $\{v\} = [U]\{x\}$

i.e

$$\begin{Bmatrix} v_1 \\ v_2 \\ v_3 \\ \vdots \\ v_n \end{Bmatrix} = \begin{bmatrix} l_{11} & l_{21} & l_{31} & \cdots & l_{n1} \\ 0 & l_{22} & l_{32} & \cdots & l_{n2} \\ 0 & 0 & l_{33} & \cdots & l_{n3} \\ \vdots & \vdots & \vdots & \vdots & \ddots & \vdots \\ 0 & 0 & 0 & \cdots & l_{nn} \end{bmatrix} \begin{Bmatrix} x_1 \\ x_2 \\ x_3 \\ \vdots \\ x_n \end{Bmatrix}$$

from which, elements of $\{ \delta \}$ can be obtained as

$$x_n = \frac{v_n}{l_{nn}}$$

$$x_{n-1} = \frac{\left(v_{n-1} - l_{n, \ n-1} \ x_n\right)}{l_{n-1, \ n-1}}$$

$$x_{n-2} = \frac{\left(v_{n-2} - l_{n-1, \ n-2} \ x_{n-1} - l_{n, n-2} \cdot x_n\right)}{l_{n-2, \ n-2}} \qquad \text{and so on}$$

(iv) *Gauss Jordan method* is also similar to Gauss elimination method except that back substitution operation is avoided by transforming coefficient matrix [K] into a unity (diagonal) matrix, instead of an upper triangular matrix. This increases number of operations and hence is not economical. However, this method is popular for inversion of matrix [K] which can then be used with multiple right side vectors. In such applications, this method is more economical.

2.5.3 ITERATIVE METHODS

There are many different approaches, but two popular methods are briefly explained here.

(a) *Jacobi iteration or Method of simultaneous corrections :* In this method also, each row is first divided by the corresponding diagonal element to make diagonal elements equal to 1.

Then, $\{x\}^{(m+1)} = \{P\} + ([I] - [K]) \{x\}^{(m)}$

This form is used for successive improvement of the solution vector from an initial approximation $\{x\}^{(0)}$. No component of $\{x\}^{(m)}$ is replaced with new ones, until all components are computed. All components are simultaneously changed at the beginning of each step. For better understanding, matrix notation is avoided in the example.

Example 2.9

Solve the following simultaneous equations by Jacobi iteration method.

$$20x + y - 2z = 17$$
$$3x + 20y - z = -18$$
$$2x - 3y + 20z = 25$$

Solution

These equations are rewritten as

$$x^i = (17 - y^{i-1} + 2z^{i-1}) / 20$$
$$y^i = (-18 - 3x^{i-1} + z^{i-1}) / 20$$
$$z^i = (25 - 2x^{i-1} + 3y^{i-1}) / 20$$

The values obtained after each step are tabulated below. Starting assumption refers to $i = 0$

$i \to$	*0*	*1*	*2*	*3*	*4*	*5*	*6*
x	0	0.85	1.02	1.0134	1.0009	1.0000	1.0000
y	0	−0.9	−0.965	−0.9954	−1.0018	−1.0002	−1.0000
z	0	1.25	1.1515	1.0032	0.9993	0.9996	1.0000

Exact solution is $x = 1$, $y = -1$ and $z = 1$

(b) *Gauss Siedel iteration* or **Method of successive corrections** : In this method, each row is first divided by the corresponding diagonal element to make diagonal elements equal to 1. Coefficient matrix is assumed as sum of three matrices, [I], [L] and [U] where [I] is the unity matrix, [L] is a lower triangular matrix with diagonal elements equal to 0 and [U] is an upper triangular matrix with diagonal elements equal to 0.

Then, $[K] \{x\} = ([I] + [L] + [U])\{x\} = \{P\}$ or $\{x\} = \{P\} - [L] \{x\} - [U] \{x\}$

This form is used for successive improvement of the solution vector from an initial approximation $\{x\}^{(0)}$ with

$$\{x\}^{(m+1)} = \{P\} - [L] \{x\}^{(m)} - [U] \{x\}^{(m)}$$

replacing old terms of solution vector $\{x\}^{(m)}$ with new ones, as soon as they are computed.

For better understanding, matrix notation is avoided in the example.

Example 2.10

Solve the following simultaneous equations by Gauss Siedel iteration method.

$$20x + y - 2z = 17$$
$$3x + 20y - z = -18$$
$$2x - 3y + 20z = 25$$

Solution

These equations are rewritten as

$$x^i = (17 - y^{i-1} + 2z^{i-1}) / 20$$
$$y^i = (-18 - 3x^i + z^{i-1}) / 20$$
$$z^i = (25 - 2x^i + 3y^i) / 20$$

Note the change in the superscripts (iteration no), compared to the Jacobi iteration method.

The values obtained after each step are tabulated below. Starting assumption refers to $i = 0$

$i \rightarrow$	*0*	*1*	*2*	*3*
x	0	0.85	1.0025	1.0000
y	0	−1.0275	−0.9998	−1.0000
z	0	1.0109	0.9998	1.0000

Thus, successive correction (Gauss Siedel iteration) method converges much faster than simultaneous correction (Jacobi's iteration) method.

2.6   EIGEN VALUES AND EIGEN VECTORS

For every square matrix [A], there exist λ and $\{u\}$ such that $[A]\{u\} = \lambda\{u\}$ or

$$([A] - \lambda[I]) \{u\} = \{0\} \qquad\qquad .....(2.22)$$

For a non-trivial solution, $\{u\} \neq \{0\}$,

$$\left| [A] - \lambda[I] \right| = 0 \text{ is called the } \textbf{characteristic equation} \quad .....(2.23)$$

λ_i s are called eigen values. Eigen values may be real or complex. Most of the engineering problems will have real eigen values. The system of n independent equations represented by $([A] - \lambda[I]) \{u\} = \{0\}$ will have n eigen values, where n is the order of the square matrix [A]. Some of them may be repeated. $\{u_i\}$ associated with each λ_i is called an eigen vector and is calculated from the system of equations $([A] - \lambda_i[I]) \{u_i\} = \{0\}$.

[A] is a *positive definite matrix* if all its λ_i are +ve

or if $\{x\}^T[A]\{x\} > 0$ for any non-zero vector $\{x\}$

[A] is a positive semi-definite matrix if all its λ_i are $\geq$ zero

[A] is an indefinite matrix if its λ_i are −ve, zero or +ve

[A] is singular if and only if one or more of its eigen values are zero   .....(2.24)

Eigen values and eigenvectors of an (n × n) matrix can be obtained by one of the two approaches.

- Solving n^{th} order characteristic equation for the n eigen values and solving the system of n simultaneous equations with each eigen value for the corresponding eigenvector.
- Solving the system by iterative method through successive approximations of the eigen values and corresponding eigenvectors

Since eigen values in a dynamic system represent the natural frequencies and only the first few eigen values representing the dominant modes of vibrations are usually required, the latter approach is faster and is more commonly used. Both these techniques are explained here through simple examples.

Example 2.11

Determine the eigen values for the equation of motion given below

$$\left(\begin{bmatrix} 2 & 1 & 1 \\ 1 & 2 & 1 \\ 1 & 1 & 2 \end{bmatrix} - \lambda_1 \begin{bmatrix} 4 & 0 & 0 \\ 0 & 4 & 0 \\ 0 & 0 & 4 \end{bmatrix} \right) \{ \phi_i \} = 0$$

where λ_i and ϕ_i are eigen values and eigen vectors respectively

Solution

Characteristic equation is given by

$$\text{Det} \left(\begin{bmatrix} 2 & 1 & 1 \\ 1 & 2 & 1 \\ 1 & 1 & 2 \end{bmatrix} - \lambda_1 \begin{bmatrix} 4 & 0 & 0 \\ 0 & 4 & 0 \\ 0 & 0 & 4 \end{bmatrix} \right) = 0$$

or

$$\text{Det} \begin{bmatrix} 2-4\lambda_i & 1 & 1 \\ 1 & 2-4\lambda_i & 1 \\ 1 & 1 & 2-4\lambda_i \end{bmatrix} = 0$$

$$(2 - 4\lambda_i) [(2 - 4\lambda_i)^2 - 1] - [(2 - 4\lambda_i) - 1] + [1 - (2 - 4\lambda_i)]$$

$$= (2 - 4\lambda_i)^3 - (2 - 4\lambda_i) + 2 = 0$$

or

$$-64 \lambda_i^3 + 96 \lambda_i^2 - 36 \lambda_i + 4 = -4(\lambda - 1)(4\lambda - 1)^2 = 0$$

This cubic equation has three real roots given by $\lambda = 1, \dfrac{1}{4}, \dfrac{1}{4}$

Example 2.12

Determine the eigen values and eigen vectors of the equation

$\{[A] - \lambda[I]\}\ \{q\} = 0$ where λ = eigen value and q = eigen vector, if

$$A = \begin{bmatrix} 2 & 2 & 1 \\ 1 & 3 & 1 \\ 1 & 2 & 2 \end{bmatrix}$$

Solution

Characteristic equation is given by

$$\text{Det}\left(\begin{bmatrix} 2 & 1 & 1 \\ 1 & 2 & 1 \\ 1 & 1 & 2 \end{bmatrix} - \lambda \begin{bmatrix} 1 & 0 & 0 \\ 0 & 1 & 0 \\ 0 & 0 & 1 \end{bmatrix}\right) = \begin{vmatrix} 2-\lambda & 2 & 1 \\ 1 & 3-\lambda & 1 \\ 1 & 2 & 2-\lambda \end{vmatrix} = 0$$

$$(2-\lambda)\,[(3-\lambda)(2-\lambda) - (1)(2)] - (2)\,[(1)(2-\lambda) - (1)(1)]$$

$$+ (1)\ [\,(1).(2) - (3-\lambda)(1)\,] = 0$$

i.e., $\lambda^3 - 7\lambda^2 + 11\lambda - 5 = 0$

or $(\lambda - 5)(\lambda^2 - 2\lambda + 1) = (\lambda - 5)(\lambda - 1)(\lambda - 1) = 0$

Therefore, eigen values of the matrix are 5, 1 and 1

Eigen vectors corresponding to each of these eigen values are calculated by substituting the eigen value in the three simultaneous homogeneous equations.

Corresponding to $\lambda = 5$, $(2 - 5)x_1 + 2x_2 + x_3 = 0$

$$x_1 + (3 - 5)x_2 + x_3 = 0$$

$$x_1 + 2x_2 + (2 - 5)x_3 = 0$$

which gives $\{x\} = \begin{Bmatrix} 1 \\ 1 \\ 1 \end{Bmatrix}$

This eigenvector indicates rigid body mode of vibration.

Corresponding to $\lambda = 1$, $(2-1)x_1 + 2x_2 + x_3 = 0$

$$x_1 + (3-1)x_2 + x_3 = 0$$

$$x_1 + 2x_2 + (2-1)x_3 = 0$$

Since the 2nd and 3rd equations are identical, we have effectively two independent equations in *three unknowns*. Therefore, a unique solution does not exist. These equations are satisfied by

$$\{x\} = \begin{Bmatrix} 1 \\ 0 \\ -1 \end{Bmatrix} \text{ or } \begin{Bmatrix} 2 \\ -1 \\ 0 \end{Bmatrix}$$

In many practical applications involving many degrees of freedom, only the fundamental mode is dominant or important. **Iterative methods** such as Rayleigh's Power method, House holder's tri-diagonalisation method are very commonly used for computing few dominant eigen values and eigen vectors.

Rayleigh's Power method: This iterative method is used for calculating fundamental or largest eigen value

$[A] \{x\} = \lambda \{x\}$ is rewritten for iterations as $[A] \{x\}^{(i-1)} = \lambda^{i} \{x\}^{(i)}$

Example 2.13

Calculate largest eigen value for the matrix [A] by Rayleigh's Power method

where $[A] = \begin{bmatrix} 2 & -1 & 0 \\ -1 & 2 & -1 \\ 0 & -1 & 2 \end{bmatrix}$

Solution

Let the initial assumption be $\{x\}^{(0)} = [\,1 \quad 0 \quad 0\,]^{T}$

Then, $[A]\{x\}^{(0)} = \begin{bmatrix} 2 & -1 & 0 \\ -1 & 2 & -1 \\ 0 & -1 & 2 \end{bmatrix} \begin{Bmatrix} 1 \\ 0 \\ 1 \end{Bmatrix} = \begin{Bmatrix} 2 \\ -1 \\ 0 \end{Bmatrix} = 2 \begin{Bmatrix} 1 \\ -0.5 \\ 0 \end{Bmatrix} = \lambda^{(1)} \{x\}^{(1)}$

$[A]\{x\}^{(1)} = \begin{bmatrix} 2 & -1 & 0 \\ -1 & 2 & -1 \\ 0 & -1 & 2 \end{bmatrix} \begin{Bmatrix} 1 \\ -0.5 \\ 0 \end{Bmatrix} = \begin{Bmatrix} 2.5 \\ -2 \\ 0.5 \end{Bmatrix} = 2.5 \begin{Bmatrix} 1 \\ -0.8 \\ 0.2 \end{Bmatrix} = \lambda^{(2)} \{x\}^{(2)},$

Eigen values and eigen vectors obtained after successive iterations are tabulated below.

$i\rightarrow$	0	1	2	3	4	5	6	7
	1	1	1	1	0.87	0.80	0.76	0.74
$\{x\}$	0	−0.5	−0.8	−1	−1	−1	−1	−1
	0	0	0.2	0.43	0.54	0.61	0.65	0.67
λ	−	2	2.5	2.8	3.43	3.41	3.41	3.41

2.7 MATRIX INVERSION THROUGH CHARACTERISTIC EQUATION

Cayley-Hamilton theorem states that that every square matrix satisfies its own characteristic equation. Let characteristic equation of square matrix [A] be given by

$$D(\lambda) = \lambda^n + C_{n-1}\lambda^{n-1} + C_{n-2}\lambda^{n-2} + C_{n-3}\lambda^{n-3} + \ldots\ldots + C_1\lambda + C_0 = 0$$

Then, according to this theorem,

$$A^n + C_{n-1}A^{n-1} + C_{n-2}A^{n-2} + C_{n-3}A^{n-3} + \ldots\ldots + C_1 A + C_0 = 0$$

Multiplying throughout by

$$A^{-1}, \quad A^{n-1} + C_{n-1}A^{n-2} + C_{n-2}A^{n-3} + C_{n-3}A^{n-4} + \ldots + C_1 I + C_0 A^{-1} = 0$$

or $A^{-1} = -[A^{n-1} + C_{n-1}A^{n-2} + C_{n-2}A^{n-3} + C_{n-3}A^{n-4} + \ldots + C_1 I] / C_0$

Example 2.14

Obtain inverse of the following matrix using Cayley-Hamilton theorem

$$A = \begin{bmatrix} 1 & 2 & 3 \\ 2 & 4 & 5 \\ 3 & 5 & 6 \end{bmatrix}$$

Solution

Characteristic equation is given by

$$\text{Det} \begin{bmatrix} 1-\lambda & 2 & 3 \\ 2 & 4-\lambda & 5 \\ 3 & 5 & 6-\lambda \end{bmatrix} = 0$$

i.e., $\lambda^3 - 11\lambda^2 - 4\lambda + 1 = 0$ or $A^3 - 11A^2 - 4A + I = 0$

Multiplying by A^{-1},

we get $A^2 - 11A - 4I + A^{-1} = 0$ or $A^{-1} = -[A^2 - 11A - 4I]$

$$A^2 = [A][A] = \begin{bmatrix} 1 & 2 & 3 \\ 2 & 4 & 5 \\ 3 & 5 & 6 \end{bmatrix} \begin{bmatrix} 1 & 2 & 3 \\ 2 & 4 & 5 \\ 3 & 5 & 6 \end{bmatrix} = \begin{bmatrix} 14 & 25 & 31 \\ 25 & 45 & 56 \\ 31 & 56 & 70 \end{bmatrix}$$

Then,

$$A^{-1} = -[A^2 - 11A - 4I] = -\begin{bmatrix} 14 & 25 & 31 \\ 25 & 45 & 56 \\ 31 & 56 & 70 \end{bmatrix} + 11\begin{bmatrix} 1 & 2 & 3 \\ 2 & 4 & 5 \\ 3 & 5 & 6 \end{bmatrix} + 4\begin{bmatrix} 1 & 0 & 0 \\ 0 & 1 & 0 \\ 0 & 0 & 1 \end{bmatrix}$$

$$= \begin{bmatrix} 1 & -3 & 2 \\ -3 & 3 & -1 \\ 2 & -1 & 0 \end{bmatrix}$$

Note : Reader is advised to verify the solutions obtained by these various methods with the standard check $[A] [A]^{-1} = [I]$, since numerical errors are frequently encountered while practicing these methods.

2.8 SUMMARY

- A matrix is a group of (m × n) numbers (scalars or vectors) arranged in 'm' number of rows and 'n' number of columns. FEM deals with solution to a large number of simultaneous equations, which can be expressed more conveniently in matrix form as $[K] \{x\} = \{P\}$, where $[K]$ is the stiffness matrix of the component which is usually square, symmetric and non-singular.

- This system of equations can be solved for $\{x\}$ when $[K]$ is a square, non-singular matrix, in the form $\{x\} = [K]^{-1}\{P\}$. Direct methods such as Gauss elimination method or factorisation method and iteration method like Gauss Siedel method are commonly used techniques for the solution of the system of equations.

- For every square matrix $[A]$, there exist λ and $\{u\}$ such that $[K]\{u\} = \lambda\{u\}.\lambda$ is called the eigen value and $\{u\}$ is the corresponding eigen vector. Stiffness matrix is usually positive definite i.e., $\{u\}^T[K] \{u\} > 0$ for all non-zero $\{u\}$. Eigen values represent the natural frequencies and eigen vectors represent natural modes of a dynamic system. Iterative methods such as Rayleigh's Power method, House holder's tri-diagonalisation method are very commonly used for computing the first few dominant eigen values and eigenvectors.

CHAPTER 3

THEORY OF ELASTICITY

A brief review of theory of elasticity, with specific reference to applications of FEM, is presented here for a clear understanding of the subsequent chapters. For a more detailed explanation, the reader may refer to other standard books on this subject.

Every physical component is a three-dimensional solid. However, based on the relative dimensions along three coordinate directions and nature of applied loads / boundary conditions, they are classified as 1-D, 2-D or 3-D components. This idealisation helps in analysing the component quickly and at lower cost.

3.1 DEGREES OF FREEDOM

The direction in which a point in a structure is free to move is defined as its degree of freedom (DOF). In general, any point in a component can move along an arbitrary direction in space and rotate about an arbitrary direction, depending upon the loads applied on the component. Since specifying this arbitrary direction through angles is tedious, the movement and rotation at any point are identified by their components in the chosen coordinate system. Thus, in Cartesian coordinate system, a point can at best have translation identified by its three components *along* the three coordinate directions X, Y and Z; and rotation identified by its three components *about* the three coordinate directions X, Y and Z. Depending on the way a member is assembled in a structure, some or all of its DOF at a point may be fixed or free. Truss members are designed for only axial loads and rotational DOFs are not relevant whereas beam is designed for bending loads which result in displacement at each point normal to the axis. Its derivative or slope is independently constrained and hence is considered as an independent DOF. Similarly, in plates subjected to in-plane loads, rotational DOFs are not significant and need not be considered as independent DOFs whereas in plates and shells subjected to bending loads, rotational DOF are significant and need to be treated as independent DOFs.

3.2 RIGID BODY MOTION

It is the motion of entire component or part under the influence of external applied loads. Such a rigid body motion, with no relative deformation between any two points in the component, cannot induce stresses or strains in a component. Since design or analysis of a component involves calculation of stresses and strains in a component, any static part with 'n' degrees of freedom can not be solved unless it is restrained from moving as a rigid body by constraining it at least at one point along each DOF. The number of DOFs to be constrained depends on the type of component. For example, a truss has to be constrained for rigid body motion along X and Y axes; a plane frame (in X-Y plane) has to be constrained along X and Y axes as well as for rotation about Z axis; a thick shell has to be constrained along X, Y and Z axes as well as for rotation about X, Y and Z axes. In the conventional analysis by closed form solution to the differential equation, the rigid body motion is constrained by using relevant boundary conditions for evaluating constants of integration.

3.3 DISCRETE STRUCTURES

Structures such as trusses and frames, which have many identifiable members, connected only at their end points or nodes are called discrete structures. Each member of the structure is considered as a one-dimensional (1-D) element along its length, identified by its end point coordinates. Their lateral dimensions are reflected in element properties like area of cross section in trusses and moment of inertia and depth of section in beams.

3.4 CONTINUUM STRUCTURES

Structures such as plates, thin shells, thick shells, solids, which do not have distinctly identifiable members, can be modeled by an arbitrary number of elements of different shapes viz., triangles and quadrilaterals in 2-D structures and tetrahedron and brick elements in 3-D structures. These are called continuum structures. In these structures, adjacent elements have a common boundary surface (or line, if stress variation across thickness is neglected as in the case of plates). *The finite element model represents true situation only when displacements and their significant derivates of adjacent elements are same along their common boundary.*

3.5   MATERIAL PROPERTIES

FEM is ideally suited for analysing structures with varying material properties. The variation of properties such as Modulus of elasticity (E), Modulus of rigidity (G) and coefficient of linear thermal expansion (α) may be:

(a)   constant (linear stress-strain relationship) or variable (non-linear stress-strain relationship) over the range of load.

(b)   same in all directions (isotropic) or vary in different directions (anisotropic or orthotropic).

(c)   constant over the temperature range or vary with temperature, particularly when the temperature range over the component is large.

These variations may be inherent in the material or induced by the manufacturing processes like rolling, casting. Treatment of non-homogeneous material, with varying properties at different locations of a component, is difficult and is also very unusual. In most cases, variation of material properties with the direction and with temperature may not be significant and hence neglected. So, an isotropic, homogeneous material with constant (temperature-independent) properties is most often used in the analysis of a component.

3.6   LINEAR ANALYSIS

It is based on linear stress-strain relationship (Hooke's law) and is usually permitted when stress at any point in the component is below the elastic limit or yield stress. In this analysis, linear superposition of results obtained for individual loads on a component is valid in order to obtain stresses due to any combination of these loads acting simultaneously. In some designs, it is necessary to check for many combinations of loads such as pressure and thermal at different times of a start-up transient of a steam turbine. In such a case, analysing for unit pressure; multiplying the stress results with the pressure corresponding to that particular time of the transient and adding to the stresses due to temperature distribution will be economical.

3.7   NON-LINEAR ANALYSIS

In many cases, the mathematical formulations are based on small deflection theory. A component with large deflections due to loads, such as aircraft wing, comes under the category of 'geometric non-linearity'. In some aerospace applications, where the component is designed for single use, stress level above yield point, where stress-strain relationship is non-linear, may be permitted. In some other cases involving non-metallic components, material may exhibit non-linear stress-strain behaviour in the operating load range. These two cases come under the category of 'material non-linearity'. In both these cases, analysis carried out by applying the load in small steps.

(a) **_Geometric non-linearity_** : In these problems, geometry of the component is redefined after every load step by adding the displacements at various nodes to the nodal coordinates for defining the true geometry to be used for the next load step.

(b) **_Material non-linearity_** : In these problems, total load on the component is applied in small steps and non-linear stress-strain relationship in the material, usually represented by the value of Young's modulii or Modulii of elasticity (normal stress/normal strain) E_X, E_Y and E_Z in different directions, is considered as linear in each load step. These values are suitably modified after each load step, till the entire load range is covered. Here, normal stress and normal strain can be tensile (+ve) or compressive (–ve) and E has the same units as stress, since strain is non-dimensional.

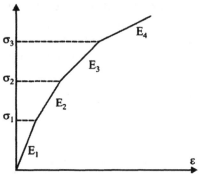

FIGURE 3.1 Stress-strain diagram of a non-linear material

3.8 STIFFNESS AND FLEXIBILITY

Loads and displacements in an element are related through stiffness and flexibility coefficients. Stiffness coefficient (K) is the force required to produce unit displacement, while Flexibility coefficient (F) is the displacement produced by a unit force. They are usually defined for 1-D elements such as truss elements or spring elements by

$$P = K.u \qquad \text{where, } K = \frac{AE}{L} \text{ is the stiffness coefficient}$$

$$\text{or} \quad u = F.P \qquad \text{where, } F = \frac{L}{AE} \text{ is the flexibility coefficient}$$

Since $\sigma = \dfrac{P}{A} = E\varepsilon = E\left(\dfrac{u}{L}\right)$ where, u is the change in length and, therefore, $P = \left(\dfrac{AE}{L}\right).u$

However, a more general definition of stiffness coefficient k_{ij} is the force required at node 'i' to produce unit displacement at node 'j'. i.e., $P_i = k_{ij}\, u_j$.

An important feature of the *stiffness coefficient* is that it **has different units with reference to different loads and different displacements**. For example, units of k_{ij} relating load at 'i' to displacement at 'j' in a truss element are in 'N/mm' whereas units of k_{ij} relating moment at 'i' to slope at 'j' in a beam element are in 'N mm'.

Stiffness coefficients k_{ij} connecting loads at 'n' number of nodes with displacements at these 'n' number of nodes thus form a square matrix of order n × n, represented by [K]. In structures having linear force-deflection relationship, the flexibility and stiffness coefficients have the property $k_{ij} = k_{ji}$ and $f_{ij} = f_{ji}$. This is called *Maxwell's reciprocity relationship*. This makes the flexibility and stiffness matrices symmetric.

3.9   PRINCIPLE OF MINIMUM POTENTIAL ENERGY

Every component subjected to external applied loads reaches stable equilibrium, when its potential energy or the difference between work done by external forces and internal strain energy due to stresses developed is zero. It can also be expressed as – During any arbitrary kinematically consistent virtual displacement from the equilibrium state, satisfying constraints prescribed for the body, potential energy equals tozero or the work done by the external forces equals the increment in strain energy

i.e., $\qquad \delta I = \delta W_{ext} - \delta U = 0$

3.10   STRESS AND STRAIN AT A POINT

Stress is the internal reaction in a component subjected to external forces. Thus, stress exists only when external force is applied on a component and varies from point to point. Stress at any point in a component is defined as a tensor with three components on each of the six faces of an infinitesimal cube around that point, as shown in the Figure 3.2.

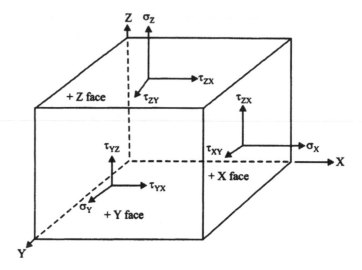

FIGURE 3.2 Stress at a point

Stress components are identified by two subscripts, the first subscript representing the direction of outward normal of the plane or face of the cube and the second subscript representing the direction of stress on that plane. Thus σ_{xx} (or σ_x) represents normal stress on the plane whose outward normal is along +ve X-axis (identified as + X face) while σ_{xy} (or τ_{xy}) and σ_{xz} (or τ_{xz}) represent shear stresses along Y and Z directions on the same plane. Stresses are similarly defined on the other five faces of the cube.

For the equilibrium of this small element along X, Y and Z directions, it is seen that σ_{xx}, σ_{xy} and σ_{xz} on the +X face are equal to σ_{xx}, σ_{xy} and σ_{xz} on the −X face. Similarly on the Y and Z faces. So, out of the 18 stress components on the 6 faces of the cube, only the following 9 stress components are considered independent. These are represented as

$$\sigma_x \quad \tau_{xy} \quad \tau_{xz} \qquad \tau_{yx} \quad \sigma_y \quad \tau_{yz} \qquad \tau_{zx} \quad \tau_{zy} \quad \sigma_z$$

Also, for the equilibrium of the element for moments about X, Y and Z directions, $\tau_{xy} = \tau_{yx}$, $\tau_{xz} = \tau_{zx}$ and $\tau_{yz} = \tau_{zy}$ being the complementary shear stresses on two perpendicular faces. Thus, only three normal stress components σ_x, σ_y, σ_z and three shear stress components τ_{xy}, τ_{yz} and τ_{zx} are identified at each point in a component. In the following chapters, these components are written as a stress vector $\{\varepsilon\}$ for convenience of matrix operations.

Normal stress on any plane whose normal is N (with components N_x, N_y and N_z along cartesian x, y and z directions) can be obtained from the stress components available in cartesian coordinate system by the following relationship.

$$\sigma_n = T_X N_X + T_Y N_Y + T_Z N_Z$$

where,
$$T_X = \sigma_X N_X + \tau_{XY} N_Y + \tau_{XZ} N_Z$$

$$T_Y = \tau_{XY} N_X + \sigma_Y N_Y + \tau_{YZ} N_Z$$

$$T_Z = \tau_{XZ} N_X + \tau_{YZ} N_Y + \sigma_Z N_Z$$

In a similar way, strain at a point is defined as a tensor of 18 components, out of which six components consisting of normal strains ε_x, ε_y and ε_z and shear strains γ_{xy}, γ_{yz} and γ_{zx} are considered independent. Normal strain along a direction is defined as the change in length per unit length along that direction while shear strain is defined as the change in the included angle as shown in Fig. 3.3. In the following chapters, these components are written as a strain vector $\{\varepsilon\}$ for convenience of matrix operations.

In the case of 2-D and 3-D elements, general relations between displacements and strains and between strains and stresses, as obtained in the theory of elasticity, are used for calculating element stiffness matrices. These relations are given later with the following notation.

Notation :   u, v, w are the displacements along x, y and z directions

ε_x, ε_y, ε_z are the normal strains

γ_{xy}, γ_{yz}, γ_{zx} are the shear strains

σ_x, σ_y, σ_z are the normal stresses

τ_{xy}, τ_{yz}, τ_{zx} are the shear stresses

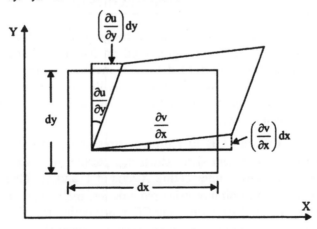

FIGURE 3.3 Shear strain at a point

3.11 PRINCIPAL STRESSES

Depending on the type of structure and the load applied, three normal stresses and three shear stresses can exist at any point in the structure. These stress components are calculated with reference to the coordinate system used. However, there exists a plane along which shear stress is zero and the corresponding normal stresses (maximum and minimum) are called principal normal stresses or principal stresses. Along some direction, inclined to the two principal stresses, there exists maximum or principal shear stress. These are of interest to any designer as the component has to be designed to limit these stresses to the allowable limits of the material. They can be calculated, for a two-dimensional stress state, from

$$\sigma_1 = \frac{\left(\sigma_x + \sigma_y\right)}{2} + \frac{\sqrt{\left(\sigma_x - \sigma_y\right)^2 + 4\tau^2}}{2}$$

$$\sigma_2 = \frac{\left(\sigma_x + \sigma_y\right)}{2} - \frac{\sqrt{\left(\sigma_x - \sigma_y\right)^2 + 4\tau^2}}{2}$$

$$\tau_{max} = \frac{\left(\sigma_1 - \sigma_2\right)}{2} = \frac{\sqrt{\left(\sigma_x - \sigma_y\right)^2 + 4\tau^2}}{2}$$

From uni-axial tensile test, where $\sigma_X = 0$ and $\tau = 0$,

we get $\qquad \sigma_1 = \sigma_{max} = \sigma_Y; \qquad \sigma_2 = 0 \quad$ and $\quad \tau_{max} = \frac{\sigma_Y}{2} = 0.5\,\sigma_{max}$

3.12 MOHR'S CIRCLE FOR REPRESENTATION OF 2-D STRESSES

The principal stresses can also be obtained by graphical method using Mohr's circle. Here, normal stresses are represented on X-axis and shear stresses on Y-axis. Principal or maximum normal stress is inclined to the given stress state by an angle θ, while maximum shear stress is the maximum ordinate of the circle and exists on a plane inclined at $(45 - \theta)°$ from the given stress state or at 45° from the Principal stresses.

Procedure : Plot OA and OB to represent stresses σ_x and σ_y along X-axis. Plot AC and BD, to represent shear stress τ_{xy}, parallel to Y-axis. With CD as diameter, construct a circle intersecting X-axis at F and G. Then OF and OG indicate maximum and minimum principal stresses, inclined at an angle θ (half of angle between EC and EA) with line CD representing the given stress state. Length EH, radius of the circle, indicates maximum shear stress.

It can be observed that sum of normal stresses on any plane is constant and is called 1st stress invariant I_1

i.e., $I_1 = \sigma_x + \sigma_y = \sigma_1 + \sigma_2$ for 2-D stress case

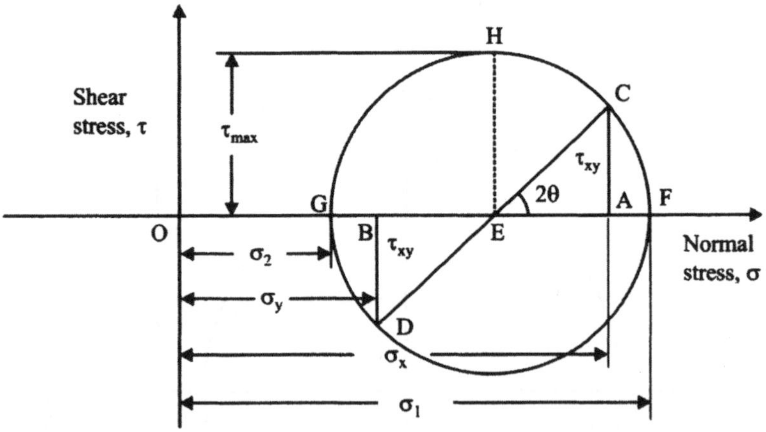

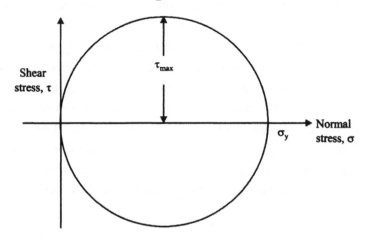

FIGURE 3.4 Mohr's circle for 2-D stress representation

Two special cases of Mohr's circle are of special importance:

(a) For uni-axial tensile test, from which material properties are usually evaluated, the load is applied along one axis (usually Y-axis) and other components of stress are all zero. Then,

$\sigma_1 = \sigma_y$; $\sigma_2 = 0$ and $\tau_{max} = \dfrac{\sigma_y}{2}$ as can be seen from the Figure 3.5.

FIGURE 3.5 Mohr's circle for uni-axial tensile test

(b) In the earlier cases, it is seen that normal stress corresponding to maximum shear stress is not zero. If the normal stress, associated with maximum shear stress, is zero, then it is called *pure shear state*. It is possible when a component is subjected to torsion only. In this case of pure shear, on a plane inclined at 45° to the plane of maximum shear stress, radius of the circle is

$\sigma_1 = -\sigma_2 = \tau_{max}$ as can be seen from Fig. 3.6.

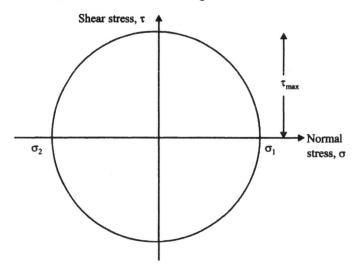

FIGURE 3.6 Mohr's circle for pure shear

This can also be understood by considering stresses acting at a point, identified by a small cube around it, when the component is subjected to torsion. Torsion is represented by a couple formed by two equal and opposite shear forces acting on opposite faces of the cube, as shown in Fig. 3.7 (a). An equal and opposite couple is automatically formed, if the component is in static equilibrium, as shown in Fig. 3.7 (b). The shear forces resulting from this couple are called *complementary shear forces (stresses)*. If two different free bodies of half this cube about its diagonals are considered, resultant of the two shear forces on adjacent surfaces will result in tensile stress on one diagonal (BD) and a compressive stress on the other diagonal (AC), as shown in Fig. 3.7 (c).

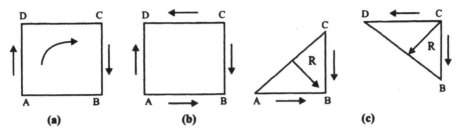

FIGURE 3.7 Representation of shear, complementary shear and principal normal stresses

If 'q' is the shear stress on the face of the cube and 's' is the side of the cube, then

Resultant force on the diagonal, $R = 2.q.s^2.\cos\theta$.

Stress on this diagonal, $\sigma = \dfrac{R}{A} = \dfrac{2.q.s^2.\cos\theta}{2.s^2.\cos\theta} = q$

Thus, both these stresses (maximum or principal normal stresses) are equal in magnitude to the shear stress on the surfaces and are inclined at $45|°$ to the shear stresses, as already seen in the corresponding Mohr's circle.

Maximum shear stress theory, the most conservative and commonly used theory of failure, suggests that a component fails when the maximum shear stress at any point in a component exceeds the allowable maximum shear stress value of the material.

3.13 VONMISES STRESS

VonMises stress or equivalent stress is related to the three principal stresses at any point. It is used in Maximum distortion energy theory (which states that a component fails only due to distortion in shape and is independent of volumetric expansion or contraction). Equivalent stress is given by

$$\sigma_{eq} = \sqrt{\frac{[(\sigma_1 - \sigma_2)^2 + (\sigma_2 - \sigma_3)^2 + (\sigma_3 - \sigma_1)^2]}{2}} \le \sigma_y$$

It is also represented in terms of 1st and 2nd stress invariants (I_1 and I_2) as

$$\sigma_{eq} = \sqrt{I_1^2 - 3I_2}$$

where, $I_1 = \sigma_x + \sigma_y + \sigma_z = \sigma_1 + \sigma_2 + \sigma_3$

and $I_2 = \sigma_x\sigma_y + \sigma_y\sigma_z + \sigma_z\sigma_x - \tau_{xy}^2 - \tau_{yz}^2 - \tau_{zx}^2 = \sigma_1\sigma_2 + \sigma_2\sigma_3 + \sigma_3\sigma_1$

In 2-D *plane stress case*, $\sigma_z = 0$ and $\sigma_3 = 0$

$$I_1 = \sigma_x + \sigma_y = \sigma_1 + \sigma_2$$

$$I_2 = \sigma_x \sigma_y - \tau_{xy}^2 = \sigma_1 \sigma_2$$

In 2-D *plane strain case*, $\sigma_z = \nu (\sigma_x + \sigma_y)$

$$I_1 = \sigma_x + \sigma_y + \sigma_z = \sigma_1 + \sigma_2 + \sigma_3$$

and

$$I_2 = \sigma_x \sigma_y + \sigma_y \sigma_z + \sigma_z \sigma_x - \tau_{xy}^2 = \sigma_1 \sigma_2 + \sigma_2 \sigma_3 + \sigma_3 \sigma_1$$

Comparing distortion energy of specimen in uni-axial tensile test,

where $\sigma_1 = \sigma_Y$; $\sigma_2 = \sigma_3 = 0$,

with pure shear state where $\tau_{max} = \sigma_1 = \sigma_2$,

we get, $\tau_{max} \le \dfrac{\sigma_Y}{\sqrt{3}} = 0.577 \; \sigma_Y$

This stress value is used in the ***distortion energy theory of failure*** and is very popular.

3.14 THEORY OF ELASTICITY

The definitions of stress, strain and the relationships between displacement, strain and stress are explained in Strength of materials with special reference to 1-D structures. Before analysing continuum structures, a more general understanding of these concepts is essential. In the following, a brief discussion of these concepts is presented. Reader is advised to go through any book on theory of elasticity for a more detailed presentation of these concepts.

(a)   Poisson's ratio

Normal strain (tensile or compressive) due to an applied load along the direction of load is called longitudinal strain. In most engineering materials, increase of size in one direction is associated with reduction of size in the other two directions, to minimise change in the volume of the component. Thus tensile longitudinal strain is associated with compressive lateral strains and vice versa. Ratio of lateral strain to longitudinal strain is found to be a constant for each material and is called Poisson's ratio (usually represented by ν or $1/m$). With the usual notation of tensile strains as +ve and compressive strains as −ve, this ratio is always negative. However, this ratio is given a positive value and the negative effect is taken care of in the corresponding equations. Its value ranges theoretically from 0 to 1, while it varies from $1/3$ to $1/4$ for most of the engineering materials. $\nu = \frac{1}{2}$ indicates a perfectly plastic material.

Poisson's ratio, ν = Lateral strain / Longitudinal strain

(b) **Rigidity modulus, Bulk modulus**

Shear modulus or Rigidity modulus, G = shear stress / shear strain

and Bulk modulus, K = Normal stress / Volumetric strain

where,

$$\text{volumetric strain} = \frac{\delta v}{v} \simeq \varepsilon_x + \varepsilon_y + \varepsilon_z$$

(neglecting higher order small terms)

$$= \frac{\left(\sigma_x - v\,\sigma_y - v\,\sigma_z\right)}{E} + \frac{\left(\sigma_y - v\,\sigma_z - v\,\sigma_x\right)}{E} + \frac{\left(\sigma_z - v\,\sigma_x - v\,\sigma_y\right)}{E}$$

$$= \left(\sigma_x + \sigma_y + \sigma_z\right)\frac{\left(1 - 2v\right)}{E}$$

when same load is acting along x, y and z directions

$$= \sigma_x \frac{\left(1 - 2v\right)}{E} \text{ when load is acting along x direction only}$$

Rigidity modulus and bulk modulus also have the same units as stress (N/m^2 or Pa)

Bulk modulus has very limited applications in structural analysis. For any given material, modulus of elasticity and modulus of rigidity reduce at higher temperatures.

These material constants (modulii) are mutually related by the following expression

$$E = 2G\,(1 + v) = 3K\,(1 - 2v)$$

(c) **Strain-Displacement Relations**

Strains can also be expressed as functions of displacement components at a point in the three Cartesian coordinate directions. If u, v and w (all functions of location of point, represented by its X, Y, Z coordinates) and represent the displacement components along X, Y and Z directions, then

$$\varepsilon_x = \frac{\partial u}{\partial x} \qquad \varepsilon_y = \frac{\partial v}{\partial y} \qquad \varepsilon_z = \frac{\partial w}{\partial z}$$

$$\gamma_{xy} = \frac{\partial u}{\partial y} + \frac{\partial v}{\partial x} \qquad \gamma_{yz} = \frac{\partial v}{\partial z} + \frac{\partial w}{\partial y} \qquad \gamma_{zx} = \frac{\partial w}{\partial x} + \frac{\partial u}{\partial z} \qquad .....(3.1)$$

(d) **Thermal stress**

Thermal strains do not induce any stresses unless thermal expansion is constrained.

For example stress in a uniform bar subjected to temperature rise by ΔT, is dependent on the end condition as shown below. Let α be the coefficient of linear thermal expansion. Total elongation of a bar of length L due to increase in its temperature by ΔT is $L \propto \Delta T$. Then, stress in the bar depends on the constraint (boundary condition) for its expansion, as shown Fig. 3.8. In each case,

$$\text{stress } \sigma = E.\varepsilon = \left(\frac{E}{L}\right) \times \text{Restrained part of expansion.}$$

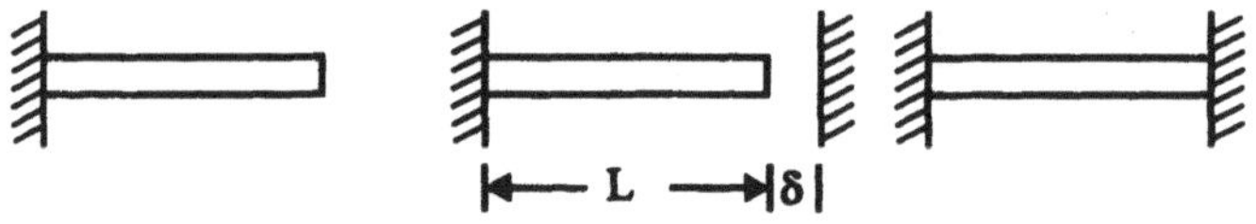

Case (a) : Unconstrained	Case (b) : Partially constrained $(\delta < L.\alpha.\Delta T)$	Case (c) : Fully constrained
Stress, $\sigma = 0$	$\sigma = \dfrac{E(L.\alpha.\Delta T - \delta)}{L}$	$\sigma = E.\alpha.\Delta T$

FIGURE 3.8

(e)   **Stress-strain Relations** or Constitutive equations (from generalised Hooke's law)

• For linearly elastic and isotropic material

$$\varepsilon_x = \frac{\left(\sigma_x - \nu\sigma_y - \nu\sigma_z\right)}{E}; \qquad \gamma_{xy} = \frac{\tau_{xy}}{G}$$

$$\varepsilon_y = \frac{\left(-\nu\sigma_x + \sigma_y - \nu\sigma_z\right)}{E}; \qquad \gamma_{yz} = \frac{\tau_{yz}}{G} \qquad .....(3.2)$$

$$\varepsilon_z = \frac{\left(-\nu\sigma_x - \nu\sigma_y + \sigma_z\right)}{E}; \qquad \gamma_{zx} = \frac{\tau_{zx}}{G}$$

where $G = \dfrac{E}{2(1+\nu)}$ is the shear modulus or rigidity modulus

Sum of the three equations gives,

$$\varepsilon_x + \varepsilon_y + \varepsilon_z = (1 - 2\nu)\left(\frac{\sigma_x + \sigma_y + \sigma_z}{E}\right)$$

These equations can also be written in terms of stresses as functions of strains.

$$\sigma_x = \frac{E[(1-v)\varepsilon_x + v\varepsilon_y + v\varepsilon_z]}{(1+v)(1-2v)} \quad ; \quad \tau_{xy} = \frac{E\gamma_{xy}}{2(1+v)}$$

$$\sigma_y = \frac{E[vv_x + (1-v)\varepsilon_y + v\varepsilon_z]}{(1+v)(1-2v)} \quad ; \quad \tau_{yz} = \frac{E\gamma_{yz}}{2(1+v)}$$

$$\sigma_z = \frac{E[v[_x + v\varepsilon_y + (1-v)\varepsilon_z]}{(1+v)(1-2v)} \quad ; \quad \tau_{zx} = \frac{E}{2(1+v)}\gamma_{zx} \quad .....(3.3)$$

These equations are expressed more conveniently in matrix notation as

$$\{\sigma\} = [\,D\,]\,\{\varepsilon\} \qquad\qquad .....(3.4)$$

In general, $\{\sigma\} = [D]\,(\{\varepsilon\} - \{\varepsilon_0\})$ \qquad\qquad .....(3.5)

where, $\{\varepsilon_0\} = [\alpha\Delta T,\ \alpha\Delta T,\ \alpha\Delta T,\ 0,\ 0,\ 0]^T$ is the initial or stress-free strain vector

and ΔT is the change in temperature of the component

since thermal expansion produces only normal strain (with no shear strain and no Poisson's effect) and thermal strains do not induce any stresses unless thermal expansion is constrained

The 3-D stress-strain relations are simplified below for 1-D and 2-D cases.

1-D Case :

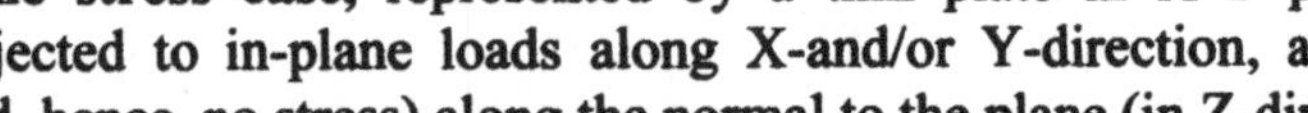

FIGURE 3.9

$$\sigma = E\,\varepsilon \qquad\qquad .....(3.6)$$

2-D cases :

(i) **Plane stress case,** represented by a thin plate in X-Y plane, plane subjected to in-plane loads along X-and/or Y-direction, and no load (and, hence, no stress) along the normal to the plane (in Z-direction).

(a) Subjected to no load (and, hence, no stress) along

i.e., $(\sigma_z = 0,\ \varepsilon_z \neq 0)$ \qquad\qquad .....(3.7)

$$[D] = \frac{E}{(1-v^2)}\begin{bmatrix} 1 & v & 0 \\ v & 1 & 0 \\ 0 & 0 & \dfrac{(1-v)}{2} \end{bmatrix} \qquad\qquad .....(3.8)$$

$$\{\varepsilon_0\} = [\,\alpha\Delta T \quad \alpha\Delta T \quad 0\,]^T \qquad\qquad .....(3.9)$$

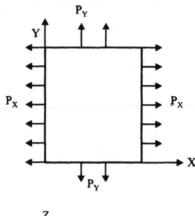

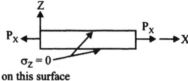

$\sigma_z = 0$ —
on this surface

FIGURE 3.10

(ii) **Plane strain case,** represented by a thin plate in X-Y plane, which is constrained along the normal to the plane in Z-direction (i.e., no strain along the normal). It is an subjected to in-plane loads along X-and/or Y-directions, and no load (and, hence, no stress) along the normal to the which is constrained along the normal to the plane, has no strain along the normal. It is an approximation of a 3-dimensional solid of vary large dimension along Z compared to its dimensions along X and Y and loaded in X-Y plane. Example : A hydro dam between two hills, which can be considered as a set of slices or plates in the flow direction (X). Each slice in X-Y plane is modeled by plane strain elements.

Here, $(\sigma_z \neq 0, \varepsilon_z = 0)$ .....(3.10)

$$[D] = \frac{E}{(1+v)(1-2v)} \begin{bmatrix} 1-v & v & 0 \\ v & 1-v & 0 \\ 0 & 0 & \dfrac{(1-2v)}{2} \end{bmatrix}$$ .....(3.11)

and $\{\varepsilon_0\} = (1+v) \begin{bmatrix} \alpha\Delta T & \alpha\Delta T & 0 \end{bmatrix}^T$ .....(3.12)

This is obtained by using $\varepsilon_z = \dfrac{\left(-v\sigma_x - v\sigma_y + \sigma_z\right)}{E} = 0$ to represent σ_z in terms of σ_x and σ_y as $\sigma_z = v(\sigma_X - \sigma_Y)$ and substituting for σ_z in the relations for ε_x and ε_y.

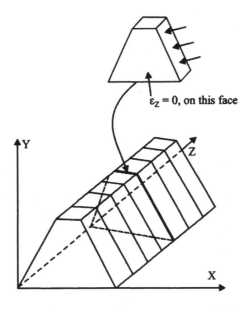

$\varepsilon_z = 0$, on this face

FIGURE 3.11

Another example of plane strain is a segment of an axi-symmetric solid which is self constrained in the circumferential or hoop direction.

- **Orthotropic materials**

3-D case

Eq. 3.2 is modified to account for variation of E and ν along material axis (1, 2,3) as

$$\begin{Bmatrix} \varepsilon_1 \\ \varepsilon_2 \\ \varepsilon_3 \\ \gamma_{12} \\ \gamma_{23} \\ \gamma_{31} \end{Bmatrix} = \begin{bmatrix} \dfrac{1}{E_1} & \dfrac{-\nu_{21}}{E_2} & \dfrac{-\nu_{31}}{E_3} & 0 & 0 & 0 \\[2mm] \dfrac{-\nu_{12}}{E_1} & \dfrac{1}{E_2} & \dfrac{-\nu_{32}}{E_3} & 0 & 0 & 0 \\[2mm] \dfrac{-\nu_{13}}{E_1} & \dfrac{-\nu_{23}}{E_2} & \dfrac{1}{E_3} & 0 & 0 & 0 \\[2mm] 0 & 0 & 0 & \dfrac{1}{G_{12}} & 0 & 0 \\[2mm] 0 & 0 & 0 & 0 & \dfrac{1}{G_{23}} & 0 \\[2mm] 0 & 0 & 0 & 0 & 0 & \dfrac{1}{G_{31}} \end{bmatrix} \begin{Bmatrix} \sigma_1 \\ \sigma_2 \\ \sigma_3 \\ \tau_{12} \\ \tau_{23} \\ \tau_{31} \end{Bmatrix} \quad .....(3.13)$$

where, $E_1 \nu_{21} = E_2 \nu_{12}$; $E_2 \nu_{32} = E_3 \nu_{23}$ and $E_3 \nu_{13} = E_1 \nu_{31}$

2-D Plane stress case

Eq. (3.8) is modified to account for variation of E and ν along material axes (1-2) as

$$[D] = \frac{1}{(1 - v_{12}v_{21})} \begin{bmatrix} E_1 & v_{21}E_1 & 0 \\ v_{12}E_2 & E_2 & 0 \\ 0 & 0 & G(1 - v_{12}v_{21}) \end{bmatrix} \qquad .....(3.14)$$

where, $v_{12}E_2 = v_{21}E_1$

When an orthotropic plate is loaded parallel to its material axes, it results only in normal strains. If the material axes (1, 2) are oriented at an angle θ w.r.t. global (x,y) axes, then

$$[D] = [T]^T[D][T] \qquad\qquad .....(3.15)$$

where

$$[T] = \begin{bmatrix} \cos^2\theta & \sin^2\theta & 2\sin\theta\cos\theta \\ \sin^2\theta & \cos^2\theta & -2\sin\theta\cos\theta \\ 2\sin\theta\cos\theta & -2\sin\theta\cos\theta & \cos^2\theta - \sin^2\theta \end{bmatrix} \qquad .....(3.16)$$

(f)   Compatibility equations

While in 1-D elements, only one stress component and one strain component are used in the strain energy calculation, stresses and strains in 2-D and 3-D elements have more components due to the effect of Poisson's ratio even for a simple loading. While six strain components can be obtained from the three displacement components by partial differentiation, the reverse requirement of *calculating three displacement components is possible only when the six strain components are inter-related* through the following three additional conditions, called compatibility equations.

$$\frac{\partial^2\varepsilon_x}{\partial y^2} + \frac{\partial^2\varepsilon_y}{\partial x^2} = \frac{\partial^2\gamma_{xy}}{\partial x \partial y}$$

$$\frac{\partial^2\varepsilon_y}{\partial z^2} + \frac{\partial^2\varepsilon_z}{\partial y^2} = \frac{\partial^2\gamma_{yz}}{\partial y \partial z}$$

$$\frac{\partial^2\varepsilon_z}{\partial x^2} + \frac{\partial^2\varepsilon_x}{\partial z^2} = \frac{\partial^2\gamma_{zx}}{\partial z \partial x} \qquad .....(3.17)$$

(g)   Equilibrium equations

Stress at a point in a component is described by the stress tensor – one normal stress component and two shear stress components on each of the six faces of a cube around that point. For equilibrium of this cube, these eighteen stress components should satisfy the following equilibrium conditions, where F_x, F_y and F_z are the forces acting on the cube along X, Y, and Z axes.

$$\frac{\partial \sigma_x}{\partial x} + \frac{\partial \tau_{xy}}{\partial y} = \frac{\partial \tau_{xz}}{\partial z} = F_x; \qquad \tau_{xy} = \tau_{yx}$$

$$\frac{\partial \tau_{yx}}{\partial x} + \frac{\partial \sigma_y}{\partial y} + \frac{\partial \tau_{yz}}{\partial z} = F_y; \qquad \tau_{yz} = \tau_{zy} \qquad \qquad .....(3.18)$$

$$\frac{\partial \tau_{zx}}{\partial x} + \frac{\partial \tau_{zy}}{\partial y} + \frac{\partial \sigma_z}{\partial z} = F_z; \qquad \tau_{zx} = \tau_{xz}$$

A few problems are included here, based on the equations presented above.

Example 3.1

If a displacement field is described by $u = -5x^2 - 6xy + 3y^2$; $v = 4x^2 + 9xy + 4y^2$ determine ε_x, ε_y and γ_{xy} at point $x = 1$ and $y = -1$

Solution

$$\varepsilon_x = \frac{\partial u}{\partial x} = -10x - 6y + 0 = -10 \times 1 - (-6)(-1) = -4$$

$$\varepsilon_y = \frac{\partial v}{\partial y} = 0 + 9x + 8y = 9 \times 1 + 8(-1) = +1$$

$$\gamma_{xy} = \frac{\partial u}{\partial y} + \frac{\partial v}{\partial x} = (0 - 6x + 6y) + (8x + 9y + 0) = 2x + 15y$$

$$= 2 \times 1 + 15(-1) = -13$$

Example 3.2

If a displacement field is described by $u = -4x^2 - 12xy + 3y^2$;

$v = -3x^2 + 4xy + 2y^2$ determine ε_x, ε_y and γ_{xy} at point $x = 1$ and $y = -1$

Solution

$$\varepsilon_x = \frac{\partial u}{\partial x} = -8x - 12y + 0 = -8 \times 1 - 12 \times (-1) = +4$$

$$\varepsilon_y = \frac{\partial v}{\partial y} = 0 + 4x + 4y = 4 \times 1 + 4 \times (-1) = 0$$

$$\gamma_{xy} = \frac{\partial u}{\partial y} + \frac{\partial v}{\partial x} = (0 - 12x + 6y) + (6x + 4y + 0)$$

$$= -6x + 10y = -6 \times 1 + 10(-1) = -16$$

Example 3.3

If a displacement field is described by $u = 4x^2 - 4xy + 6y^2$; $v = -2x^2 - 8xy + 3y^2$ determine ε_x, ε_y and γ_{xy} at point $x = -1$ and $y = -1$

Solution

$$\varepsilon_x = \frac{\partial u}{\partial x} = 8x - 4y + 0 = 8(-1) - 4(-1) = -4$$

$$\varepsilon_y = \frac{\partial v}{\partial y} = 0 - 8x + 6y = -8 \times (-1) + 6 \times (-1) = +2$$

$$\gamma_{xy} = \frac{\partial u}{\partial y} + \frac{\partial v}{\partial x} = (0 - 4x + 12y) + (4x - 8y + 0) = 4y = 4(-1) = -4$$

Example 3.4

If a displacement field is described by $u = 3x^2 - 2xy + 6y^2$; $v = 4x^2 + 6xy - 8y^2$ determine ε_x, ε_y and γ_{xy} at point $x = -1$ and $y = 1$

Solution

$$\varepsilon_x = \frac{\partial u}{\partial x} = 6x - 2y + 0 = 6(-1) - 2 \times 1 = -8$$

$$\varepsilon_y = \frac{\partial v}{\partial y} = 0 + 6x - 16y = 6(-1) - 16 \times 1 = -22$$

$$\gamma_{xy} = \frac{\partial u}{\partial y} + \frac{\partial v}{\partial x} = (0 - 2x + 12y) + (8x + 6y - 0) = 6x + 18y$$

$$= 6(-1) + 18 \times 1 = 12$$

Example 3.5

If $u = 2x^2 + 3y$, $v = 3y + y^3$ find the normal and shear strains

Solution

Normal strains are

$$\varepsilon_x = \frac{\partial u}{\partial x} = 4x + 0 = 4x$$

$$\varepsilon_y = \frac{\partial v}{\partial y} = 3 + 3y^2 = 3(1 + y^2)$$

Shear strain is

$$\gamma_{xy} = \frac{\partial u}{\partial y} + \frac{\partial v}{\partial x} = (0 + 3) + (0 + 0) = 3$$

Example 3.6

A long rod is subjected to loading and a temperature increase of 30 °C. The total strain at a point is measured to be 1.2×10^{-5}. If E = 200 GPa and

$\alpha = 12 \times 10^{-6}/$ °C, determine the stress at the point.

Solution

$$\sigma = E\,(\varepsilon_{Total} - \varepsilon_{Thermal}) = E\,(\varepsilon_{Total} - \alpha.\Delta T)$$

$$= 200 \times 10^9\,(1.2 \times 10^{-5} - 12 \times 10^{-6} \times 30) = 200 \times 10^9\,(-34.8 \times 10^{-5})$$

$$= -69.6 \times 10^6\ \text{N/m}^2\ \text{or} - 69.6\ \text{N/mm}^2$$

Example 3.7

Consider the rod shown in Fig. 3.12 where the strain at any point x is given by $\varepsilon_x = 1 + 2x^2$. Find the tip displacement δ.

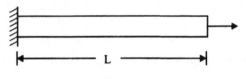

FIGURE 3.12

Solution

$$\delta_L = \int \varepsilon_x \, dx = \int_0^L \left(1 + 2x^2\right) dx$$

$$= \left[\frac{x + 2x^3}{3} \right]_0^L = \frac{L + 2L^3}{3}$$

Example 3.8

In a plane strain problem, if $\sigma_x = 150$ N/mm^2, $\sigma_y = -100$ N/mm^2, $E = 200$ KN/mm^2 and $v = 0.3$, determine the value of stress σ_z

Solution

In a plane strain problem, strain $\varepsilon_z = \dfrac{\left(-v\sigma_x - v\sigma_y + \sigma_z\right)}{E} = 0$

Therefore, $\sigma_z = v\sigma_x + v\,\sigma_y = 0.3 \times 150 + 0.3 \times (-100) = 15$ N/mm^2

Note that the value of E is not required, but given only to mislead students.

3.15 SUMMARY

- Based on the relative dimensions, a component may be idealised by 1-D, 2-D or 3-D elements. At every point in a component, stress and strain are expressed by a normal component and two shear components on each of the six faces of a small cube around that point. Out of 18 such vector components, from equilibrium considerations, only three normal components and three shear components in any orthogonal coordinate system, are independent and are represented by a column matrix of order (n × 1). Combined effect of these various components will be different on planes, inclined to the coordinate axes. Principal stresses are the maximum values on any such plane. These values are significant while designing a component and can be also obtained by graphical method, called Mohr's circle.

- The flexibility of movement of any point in the component is identified by degrees of freedom (DOF). With reference to the orthogonal coordinate system, the every point in a component can have 1-6 DOFs, covering 3 translations along the three axes and 3 rotations about the three axes.

- Components, which can be modeled by a combination of 1-D elements, are also called discrete structures. These include truss element for axial loads (1 DOF/node, along its axis), beam element for bending loads (2 DOFs/node in each of its two planes of bending, deflection and slope normal to axis) and torsion element for torque load (1 DOF/node, rotation about its axis). These four modes of deformation are mutually independent and are called uncoupled DOF. A generalised beam element has 6 DOFs/node and is subjected to all these loads.

- Different elements of a discrete structure are joined at their end points only and hence only modulus of elasticity of the material is relevant (Poisson's ratio is not relevant in their analysis). Normal stress and normal strain are related by modulus of elasticity (E) while shear stress and shear strain are related by modulus of rigidity (G).

- Components, which are modeled by 2-D or 3-D elements, are called continuum structures. These elements have surface contact along the common boundary and hence Poisson's ratio is also relevant. Isotropic and homogeneous material is commonly used, within its linear elastic range. Stress strain relationship is a matrix of order 3×3 for 2-D elements and 6×6 for 3-D elements. 2-D elements may be used in plane stress condition or plane strain condition, depending on the particular component.

- 2-D and 3-D elements may be used with translational DOFs only (2-D plane stress, 2-D plane strain or 3-D thick solid) or along with rotational degrees of freedom (2-D plate bending, 2-D thin shell and 3-D thick shell) depending on the nature of applied loads.

CHAPTER **4**

DISCRETE (1-D) ELEMENTS

A discrete structure is assembled from a number of easily identifiable 1-D elements like spars, beams. Nodes are chosen at the junctions of two or more discrete members, at junctions of two different materials, at points of change of cross section or at points of load application. In the 1-D element, axial dimension is very large compared to the cross section and load is assumed to act uniformly over the entire cross-section. So, the displacement is taken as a function of x, along the axis of the member. Stress and strain are also uniform over the entire cross section. The solution obtained in most of these cases, is exact.

4.1 DEGREES OF FREEDOM OF DIFFERENT ELEMENTS

Based on the relative dimensions of the element, the individual elements can be broadly classified as 1-D, 2-D and 3-D elements. The load-displacement relationships of these elements depend on the nature of loads (axial/in-plane loads, torsion or bending loads) and are calculated using variational principle. Some such elements and their degrees of freedom at each node in element (or local) coordinate system are given below.

	Axial/In-plane loads	**Bending (Normal loads and/or moments)**
1-D	Spar or Truss (1 DOF/node)	Beam (2 DOF/node for bending in 1-plane)
2-D	Plane stress/Plane strain/	Plate bending (3 DOF/node)
	Axisymmetric (2 DOF/node)	Thin shell (6 DOF/node)
3-D	Solid (3 DOF/node)	Thick shell (6 DOF/node)

85

4.2 CALCULATION OF STIFFNESS MATRIX BY DIRECT METHOD

1-D elements are broadly classified based on the load applied on them as spar or truss element for axial load, torsion element for torque load and beam element for bending in one or two planes through neutral axis/plane. Stiffness matrix of each of these elements can be derived based on the well known relations in strength of materials.

(a) Truss element

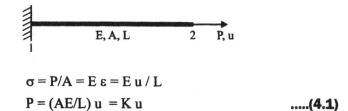

$$\sigma = P/A = E\,\varepsilon = E\,u\,/\,L$$

or $P = (AE/L)\,u\ = K\,u$ .....(4.1)

is the familiar relation for axial load carrying element fixed at one end and load applied at the free end with displacement 'u' at the free end 2.

In general, if loads P_1 and P_2 are applied at the two ends of an element

resulting in displacements u_1 and u_2 at these two ends, stress is proportional to $(u_2 - u_1)$

Then, $-P_1 = P_2 = \left(\dfrac{AE}{L}\right)(u_2 - u_1)$ .....(4.2)

In matrix notation, $\begin{Bmatrix} P_1 \\ P_2 \end{Bmatrix} = [K]\begin{Bmatrix} u_1 \\ u_2 \end{Bmatrix}$

where $[K] = \left(\dfrac{AE}{L}\right)\begin{bmatrix} 1 & -1 \\ -1 & 1 \end{bmatrix}$ .....(4.3)

(b) Torsion element

From strength of materials, $\quad \dfrac{T}{J} = \dfrac{G\theta}{L}$

$$\text{or} \quad T = \left(\dfrac{GJ}{L}\right)\theta = K\,\theta$$

is the familiar relation for torsion load carrying element fixed at one end and torque applied at the free end with rotation 'θ' at the free end 2.

For a general element on which torque loads T_1 and T_2 are applied at the two ends and corresponding rotations are θ_1 and θ_2, torque is proportional to $(\theta_2 - \theta_1)$.

Then, $\qquad\qquad\qquad -T_1 = T_2 = (GJ/L)\ (\theta_2 - \theta_1)$

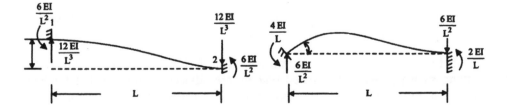

In matrix notation, $\quad \begin{Bmatrix} T_1 \\ T_2 \end{Bmatrix} = [K]\begin{Bmatrix} \theta_1 \\ \theta_2 \end{Bmatrix}$

where, $\qquad\qquad [K] = \left(\dfrac{GJ}{L}\right)\begin{bmatrix} 1 & -1 \\ -1 & 1 \end{bmatrix}$ $\qquad$ .....(4.4)

(c)   Beam element

From simple beam theory, forces and moments required at the two ends of a beam in X-Y plane to give,

(i) $v_1 = 1$ and $v_2 = \theta_1 = \theta_2 = 0$ $\qquad$ and

(ii) $\theta_1 = 1$ and $v_1 = v_2 = \theta_2 = 0$ are given in the figures with upward forces, counterclockwise moments, and corresponding displacements and rotations as +ve.

Similar values can be obtained for the conditions

(i) $v_2 = 1$ and $v_1 = \theta_1 = \theta_2 = 0$ and

(ii) $\theta_2 = 1$ and $v_1 = v_2 = \theta_1 = 0$

Arranging these values in matrix form,

$$\begin{Bmatrix} P_1 \\ M_1 \\ P_2 \\ M_2 \end{Bmatrix} = [K] \begin{Bmatrix} v_1 \\ \theta_1 \\ v_2 \\ \theta_2 \end{Bmatrix}$$

where, $\quad [K] = \dfrac{EI}{L^3} \begin{bmatrix} 12 & 6L & -12 & 6L \\ 6L & 4L^2 & -6L & 2L^2 \\ -12 & -6L & 12 & -6L \\ 6L & 2L^2 & -6L & 4L^2 \end{bmatrix}$

4.3 CALCULATION OF STIFFNESS MATRIX BY VARIATIONAL PRINCIPLE

Stiffness matrix of each element is calculated, using the principle of minimum potential energy which states that "Every component, subjected to some external loads, reaches a stable equilibrium condition when its potential energy is minimum". So, the problem lies in identifying the set of displacements at various points in the component which ensures that the potential energy of the component is minimum.

This is analogous to the problem in *variational calculus* of finding a stationary value y(x) such that the functional (function of functions)

$$I = \int_{x_1}^{x_2} F\left(x, y, \frac{dy}{dx}\right) dx \qquad \qquad .....(4.6)$$

is rendered stationary. Integral I is stationary when its first variation vanishes

i.e., $\quad \delta I = 0$ .....(4.7)

There are many different approaches such as Euler-Lagrange method, for solving such problems. Interested readers are advised to refer to the relevant books in mathematics for a more detailed presentation of the variational calculus.

In *Rayleigh-Ritz method*, which is used in FEM, a mathematical expression in the form of a power series in x is assumed for the unknown function y(x).

Then, eq. (4.7) becomes

$$\frac{\partial I}{\partial a_i} = 0 \qquad \text{for } i = 0, 1, \dots n \qquad \text{.....(4.8)}$$

where $a_0, a_i, \dots a_n$ are the coefficients of the assumed power series.

Finite element method is based on the variational principle where the functional I is the potential energy of the system with nodal displacements as the independent variables y(x) and strains as the functions $\frac{dy}{dx}$ of the independent variables. This method leads to an approximate solution. Potential energy is an *extensive property* i.e. the energy of the entire component is the sum of the energy of its individual sub regions (or elements). Hence eq.(4.7) can be written as

$$\delta I = \Sigma \, \delta I_e = 0$$

Since the number and size of the elements are arbitrary, this relation is satisfied only when

$$\delta I_e = 0 \qquad \text{.....(4.9)}$$

In using this method, y(x) must be kinematically admissible i.e., y(x) must be selected so that it satisfies the displacement boundary conditions prescribed for the problem. Choice of a function for the entire component satisfying this condition becomes difficult for complex problems. Finite element method overcomes this difficulty by relating the primary unknown function to the individual element, rather than to the total problem. Hence, geometry of the overall component and the system boundary conditions are of no concern when choosing the function.

For the individual element,

$$\delta I_e = \delta W_{ext} - \delta U = \{\partial_e\}^T \{P_e\} - \int_v \{\delta \varepsilon_e\}^T \{\sigma\} \, dv \qquad \text{.....(4.10)}$$

$$= \{\delta u_e\}^T \left(\{P_e\} - [K_e]\{u_e\} \right) = 0$$

where, $[K_e]$ is the stiffness matrix of the element

Since $\{\delta u_e\}$ represents the arbitrary nodal values of displacements, they can not be identically zero in a loaded component.

$$\therefore \qquad \{P_e\} - [K_e]\{u_e\} = 0 \quad \text{or} \quad \{P_e\} = [K_e]\{u_e\} \qquad \text{.....(4.11)}$$

In the displacement method, calculation of stiffness matrix for an element starts with an assumed displacement function over the element in each degree of freedom, usually in the form of a polynomial. By substituting nodal coordinate values in these polynomials, the unknown constants in the polynomial can be evaluated in terms of nodal displacements. This condition necessitates choosing displacement polynomial with as many coefficients as the number of nodal DOF for that element. This procedure is explained in more detail in the remaining sections of this chapter.

Displacement function for the element

The function shall be continuous over the entire element with no singularities and easily differentiable to obtain strains for calculation of potential energy. The polynomial should be symmetric in terms of the global coordinate axes, to ensure geometric isotropy.

Strains in the element are obtained as derivatives of the displacement polynomial, and are thus expressed in terms of the nodal displacements. Stresses are expressed in terms of strains, using the appropriate stress-strain relationship, given earlier. By equating work done by the external forces to the change in internal strain energy of the element and applying variational principle, load-displacement relationships of the element in terms of stiffness coefficients are obtained. They represent a system of simultaneous equations in terms of nodal loads and nodal displacements.

By using suitable transformation matrix, this stiffness matrix derived in local coordinate system of the element, is transformed to a *global coordinate system* which is common to all the elements. The stiffness matrices of all elements are then added together such that the stiffness coefficient at a common node is the sum of the stiffness coefficients at that node of all the elements joining at that node. This assembled stiffness matrix is square, symmetric, singular and positive definite. This method is described in detailed here for a truss element.

A truss element is subjected to axial load only and therefore has one degree of freedom (axial displacement) per node. Variation of axial displacement u(x) between the two end nodes is represented by a linear relationship in the form of a polynomial with two constants.

$$\text{Let} \quad u(x) = a_1 + a_2 x = \begin{bmatrix} 1 & x \end{bmatrix} \begin{Bmatrix} a_1 \\ a_2 \end{Bmatrix} = \{f(x)\}^T \{a\} \qquad .....(4.12)$$

Choosing node I as the origin of local coordinate system for this element with X-axis along the axis of the element and substituting the values of x for the two end points of the truss element, (x = 0 at node i and x = L at node j), nodal displacement vector $\{u_e\}$ or $[u_i \ u_j]^T$ can be written as

$$\begin{Bmatrix} u_i \\ u_j \end{Bmatrix} = \begin{bmatrix} 1 & 0 \\ 1 & L \end{bmatrix} \begin{Bmatrix} a_1 \\ a_2 \end{Bmatrix}$$

or $\qquad \{u_e\} = [G]\{a\}$ and $\{a\} = [G]^{-1}\{u_e\}$ $\qquad$ .....(4.13)

Solving for the coefficients $\{a\}$ from eq. (4.13) and substituting in eq.(4.12),

$$u(x) = \{f(x)^T\}\,[G]^{-1}\{u_e\}$$

$$= \begin{bmatrix} 1 & x \end{bmatrix} \begin{bmatrix} 1 & 0 \\ \dfrac{-1}{L} & \dfrac{1}{L} \end{bmatrix} \begin{Bmatrix} u_i \\ u_j \end{Bmatrix}$$

$$= \begin{bmatrix} 1 - \dfrac{x}{L} & \dfrac{x}{L} \end{bmatrix} \{u_e\} = [N]^T\{u_e\} \qquad \text{.....(4.14)}$$

Strain, $\quad \{\varepsilon\} = \dfrac{du}{dx} = \{f'(x)\}^T [G]^{-1}\{u_e\} = [B]\{u_e\}$ $\qquad$ .....(4.15)

where, $\quad [B] = \{f'(x)\}^T [G]^{-1} = \begin{bmatrix} 0 & 1 \end{bmatrix} \begin{bmatrix} 1 & 0 \\ \dfrac{-1}{L} & \dfrac{1}{L} \end{bmatrix} = \begin{bmatrix} -1/L & 1/L \end{bmatrix}$ $\quad$ .....(4.16)

and $\qquad \{\delta\varepsilon\} = [B]\{\delta u_e\}; \qquad \{\delta\varepsilon\}^T = \{\delta u_e\}^T [B]^T$

Stress, $\qquad \{\sigma\} = [D]\{\varepsilon\} = [D][B]\{u_e\}$.

Here [D] = E, since only one axial stress component is relevant for a 1-D truss element

From eq. (4.10),

$$\{\delta u_e\}^T \{P_e\} - \int_v \{\delta\varepsilon_e\}^T \{\sigma\}\,dv = 0$$

$$\{\delta u_e\}^T \{P_e\} - \int_v \{\delta u_e\}^T [B]^T [D][B]\{u_e\}\,dv = 0$$

$$\{\delta u_e\}^T \left[\{P_e\} - \int_v \{B\}^T [D][B]\,dV \{u_e\} \right] = 0$$

$$\{\delta u_e\}^T \left(\{P_e\} - [K_e]\{u_e\} \right) = 0$$

Since $\{\delta u_e\}$ cannot be zero,

$$\{P_e\} = [K_e]\{u_e\} \qquad .....(4.17)$$

where, $$[K_e] = \int_v [B]^T[D][B]dv = \iiint [B]^T E[B]dx\,dy\,dz$$

$$= AE \int_v [B]^T[B]dx$$

since, $[B]$ is not a function of y or z and $\iint dy\,dz = A$

$$= AE \int_0^L \left\{ \begin{array}{c} \dfrac{-1}{L} \\ \dfrac{1}{L} \end{array} \right\} \left[\dfrac{-1}{L} \quad \dfrac{1}{L} \right] dx = \dfrac{AE}{L} \left[\begin{array}{cc} 1 & -1 \\ -1 & 1 \end{array} \right] \qquad ..(4.18)$$

4.4 TRANSFORMATION MATRIX

Stiffness matrix and load vector of any element are initially derived in the *local coordinate system*, with its x-axis along the element, and can vary from one element to another. A *global coordinate system* is common to all the elements. If different elements have different local coordinate systems, stiffness coefficients relating nodal load vector and nodal displacement vector can not be combined together unless directions of load and displacements of different elements joining at a common node coincide i.e., sum of two vectors is equal to their algebraic sum only when the vectors are collinear. If the local coordinate system of an element is inclined to the global coordinate system at an angle θ, then transformation of load vector, displacement vector and the stiffness matrix are to be carried out before they are assembled with other elements.

For a truss element, if P_i, P_j, u_i and u_j represent axial load and displacement values in the local or element coordinate system at nodes i and j and $[K_e]$ is the 2 × 2 stiffness matrix of the element and $(P_x')_i$, $(P_y')_i$, u_i', v_i', $(P_x')_j$, $(P_y')_j$, u_j' and v_j' are the components of axial load and displacement along global x and y axes at nodes i and j then $[K'_e]$, stiffness matrix of the element in the global or structure coordinate system, is derived below.

$$P_i = (P'_x)_i \cos\theta + (P'_y)_i \sin\theta \qquad u_i = u'_i \cos\theta + v'_i \sin\theta$$

$$P_j = (P'_x)_j \cos\theta + (P'_y)_j \sin\theta \qquad u_j = u'_j \cos\theta + v'_j \sin\theta$$

These relations can be expressed in matrix form as

$$
\begin{Bmatrix} P_i \\ P_j \end{Bmatrix} = \begin{bmatrix} \cos\theta & \sin\theta & 0 & 0 \\ 0 & 0 & \cos\theta & \sin\theta \end{bmatrix} \begin{Bmatrix} (p_x')_i \\ (p_y')_i \\ (p_x')_j \\ (p_y')_j \end{Bmatrix}
$$

or $\{P_e\} = [T_e]\{P'_e\}$ and similarly, $\{u_e\} = [T_e]\{u'_e\}$ .....(4.19)

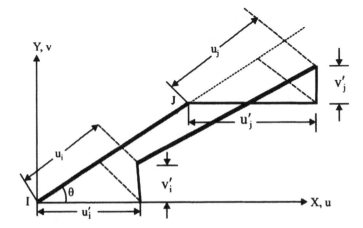

These can also be expressed as

$$\{P'_e\} = [\,T_e\,]^T\{P_e\} \quad \text{and} \quad \{u'_e\} = [\,T_e\,]^T\{u_e\}$$

$\{P_e\} = [K_e]\{u_e\}$ in local coordinate system can now be written in global coordinate system as

$$[T_e]\begin{Bmatrix} P_e{}' \end{Bmatrix} = [K_e][T_e]\begin{Bmatrix} u_e{}' \end{Bmatrix}$$

or $\{P'_e\} = [T_e]^T[K_e][T_e]\{u'_e\} = [K'_e]\{u'_e\}$

we get

$$
\left[K_e{}'\right] = \frac{AE}{L} \begin{bmatrix} \cos^2\theta & \cos\theta\sin\theta & -\cos^2\theta & -\cos\theta\sin\theta \\ \cos\theta\sin\theta & \sin^2\theta & -\cos\theta\sin\theta & -\sin^2\theta \\ -\cos^2\theta & -\cos\theta\sin\theta & \cos^2\theta & \cos\theta\sin\theta \\ -\cos\theta\sin\theta & -\sin^2\theta & \cos\theta\sin\theta & \sin^2\theta \end{bmatrix} \quad .....(4.20)
$$

By substituting $l = \dfrac{(x_2 - x_1)}{L} = \cos\theta$ and $m = \left(\dfrac{y_2 - y_1}{L}\right) = \sin\theta$

where, $L = \sqrt{(x_2 - x_1)^2 + (y_2 - y_1)^2}$

This can also be written in the form

$$\left[K_e' \right] = \frac{AE}{L} \begin{bmatrix} l^2 & lm & -l^2 & -lm \\ lm & m^2 & -lm & -m^2 \\ -l^2 & -lm & l^2 & lm \\ -lm & -m^2 & lm & m^2 \end{bmatrix} \qquad .....(4.21)$$

Transformation of truss element stiffness matrix in 3-D space

For a truss element arbitrarily oriented in 3-D space, a similar transformation matrix can also be derived and the stiffness matrix in 3-D space can be written using

$$l = \frac{(x_2 - x_1)}{L}; \quad m = \frac{(y_2 - y_1)}{L} \text{ and } n = \frac{(z_2 - z_1)}{L}$$

where, $L = \sqrt{(x_2 - x_1)^2 + (y_2 - y_1)^2 + (z_2 - z_1)^2}$

in the form

$$\left[K_e' \right] = \frac{AE}{L} \begin{bmatrix} l^2 & lm & ln & -l^2 & -lm & -ln \\ lm & m^2 & mn & -lm & -m^2 & -mn \\ ln & mn & n^2 & -ln & -mn & -n^2 \\ -l^2 & -lm & -ln & l^2 & lm & ln \\ -lm & -m^2 & -mn & lm & m^2 & mn \\ -ln & -mn & -n^2 & ln & mn & n^2 \end{bmatrix} \qquad .....(4.23)$$

4.5 ASSEMBLING ELEMENT STIFFNESS MATRICES

In a truss having three elements connecting nodes 1-2, 2-3 and 3-1, let the element stiffness matrices (each of 4 × 4) after transformation to global coordinate system be:

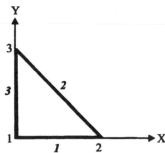

For element-1 joining nodes 1 and 2, the local coordinate system coincides with global coordinate system ($l = 1$, $m = 0$) and, so, displacement components v_1 and v_2 are zero.

Therefore,

$$
\begin{Bmatrix} (P_x)_1 \\ (P_y)_1 \\ (P_x)_2 \\ (P_y)_2 \end{Bmatrix} = \begin{bmatrix} (k_{11})_1 & 0 & (k_{13})_1 & 0 \\ 0 & 0 & 0 & 0 \\ (k_{31})_1 & 0 & (k_{33})_1 & 0 \\ 0 & 0 & 0 & 0 \end{bmatrix} \begin{Bmatrix} u_1 \\ v_1 \\ u_2 \\ v_2 \end{Bmatrix}
$$

For element-2 joining nodes 2 and 3, the member is inclined to the global axes.

Therefore,

$$
\begin{Bmatrix} (P_x)_2 \\ (P_y)_2 \\ (P_x)_3 \\ (P_y)_3 \end{Bmatrix} = \begin{bmatrix} (k_{11})_2 & (k_{12})_2 & (k_{13})_2 & (k_{14})_2 \\ (k_{21})_2 & (k_{22})_2 & (k_{23})_2 & (k_{24})_2 \\ (k_{31})_2 & (k_{32})_2 & (k_{33})_2 & (k_{34})_2 \\ (k_{41})_2 & (k_{42})_2 & (k_{43})_2 & (k_{44})_2 \end{bmatrix} \begin{Bmatrix} u_2 \\ v_2 \\ u_3 \\ v_3 \end{Bmatrix}
$$

For element-3 joining nodes 1 and 3, the local coordinate system is perpendicular to the global coordinate system ($l = 0$, $m = 1$) and so displacement components u_1 and u_3 are zero.

$$
\begin{Bmatrix} (P_x)_1 \\ (P_y)_1 \\ (P_x)_3 \\ (P_y)_3 \end{Bmatrix} = \begin{bmatrix} 0 & 0 & 0 & 0 \\ 0 & (k_{22})_3 & 0 & (k_{24})_3 \\ 0 & 0 & 0 & 0 \\ 0 & (k_{42})_3 & 0 & (k_{44})_3 \end{bmatrix} \begin{Bmatrix} u_1 \\ v_1 \\ u_3 \\ v_3 \end{Bmatrix}
$$

Then, the process of assembling element stiffness matrices involves combining the nodal stiffness values of all the elements joining at every common node so that the order of the assembled stiffness matrix equals the total number of degrees of freedom of the structure.

$$
\begin{Bmatrix} (Px)_1 \\ (P_y)_1 \\ (P_x)_2 \\ (P_y)_2 \\ (P_x)_3 \\ (P_y)_3 \end{Bmatrix} = \begin{bmatrix} (k_{11})_1 & 0 & (k_{13})_1 & 0 & 0 & 0 \\ 0 & (k_{22})_3 & 0 & 0 & 0 & (k_{24})_3 \\ (k_{31})_1 & 0 & (k_{33})_1 + (k_{11})_2 & (k_{12})_2 & (k_{13})_2 & (k_{14})_2 \\ 0 & 0 & (k_{21})_2 & (k_{22})_2 & (k_{23})_2 & (k_{24})_2 \\ 0 & 0 & (k_{31})_2 & (k_{32})_2 & (k_{33})_2 & (k_{34})_2 \\ 0 & (k_{24})_3 & (k_{41})_2 & (k_{42})_2 & (k_{43})_2 & (k_{44})_2 + (k_{44})_3 \end{bmatrix} \begin{Bmatrix} u_1 \\ v_1 \\ u_2 \\ v_2 \\ u_3 \\ v_3 \end{Bmatrix}
$$

or $\{P\} = [K] \{u\}$

which represents a set of n simultaneous equations, where n is the total number of degrees of freedom in the structure. In the case of mechanical structures, the assembled stiffness matrix is symmetric, singular and positive definite.

4.6   BOUNDARY CONDITIONS

The singularity of the matrix indicates possibility of rigid body movement of the structure in different directions and hence the possibility of many solutions for the unknown nodal displacements. Boundary conditions, in terms of fixed degrees of freedom or known values of displacements at some points of the structure, are therefore applied. In some structures, where no part of the structure is fixed, it is possible to apply different boundary conditions. Each solution gives displacements at other points in the structure, with reference to the chosen fixed points.

(a) Elimination method

The columns and rows of the stiffness matrix, displacement vector and load vector are rearranged so that the set of equations can be written as

$$\begin{Bmatrix} P_1 \\ P_2 \end{Bmatrix} = \begin{bmatrix} K_{11} & K_{12} \\ K_{21} & K_{22} \end{bmatrix} \begin{Bmatrix} q_1 \\ q_2 \end{Bmatrix} \qquad \qquad .....(4.24)$$

where q_1 is the set of unknown displacements and q_2 is the set of specified displacements.

From static equilibrium considerations, terms in the load vector $\{P_2\}$ corresponding to the fixed or specified values of degrees of freedom $\{q_2\}$ represent reactions at those degrees of freedom to balance the applied loads.

Taking all known values to the left side, first set of these equations can be rewritten as

$$\{P_1\} - [K_{12}] \{q_2\} = [K_{11}] \{q_1\} \qquad \qquad .....(4.25)$$

The reduced stiffness matrix of the structure $[K_{11}]$ is usually non-singular and can be inverted so that unknown displacements $\{q_1\}$ and the reactions $\{R\}$ can be evaluated from

$$\{q_1\} = [K_{11}]^{-1} (\{P_1\} - [K_{12}] \{q_2\}) \qquad \qquad .....(4.26)$$

$$\{R\} = \{P_2\} = \begin{bmatrix} K_{21} & K_{22} \end{bmatrix} \begin{Bmatrix} q_1 \\ q_2 \end{Bmatrix} = [K_{21}]\{q_1\} + [K_{22}]\{q_2\} \qquad .....(4.27)$$

In case of specified zero displacements $\{q_2\}$, eq.(4.24) reduces to

$$\{P_1\} = [K_{11}]\{q_1\} \quad \text{or} \quad \{P_r\} = [K_r]\{u_r\} \qquad \text{.....(4.28)}$$

This procedure is equivalent to deleting rows and columns corresponding to the fixed degrees of freedom from the assembled stiffness matrix, displacement vector and load vector. The reduced stiffness matrix $[K_r]$ is a non-singular matrix.

The unknown displacements are now obtained by using a suitable matrix inversion algorithm like Gauss elimination method or Gauss-Jordan method or Cholesky method. Eq. (4.26) and (4.27) thus simplify to

$$\{q_1\} = [K_{11}]^{-1}\{P_1\} \text{ or } \{u_r\} = [K_r]^{-1}\{P_r\} \qquad \text{.....(4.29)}$$

and $\quad \{R\} = \{P_2\} = [K_{21}]\{q_1\} \qquad \text{.....(4.30)}$

In the displacement formulation, displacements are calculated in the global coordinate system for the entire structure while the stresses are calculated in each element, in the local or element coordinate system, from the nodal displacements of that element using

$$\{\sigma_e\} = [D]\{\varepsilon\} = [D][B_e]\{u_e\}$$

In a structure with 'm' fixed degrees of freedom, assembling the complete stiffness matrix and then deleting some rows and columns will involve more computer memory as well as more time. It is therefore a common practice to ignore the rows and columns of element stiffness matrices corresponding to the fixed degrees of freedom during the assembly process, thus storing $[K_r]$ of order $(n–m) \times (n–m)$ only. In that case, calculation of reaction values corresponding to the fixed degrees of freedom requires storing appropriate terms $[K_{21}]$ in a different matrix.

(b) Penalty approach

In this technique, a quantity C is added to the diagonal term in the row corresponding to the specified displacement (n^{th} DOF) while ($C \times q_n$) is added to the force term of the corresponding equation, where C is a large stiffness value, usually max $(k_{ij}) \times 10^4$. The assembled stiffness matrix is then inverted by any one of the conventional approaches. This method gives a very small value of the order of 10^{-4} for the displacement q_n corresponding to the fixed degree of freedom. This displacement value can be reduced further by using a smaller value of C. This may give rise to numerical errors during the inversion of the stiffness matrix. Reaction is then obtained from $R_n = C \times q_n$. This approach was used in the first general purpose software (SAP-IV) developed by Prof. Wilson and his associates at the University of Southern California. This method is equivalent to adding a spring of very large stiffness value in the direction of the fixed degree of freedom. Member force calculated for the spring element indicates reaction corresponding to the fixed degree of freedom. The size of the matrix to be inverted is large in this approach of order n × n, without deletion of rows and columns corresponding to the fixed degrees of freedom and is, therefore, **_not_** usually adopted now.

(c) Multi-point constraints

There are many situations in trusses where the end supports are on inclined plane and do not coincide with the coordinate system used to describe the truss. In such cases, the displacement and force components along the coordinate axes have to be resolved along and perpendicular to the inclined plane and necessary conditions specified on them.

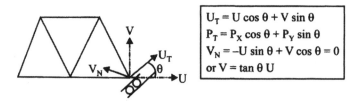

$$U_T = U \cos \theta + V \sin \theta$$
$$P_T = P_X \cos \theta + P_Y \sin \theta$$
$$V_N = -U \sin \theta + V \cos \theta = 0$$
$$\text{or } V = \tan \theta \, U$$

Some other types of multi-point constraints are those linking displacement of one node with that of another. A few of them are shown below, with node 1 as the fixed point and node 2 as the point of load application. If 'δ' is the gap, $U_3 = U_2 - \delta$ can be substituted in the load-displacement relations to reduce the number of unknowns by 1 and corresponding columns of stiffness matrix are modified. Accordingly, order of the stiffness matrix also reduces by 1.

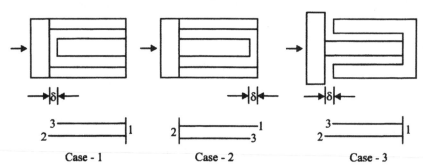

Case - 1   Case - 2   Case - 3

4.7 BEAM ELEMENT STIFFNESS MATRIX BY VARIATIONAL APPROACH

Let us consider bending of a typical beam of uniform cross section in a plane perpendicular to its axis, due to load and moment applied at its two ends. From strength of materials, it is well known that its deformed shape is a curve and can not be represented by a linear function. Considering 2nd order polynomial having three coefficients, with deflection alone as the nodal degree of freedom fails to express the three coefficients in terms of the two nodal deflections. Also, natural boundary conditions at the ends of the beam may include not only deflection but also the slope. Hence, a cubic polynomial is generally used and its four unknown coefficients are represented in terms of deflection and slope (first derivative of deflection) at each end.

If X-Y is the plane of bending, P_Y and M_Z are the loads applied while v and θ_z are the deflection and slope in the plane of bending,

Let $\quad v(x) = a_1 + a_2.x + a_3.x^2 + a_4.x^3 = [\,1 \quad x \quad x^2 \quad x^3\,]\,\{a\} \qquad .....(4.37)$

and $\quad \theta_z = \dfrac{dv}{dx} = \begin{bmatrix} 0 & 1 & 2x & 3x^2 \end{bmatrix}\{a\}$

or $\quad \begin{Bmatrix} v \\ \theta_z \end{Bmatrix} = \{u\} = \{f(x)\}^T \{a\} \qquad .....(4.38)$

Choosing node i as the origin of local coordinate system for this element with X-axis along the axis of the element and substituting $x = 0$ at node i and $X = L$ at node j, we get the nodal displacement vector,

$$\begin{Bmatrix} v_i \\ (\theta_z)_i \\ v_j \\ (\theta_z)_j \end{Bmatrix} = \begin{bmatrix} 1 & 0 & 0 & 0 \\ 0 & 1 & 0 & 0 \\ 1 & L & L^2 & L^3 \\ 0 & 1 & 2L & 3L^2 \end{bmatrix} \begin{Bmatrix} a_1 \\ a_2 \\ a_3 \\ a_4 \end{Bmatrix} \text{ or } \{u_e\} = [G]\,\{a\} \qquad .....(4.39)$$

Solving for the coefficients $\{a\}$ and substituting in eq. (4.38)

$$u = \{f(x)\}^T \{a\} = \{f(x)\}^T [G]^{-1} \{u_e\} \qquad .....(4.40)$$

From theory of bending,

strain, $\quad \varepsilon = \dfrac{\sigma}{E} = \dfrac{y}{R} = y.\dfrac{d\theta}{dx} = y\left(\dfrac{d^2v}{dx^2}\right) = [B]\{u_e\} \qquad .....(4.41)$

where, $\quad \dfrac{d^2v}{dx^2} = \begin{bmatrix} 0 & 0 & 2 & 6x \end{bmatrix}\{a\} = \begin{bmatrix} 0 & 0 & 2 & 6x \end{bmatrix}[G]^{-1}\{u_e\}$

and so, $[B] = y\begin{bmatrix} 0 & 0 & 2 & 6x \end{bmatrix}\begin{bmatrix} 1 & 0 & 0 & 0 \\ 0 & 1 & 0 & 0 \\ -3/L^2 & -2/L & 3/L^2 & -1/L \\ 2/L^3 & 1/L^2 & -2/L^3 & 1/L^2 \end{bmatrix}$

$\quad = (y/L^3)\,[\,6(2x - L) \quad 2L(3x - 2L) \quad -6(2x - L) \quad 2L(3x - L)\,] \qquad .....(4.42)$

Then, $\{P_e\} = [K_e]\ \{u_e\}$ .....(4.43)

where,

$$[K_e] = \int_v [B]^T\ E[B]dv = \iiint [B]^T\ E[B]dx\ dy\ dz$$

$$[K_e] = \left(\frac{E}{L^6}\right) \int_A y^2\ (dy\ dz) \int_L \begin{Bmatrix} 6(2x-L) \\ 2L(3x-2L) \\ -6(2x-L) \\ 2L(3x-L) \end{Bmatrix} [6(2x-L) \quad 2L(3x-2L) \quad -6(2x-L) \quad 2L(3x-L)]dx$$

$$= \frac{E\ I_z}{L^3} \begin{bmatrix} 12 & 6L & -12 & 6L \\ 6L & 4L^2 & -6L & 2L^2 \\ -12 & -6L & 12 & -6L \\ 6L & 2L^2 & -6L & 4L^2 \end{bmatrix} \qquad .....(4.44)$$

4.8 GENERAL BEAM ELEMENT

A beam in a space frame is generally subjected to axial load, torsion load and bending loads in two planes, due to the combined effect of loads acting at different locations of the space frame and in different directions, as shown in the figure.

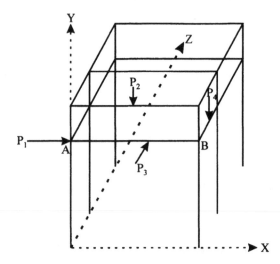

If a single beam AB in a space frame with concentrated loads P_1, P_2, P_3 and P_4 acting on some members is considered, load P_1 contributes to axial load giving rise to displacement u; load P_2 contributes to bending in X-Y plane giving rise to deflection V and slope θ_Z, load P_3 contributes to bending in X-Z plane giving rise to deflection w and slope θ_y and load P_4 contributes to torsion in AB giving rise to θ_2. Therefore, general stiffness matrix for beam AB in its local coordinate system should include response of the beam for these four types of deformations, which are mutually independent. Such DOFs are called **_uncoupled degrees of freedom_** i.e. torsion of a beam does not result in axial elongation or compression of the beam; deflection of the beam in X-Y plane does not cause any displacement of the beam in X-Z plane etc. Hence, stiffness contribution of a beam in these 6 DOFs can be placed directly, without any modifications, in the appropriate positions of the general stiffness matrix of order 12 (2 nodes x 6 DOF/node) .

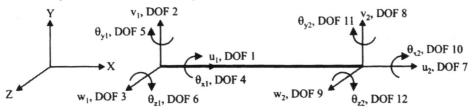

Stiffness matrix with only axial load response (relating load P_x and displacement u) is

$$
\begin{Bmatrix} P_{x1} \\ P_{y1} \\ P_{z1} \\ M_{x1} \\ M_{y1} \\ M_{z1} \\ P_{x2} \\ P_{y2} \\ P_{z2} \\ M_{x2} \\ M_{y2} \\ M_{z2} \end{Bmatrix} = \frac{1}{L}
\begin{bmatrix}
EA & 0 & 0 & 0 & 0 & 0 & -EA & 0 & 0 & 0 & 0 & 0 \\
0 & 0 & 0 & 0 & 0 & 0 & 0 & 0 & 0 & 0 & 0 & 0 \\
0 & 0 & 0 & 0 & 0 & 0 & 0 & 0 & 0 & 0 & 0 & 0 \\
0 & 0 & 0 & 0 & 0 & 0 & 0 & 0 & 0 & 0 & 0 & 0 \\
0 & 0 & 0 & 0 & 0 & 0 & 0 & 0 & 0 & 0 & 0 & 0 \\
0 & 0 & 0 & 0 & 0 & 0 & 0 & 0 & 0 & 0 & 0 & 0 \\
-EA & 0 & 0 & 0 & 0 & 0 & EA & 0 & 0 & 0 & 0 & 0 \\
0 & 0 & 0 & 0 & 0 & 0 & 0 & 0 & 0 & 0 & 0 & 0 \\
0 & 0 & 0 & 0 & 0 & 0 & 0 & 0 & 0 & 0 & 0 & 0 \\
0 & 0 & 0 & 0 & 0 & 0 & 0 & 0 & 0 & 0 & 0 & 0 \\
0 & 0 & 0 & 0 & 0 & 0 & 0 & 0 & 0 & 0 & 0 & 0 \\
0 & 0 & 0 & 0 & 0 & 0 & 0 & 0 & 0 & 0 & 0 & 0
\end{bmatrix}
\begin{Bmatrix} u_1 \\ v_1 \\ w_1 \\ \theta_{x1} \\ \theta_{y1} \\ \theta_{z1} \\ u_2 \\ v_2 \\ w_2 \\ \theta_{x2} \\ \theta_{y2} \\ \theta_{z2} \end{Bmatrix} \quad .....(4.45)
$$

Stiffness matrix with only torsion load response (relating load M_x and displacement θ_x) is

$$
\begin{Bmatrix} P_{x1} \\ P_{y1} \\ P_{z1} \\ M_{x1} \\ M_{y1} \\ M_{z1} \\ P_{x2} \\ P_{y2} \\ P_{z2} \\ M_{x2} \\ M_{y2} \\ M_{z2} \end{Bmatrix} = \frac{1}{L}
\begin{bmatrix}
0 & 0 & 0 & 0 & 0 & 0 & 0 & 0 & 0 & 0 & 0 & 0 \\
0 & 0 & 0 & 0 & 0 & 0 & 0 & 0 & 0 & 0 & 0 & 0 \\
0 & 0 & 0 & 0 & 0 & 0 & 0 & 0 & 0 & 0 & 0 & 0 \\
0 & 0 & 0 & GJ & 0 & 0 & 0 & 0 & 0 & -GJ & 0 & 0 \\
0 & 0 & 0 & 0 & 0 & 0 & 0 & 0 & 0 & 0 & 0 & 0 \\
0 & 0 & 0 & 0 & 0 & 0 & 0 & 0 & 0 & 0 & 0 & 0 \\
0 & 0 & 0 & 0 & 0 & 0 & 0 & 0 & 0 & 0 & 0 & 0 \\
0 & 0 & 0 & 0 & 0 & 0 & 0 & 0 & 0 & 0 & 0 & 0 \\
0 & 0 & 0 & 0 & 0 & 0 & 0 & 0 & 0 & 0 & 0 & 0 \\
0 & 0 & 0 & -GJ & 0 & 0 & 0 & 0 & 0 & GJ & 0 & 0 \\
0 & 0 & 0 & 0 & 0 & 0 & 0 & 0 & 0 & 0 & 0 & 0 \\
0 & 0 & 0 & 0 & 0 & 0 & 0 & 0 & 0 & 0 & 0 & 0
\end{bmatrix}
\begin{Bmatrix} u_1 \\ v_1 \\ w_1 \\ \theta_{x1} \\ \theta_{y1} \\ \theta_{z1} \\ u_2 \\ v_2 \\ w_2 \\ \theta_{x2} \\ \theta_{y2} \\ \theta_{z2} \end{Bmatrix} \qquad ..(4.46)
$$

Stiffness matrix for bending in X-Y plane (relating P_y and M_z with displacements v and θ_z) is

$$
\begin{Bmatrix} P_{x1} \\ P_{y1} \\ P_{z1} \\ M_{x1} \\ M_{y1} \\ M_{z1} \\ P_{x2} \\ P_{y2} \\ P_{z2} \\ M_{x2} \\ M_{y2} \\ M_{z2} \end{Bmatrix} = \frac{1}{L}
\begin{bmatrix}
0 & 0 & 0 & 0 & 0 & 0 & 0 & 0 & 0 & 0 & 0 & 0 \\
0 & 12EI_z/L^2 & 0 & 0 & 0 & 6EI_z/L & 0 & -12EI_z/L^2 & 0 & 0 & 0 & 6EI_z/L \\
0 & 0 & 0 & 0 & 0 & 0 & 0 & 0 & 0 & 0 & 0 & 0 \\
0 & 0 & 0 & 0 & 0 & 0 & 0 & 0 & 0 & 0 & 0 & 0 \\
0 & 0 & 0 & 0 & 00 & 0 & 0 & 0 & 0 & 0 & 0 & 0 \\
0 & 6EI_z/L & 0 & 0 & 0 & 4EI_z & 0 & -6EI_z/L & 0 & 0 & 0 & 2EI_z \\
0 & 0 & 0 & 0 & 0 & 0 & 0 & 0 & 0 & 0 & 0 & 0 \\
0 & -12EI_z/L^2 & 0 & 0 & 0 & -6EI_z/L & 0 & 12EI_z/L^2 & 0 & 0 & 0 & -6EI_z/L \\
0 & 0 & 0 & 0 & 0 & 0 & 0 & 0 & 0 & 0 & 0 & 0 \\
0 & 0 & 0 & 0 & 0 & 0 & 0 & 0 & 0 & 0 & 0 & 0 \\
0 & 0 & 0 & 0 & 0 & 0 & 0 & 0 & 0 & 0 & 0 & 0 \\
0 & 6EI_z/L & 0 & 0 & 0 & 2EI_z & 0 & -6EI_z/L & 0 & 0 & 0 & 4EI_z
\end{bmatrix}
\begin{Bmatrix} u_1 \\ v_1 \\ w_1 \\ \theta_{x1} \\ \theta_{y1} \\ \theta_{z1} \\ u_2 \\ v_2 \\ w_2 \\ \theta_{x2} \\ \theta_{y2} \\ \theta_{z2} \end{Bmatrix}
$$

$$.....(4.47)$$

Stiffness matrix for bending in X-Z plane (relating P_z and M_y with displacements w and θ_y) is

$$
\begin{Bmatrix} P_{x1} \\ P_{y1} \\ P_{z1} \\ M_{x1} \\ M_{y1} \\ M_{z1} \\ P_{x2} \\ P_{y2} \\ P_{z2} \\ M_{x2} \\ M_{y2} \\ M_{z2} \end{Bmatrix} = \frac{1}{L}
\begin{bmatrix}
0 & 0 & 0 & 0 & 0 & 0 & 0 & 0 & 0 & 0 & 0 & 0 \\
0 & 0 & 0 & 0 & 0 & 0 & 0 & 0 & 0 & 0 & 0 & 0 \\
0 & 0 & 12EI_y/L^2 & 0 & 6EI_y/L & 0 & 0 & 0 & -12EI_y/L^2 & 0 & 6EI_y/L & 0 \\
0 & 0 & 0 & 0 & 0 & 0 & 0 & 0 & 0 & 0 & 0 & 0 \\
0 & 0 & 6EI_y/L & 0 & 4EI_y & 0 & 0 & 0 & -6EI_y/L & 0 & 2EI_y & 0 \\
0 & 0 & 0 & 0 & 0 & 0 & 0 & 0 & 0 & 0 & 0 & 0 \\
0 & 0 & 0 & 0 & 0 & 0 & 0 & 0 & 0 & 0 & 0 & 0 \\
0 & 0 & 0 & 0 & 0 & 0 & 0 & 0 & 0 & 0 & 0 & 0 \\
0 & 0 & -12EI_y/L^2 & 0 & -6EI_y/L & 0 & 0 & 0 & 12EI_y/L^2 & 0 & -6EI_y/L & 0 \\
0 & 0 & 0 & 0 & 0 & 0 & 0 & 0 & 0 & 0 & 0 & 0 \\
0 & 0 & 6EI_y/L & 0 & -12EI_y/L^2 & 0 & 0 & 0 & -6EI_y/L & 0 & 4EI_y & 0 \\
0 & 0 & 0 & 0 & 0 & 0 & 0 & 0 & 0 & 0 & 0 & 0
\end{bmatrix}
\begin{Bmatrix} u_1 \\ v_1 \\ w_1 \\ \theta_{x1} \\ \theta_{y1} \\ \theta_{z1} \\ u_2 \\ v_2 \\ w_2 \\ \theta_{x2} \\ \theta_{y2} \\ \theta_{z2} \end{Bmatrix}
$$

$$.....(4.48)$$

The combined stiffness matrix of a general beam element thus includes stiffness coefficients linking loads along the three coordinate axes and moments about the three axes at each end of the beam to the corresponding displacements and rotations, and is obtained by a simple addition of the coefficients of the above four matrices.

$$
\begin{Bmatrix} P_{x1} \\ P_{y1} \\ P_{z1} \\ M_{x1} \\ M_{y1} \\ M_{z1} \\ P_{x2} \\ P_{y2} \\ P_{z2} \\ M_{x2} \\ M_{y2} \\ M_{z2} \end{Bmatrix} = \frac{1}{L}
\begin{bmatrix}
EA & 0 & 0 & 0 & 0 & 0 & -EA & 0 & 0 & 0 & 0 & 0 \\
0 & 12EI_z/L^2 & 0 & 0 & 0 & 6EI_z/L & 0 & -12EI_z/L^2 & 0 & 0 & 0 & 6EI_z/L \\
0 & 0 & 12EI_y/L^2 & 0 & 6EI_y/L & 0 & 0 & 0 & -12EI_y/L^2 & 0 & 6EI_y/L & 0 \\
0 & 0 & 0 & GJ & 0 & 0 & 0 & 0 & 0 & GJ & 0 & 0 \\
0 & 0 & 6EI_y/L & 0 & 4EI_y & 0 & 0 & 0 & -6EI_y/L & 0 & 2EI_y & 0 \\
0 & 6EI_z/L & 0 & 0 & 0 & 4EI_z & 0 & -6E_z/L & 0 & 0 & 0 & 2EI_z \\
-EA & 0 & 0 & 0 & 0 & 0 & EA & 0 & 0 & 0 & 0 & 0 \\
0 & -12EI_z/L^2 & 0 & 0 & 0 & -6EI_z/L & 0 & 12EI_z/L^2 & 0 & 0 & 0 & -6EI_z/L \\
0 & 0 & -12EI_y/L^3 & 0 & -6EI_y/L & 0 & 0 & 0 & 12EI_y/L^2 & 0 & -6EI_y/L & 0 \\
0 & 0 & 0 & -GJ & 0 & 0 & 0 & 0 & 0 & GJ & 0 & 0 \\
0 & 0 & 6EI_y/L & 0 & 2EI_y & 0 & 0 & 0 & -6EI_y/L & 0 & 4EI_y & 0 \\
0 & 6EI_z/L & 0 & 0 & 0 & 12EI_z/L^2 & 0 & -6EI_z/L & 0 & 0 & 0 & 4EI_z
\end{bmatrix}
\begin{Bmatrix} u_1 \\ v_1 \\ w_1 \\ \theta_{x1} \\ \theta_{y1} \\ \theta_{z1} \\ u_2 \\ v_2 \\ w_2 \\ \theta_{x2} \\ \theta_{y2} \\ \theta_{z2} \end{Bmatrix}
$$

$$..(4.49)$$

4.9 PIPE ELEMENT

A pipe element is essentially a one-dimensional element subjected to:

(i)   distributed load due to self weight and weight of fluid inside,

(ii)   concentrated loads in the form of pipe fittings like valves,

(iii) axial loads due to change of direction of fluid flow in pipe bends, T-joints etc. and due to restrained thermal expansion.

It is thus similar to a generalised beam element, except that stiffness of a beam member is a function of a geometric properties (I, L) and material property (E), whereas stiffness of a pipe element is a function of load (internal pressure) also. Stiffness increases due to internal fluid at high pressure (also called pressure stiffening).

For these reasons, this is listed as a different element, in many general purpose software even though the analysis is completely similar to that of a three-dimensional frame (with general beam elements).

4.10 Summary

- Finite Element Method (FEM) is based on minimum potential energy of the system, as applied to a model of the component consisting of finite number of elements connected at common nodes. In the displacement method, which is commonly used in the solution of mechanical problems, load-displacement relations (or stiffness coefficients) are calculated for each element satisfying the condition that variation of potential energy (sum of work done and strain energy) for any virtual displacement is zero.

- In 1-D elements, a polynomial expression is assumed for the variation of displacement along its axis. Axial load carrying truss element is identified by its axial displacement (1 DOF/node) at its two end points (nodes) and is modeled by a linear variation of displacement along its axis. A torsion element is similarly identified by the rotation about its axis (1 DOF/node) at its two end points (nodes) and is modeled by a linear variation of rotation along its axis. Beam element is identified by deflection normal to the axis and slope (derivative of displacement) at its two end points (2 DOF/node) and is modeled by a cubic polynomial for the deflection, in each of its two planes of bending. Stiffness matrix of a general beam element with 6 DOF per node is obtained by a combination of the above, treating them as uncoupled responses.

- Stiffness coefficients relate nodal displacement components with nodal load components, both being vectors. These vector components of different elements can be combined only when they are all oriented in the same directions. Hence, element stiffness matrix obtained in its local coordinate system, defined w.r.t. its axis, is transformed into a common or global coordinate system, using a transformation matrix (function of orientation of the member w.r.t. global coordinate system) if different elements are inclined to each other.

- The assembled stiffness matrix represents unconstrained system (with rigid body modes included) and hence can not be inverted (or has infinite solutions). Specified displacement boundary conditions are applied to avoid rigid body modes and thus obtain a unique solution for the given problem.

- One of the standard techniques for the solution of simultaneous equations will give primary unknowns (nodal displacements) in global coordinate system, since the nodes are common to many elements. Using stress-displacement relationship of each element, stresses in local coordinate system are obtained from the nodal displacements.

OBJECTIVE QUESTIONS

1. Transformation matrix ___ for all elements

   (a) is always same   (b) is different

   (c) depends on element axes   (d) depends on material

2. Transformation matrix relates ____ in element coordinate system with structural coordinate system

   (a) displacements   (b) stresses

   (c) stiffness coefficients   (d) material properties

3. Primary variable in FEM structural analysis is

   (a) displacement   (b) force   (c) stress   (d) strain

4. A singular stiffness matrix means

   (a) unstable structure

   (b) one or more DOF are unrestrained

   (c) wrong connectivity of elements

   (d) wrong solution expected

5. One possible load in structural analysis is the specified

   (a) nodal temperature   (b) stress in an element

   (c) heat flow   (d) strain in an element

6. Assembled stiffness matrix after applying boundary conditions is NOT

   (a) square   (b) symmetric   (c) banded   (d) singular

7. Determinant of assembled stiffness matrix before applying boundary conditions is

   (a) < 0 (b) $= 0$ (c) > 0 (d) depends on the problem

8. Determinant of assembled stiffness matrix after applying boundary conditions is

   (a) < 0 (b) $= 0$ (c) > 0 (d) depends on the problem

9. A pipe with internal pressure behaves _____ a hollow pipe of same section

   (a) with exactly same deflection as

   (b) with lesser bending deflection than

   (c) with more bending deflection than

   (d) with a different type of deflection

10. Any point in a structure can have maximum of __ DOF

   (a) 2   (b) 3   (c) 4   (d) 6

11. A 1-D structural element is a

   (a) truss element   (b) beam element   (c) pipe element (d) all of them

12. Meshing for 1-D elements is

   (a) essential   (b) optional

   (c) reduces input data   (d) depends on other data

13. A structure with loads at joints only is usually modeled by _____

   (a) truss elements   (b) beam elements

   (c) pipe elements   (d) any one of them

14. A frame with nodal loads only is modeled as an assembly of truss elements, if resistance to rotational degree of freedom of joints is

   (a) very small   (b) very large

   (c) not related   (d) depends on other data

15. A frame with nodal loads only is modeled as an assembly of beam elements, if resistance to rotational degree of freedom of joints is

   (a) very small   (b) very large

   (c) not related   (d) depends on other data

16. A frame with distributed loads along members is modeled by an assembly of _____ elements

   (a) truss   (b) beam   (c) pipe   (d) any one of them

17. A frame with welded joints can be approximated by truss elements

   (a) always   (b) sometimes

   (c) never   (d) depends on assumed flexibility of rotation

18. A structure assembled by multiple bolts/rivets at each joint is modeled by truss elements

    (a) always   (b) sometimes

    (c) never   (d) depends on assumed flexibility of rotation

19. Stress across any 1-D element is assumed to be constant

    (a) true for beam elements (b) true for truss elements

    (c) true for pipe element   (d) true for all 1-D elements

20. A bar is modeled as 1-D element only if its

    (a) area of cross section is small

    (b) moment of inertia is small

    (c) length is very large compared to cross section dimensions

    (d) all the above

21. A truss element in space has a stiffness matrix of order

    (a) 2×2 (b) 4×4 (c) 6×6 (d) 1×1

22. A spring element is similar to _____ element

    (a) truss   (b) beam   (c) pipe   (d) any one of them

23. A plane truss element has a stiffness matrix of order

    (a) 2×2 (b) 4×4 (c) 6×6 (d) 1×1

24. A pipe element differs from a beam element by inclusion of

    (a) cold cut   (b) internal pressure stiffening

    (c) anchors   (d) sliding supports

25. Stiffness matrix of a torsion element is of the same order as

    (a) truss element   (b) beam element

    (c) pipe element   (d) none of them

26. A spring of _____ stiffness at the supports is used for calculating support reactions in penalty approach

    (a) very small   (b) same as other connected members

    (c) very large   (d) sum of connected members

27. Penalty approach leads to _____ displacements at supports

    (a) zero   (b) very small

    (c) significant   (d) depends on stiffness of connected members

28. Penalty approach takes _____ time

    (a) more   (b) less

    (c) depends on other data   (d) no change

29. Accuracy of solution _____ with increase of number of beam elements

    (a) improves   (b) reduces

    (c) no change   (d) depends on other data

SOLVED PROBLEMS

Example 4.1

Determine the nodal displacements and element stresses by finite element formulation for the following figure. Use P=300 k N; A_1=0.5 m²; A_2=1 m²; E=200 GPa

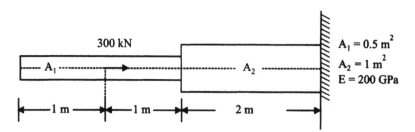

Solution

The structure is modeled with 3 axial loaded elements connected by nodes 1-2, 2-3 and 3-4 as shown below

Stiffness matrices of elements 1, 2 and 3 are given by

$$[K_1]=\begin{bmatrix} k_1 & -k_1 \\ -k_1 & k_1 \end{bmatrix} \quad [K_2]=\begin{bmatrix} k_1 & -k_1 \\ -k_1 & k_1 \end{bmatrix} \quad [K_3]=\begin{bmatrix} k_2 & -k_2 \\ -k_2 & k_2 \end{bmatrix} \quad .....(4.31)$$

where, $k_1 = A_1 E/L_1 = 0.5 \times 200 \times 10^9/1.0 = 1.0 \times 10^{11}$

and $k_2 = A_2 E/L_2 = 1 \times 200 \times 10^9/2.0 = 1.0 \times 10^{11}$

Assembled stiffness matrix is obtained by adding corresponding terms as,

$$k_1 = [K] = \begin{bmatrix} k_1 & -k_1 & 0 & 0 \\ -k_1 & k_1+k_2 & -k_1 & 0 \\ 0 & -k_1 & k_1+k_2 & -k_2 \\ 0 & 0 & -k_2 & k_2 \end{bmatrix} = 1.0 \times 10^{11} \begin{bmatrix} 1 & -1 & 0 & 0 \\ -1 & 2 & -1 & 0 \\ 0 & -1 & 2 & -1 \\ 0 & 0 & -1 & 1 \end{bmatrix}$$

.....(4.32)

Corresponding assembled nodal load vector and nodal displacement vector are

$$P = \begin{Bmatrix} 0 \\ 300,000 \\ 0 \\ R \end{Bmatrix}; q = \begin{Bmatrix} u_1 \\ u_2 \\ u_3 \\ u_4 \end{Bmatrix}$$

.....(4.33)

Thus,

$$[K]\{q\} = \{P\}$$

or

$$1.0 \times 10^{11} \begin{bmatrix} 1 & -1 & 0 & 0 \\ -1 & 2 & -1 & 0 \\ 0 & -1 & 2 & -1 \\ 0 & 0 & -1 & 1 \end{bmatrix} \begin{Bmatrix} u_1 \\ u_2 \\ u_3 \\ u_4 \end{Bmatrix} = \begin{Bmatrix} 0 \\ 300,000 \\ 0 \\ R \end{Bmatrix}$$

.....(4.34)

After applying boundary condition, $u_4=0$, the fourth row and fourth column are removed resulting in

$$1.0 \times 10^{11} \begin{bmatrix} 1 & -1 & 0 \\ -1 & 2 & -1 \\ 0 & -1 & 2 \end{bmatrix} \begin{Bmatrix} u_1 \\ u_2 \\ u_3 \end{Bmatrix} = \begin{Bmatrix} 0 \\ 300,000 \\ 0 \end{Bmatrix}$$

.....(4.35)

Solving the above set of equations gives,

$$u_1 = 6 \times 10^{-6}\,\text{m}; \qquad u_2 = 6 \times 10^{-6}\,\text{m}; \; u_3 = 3 \times 10^{-6}\,\text{m}$$ .....(4.36)

Stress in element-1,

$$\sigma_1 = E\,[B_1]\,\{q_1\} = E\,[-1/L_1 \quad 1/L_1] \begin{Bmatrix} u_1 \\ u_2 \end{Bmatrix} = 0\,\text{N/m}^2$$

Stress in element-2,

$$\sigma_2 = E\ [B_2]\ \{q_2\} = E\ [-1/L_1 \quad 1/L_1] \begin{Bmatrix} u_2 \\ u_3 \end{Bmatrix} = -6 \times 10^5\ N/m^2$$

Stress in element-3,

$$\sigma_3 = E\ [B_3]\ \{q_3\} = E\ [-1/L_2 \quad 1/L_2] \begin{Bmatrix} u_3 \\ u_4 \end{Bmatrix} = -3 \times 10^5\ N/m^2$$

Example 4.2

An axial load $P = 200 \times 10^3$ N is applied on a bar as shown. Using the penalty approach for handling boundary conditions determine nodal displacements, stress in each material and reaction forces.

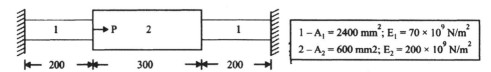

Solution

$$1 \overline{\hspace{2.5cm}} 2 \overline{\hspace{2.5cm}} 3 \overline{\hspace{2cm}} 4$$

Considering a 3-element truss model, stiffness matrices of elements 1, 2 and 3 (connected by nodes 1, 2; 2, 3 and 3, 4 respectively) are given by,

$$[K_1] = \begin{bmatrix} k_1 & -k_1 \\ -k_1 & k_1 \end{bmatrix}; \quad [K_3] = \begin{bmatrix} k_1 & -k_1 \\ -k_1 & k_1 \end{bmatrix}; \quad [K_2] = \begin{bmatrix} k_2 & -k_2 \\ -k_2 & k_2 \end{bmatrix} \quad ..(4.37)$$

where

$$k_1 = A_1 E_1 / L_1 = 2400 \times 70 \times 10^3 / 200 = 84 \times 10^4$$

and $\quad k_2 = A_2 E_2 / L_2 = 600 \times 200 \times 10^3 / 300 = 40 \times 10^4$

Assembled stiffness matrix is obtained by adding corresponding terms as,

$$[K] = \begin{bmatrix} k_1 & -k_1 & 0 & 0 \\ -k_1 & k_1+k_1 & -k_1 & 0 \\ 0 & -k_1 & k_1+k_2 & -k_2 \\ 0 & 0 & -k_2 & k_2 \end{bmatrix} = 10^4 \begin{bmatrix} 84 & -84 & 0 & 0 \\ -84 & 84+40 & -40 & 0 \\ 0 & -40 & 40+80 & -84 \\ 0 & 0 & -84 & 84 \end{bmatrix}$$

Corresponding assembled nodal load vector and nodal displacement vector are

$$P = \begin{Bmatrix} 0 \\ 200,000 \\ 0 \\ 0 \end{Bmatrix} ; \quad q = \begin{Bmatrix} u_1 \\ u_2 \\ u_3 \\ u_4 \end{Bmatrix}$$

For the penalty approach, $C = \max(k_{ij}) \times 10^4 = 124 \times 10^4$

Since the bar is fixed at nodes 1 and 4, the equations are then modified using C as,

$$\begin{Bmatrix} 0 \\ 200,000 \\ 0 \\ 0 \end{Bmatrix} = 10^4 \begin{bmatrix} 84 + 124 \times 10^4 & -84 & 0 & 0 \\ -84 & 124 & -40 & 0 \\ 0 & -40 & 124 & -84 \\ 0 & 0 & -84 & 84 + 124 \times 10^4 \end{bmatrix} \begin{Bmatrix} u_1 \\ u_2 \\ u_3 \\ u_4 \end{Bmatrix}$$

From 4th eqn. $0 = 10^4 [-84 u_3 + (84 + 124 \times 10^4) u_4]$

or $u_4 = 6.7737 \times 10^{-5} u_3$

From 3rd eqn $0 = 10^4 [-40 u_2 + 124 u_3 - 84 u_4]$

substituting for u_4 from the above,

$$u_3 = 0.3226 u_2$$

2nd eqn now becomes,

$$200,000 = 10^4 [-84 u_1 + 124 u_2 - 40 u_3]$$

or $-0.64 u_1 + 1.111 u_2 = 0.2$

1st equation gives,

$$0 = 10^4 [(84 + 124 \times 10^4) u_1 - 84 u_2]$$

From these two equations,

$$u_1 = 1.2195 \times 10^{-5} \text{mm}; \quad u_2 = 0.180034 \text{ mm}$$

Substituting in 3rd and 4th eqn.,

$$u_3 = 0.058079 \text{ mm}; \qquad u_4 = 3.9341 \times 10^{-6} \text{ mm}$$

Reactions, $R_1 = -Cu_1 = (124 \times 10^8).(1.2195 \times 10^{-5}) = -151.22 \times 10^3 \text{ N}$

$R_2 = -Cu_4 = (124 \times 10^8).(3.9341 \times 10^{-6}) = -48.78 \times 10^3 \text{ N}$

Stresses in the elements,

$$\sigma_1 = E_1\varepsilon_1 = E_1 B_1 q_{1\text{-}2}$$

$$= 70 \times 10^3 \left[\dfrac{-1}{200} \quad \dfrac{1}{200} \right] \left\{ \begin{array}{c} 1.2195 \times 10^{-5} \\ 0.180034 \end{array} \right\}$$

$$= 63.01 \text{ N/mm}^2$$

$$\sigma_2 = E_2\varepsilon_2 = E_2 B_2 q_{2\text{-}3}$$

$$= 200 \times 10^3 \left[\dfrac{-1}{300} \quad \dfrac{1}{300} \right] \left\{ \begin{array}{c} 0.180034 \\ 0.058079 \end{array} \right\}$$

$$= -81.3 \text{ N/mm}^2$$

$$\sigma_3 = E_3\varepsilon_3 = E_1 B_1 q_{3\text{-}4}$$

$$= 70 \times 10^3 \left[\dfrac{-1}{200} \quad \dfrac{1}{200} \right] \left\{ \begin{array}{c} 0.058079 \\ 3.9341 \times 10^{-6} \end{array} \right\}.$$

$$= -20.3 \text{ N/mm}^2$$

Elimination method

Since the bar is fixed at nodes 1 and 4, corresponding rows and columns of the assembled stiffness matrix are deleted, resulting in $\{P\}_R = [K]_R \{u\}_R$

or
$$\left\{ \begin{array}{c} 200,000 \\ 0 \end{array} \right\} = 10^4 \left[\begin{array}{cc} 124 & -40 \\ -40 & 124 \end{array} \right] \left\{ \begin{array}{c} u_2 \\ u_3 \end{array} \right\}$$

Solving these two simultaneous equations, we get

$$u_2 = 155 / 861 = 0.180023 \text{ mm}$$

and
$$u_3 = 50 / 861 = 0.058072 \text{ mm}$$

Reactions can now be obtained by substituting the nodal displacements in the deleted equations of the assembled stiffness matrix.

$$R_1 = 10^4 [(84 + 124 \times 10^4) \quad -84 \quad 0 \quad 0] [u_1 \quad u_2 \quad u_3 \quad u_4]^T$$

$$= -84 \times 10^4 \, u_2 \qquad = -84 \times 10^4 \, (155/861) \qquad = 151219 \text{ N}$$

$$R_4 = 10^4 [0 \quad 0 \quad -84 \quad (84 + 124 \times 10^4) \,] [u_1 \quad u_2 \quad u_3 \quad u_4]^T$$

$$= -84 \times 10^4 \, u_3 \qquad = -84 \times 10^4 \, (50/861) \qquad = 48780 \text{ N}$$

These reaction values are identical to those obtained by the penalty approach

Check : For force equilibrium of the structure,

$$R_1 + R_4 = \text{Applied load } P \approx 200 \text{ kN}$$

This equation is satisfied with the results obtained

Note that results by penalty approach match very closely with those by elimination approach.

Example 4.3

Consider the truss element with the coordinates 1 (10,10) and 2 (50,40). If the displacement vector is q=[15   10   21   43]^T mm, then determine (i) the vector q' (ii) stress in the element and (iii) stiffness matrix if E=70 GPa and A=200 mm^2

Solution :

(i) The nodal displacement vector in local coordinate system

$$\{q'\} = \begin{bmatrix} l & m & 0 & 0 \\ 0 & 0 & l & m \end{bmatrix} \{q\}$$

where $l = (x_2 - x_1)/L$ and $m = (y_2 - y_1)/L$ are the direction cosines of the element

Length of the element,

$$L = \sqrt{(x_2 - x_1)^2 + (y_2 - y_1)^2} = \sqrt{(50-10)^2 + (40-10)^2} = 50 \text{ mm}$$

$$l = \left(\frac{50-10}{50}\right) = \frac{4}{5}; \qquad m = \frac{(40-10)}{50} = \frac{3}{5}$$

$$\{q'\} = \begin{bmatrix} 4/5 & 3/5 & 0 & 0 \\ 0 & 0 & 4/5 & 3/5 \end{bmatrix} \begin{Bmatrix} 15 \\ 10 \\ 21 \\ 43 \end{Bmatrix} = \begin{Bmatrix} 90/5 \\ 213/5 \end{Bmatrix}$$

(ii)   Stress in the element, $\sigma = E \varepsilon = E \begin{bmatrix} \dfrac{-1}{L} & \dfrac{1}{L} \end{bmatrix} \{q'\}$

$$= 70,000 \begin{bmatrix} \dfrac{-1}{50} & \dfrac{1}{50} \end{bmatrix} \begin{Bmatrix} 90/5 \\ 213/5 \end{Bmatrix}$$

$$= 34.44 \times 10^3 \text{ N/mm}^2$$

(iii)   Stiffness matrix of the element,

$$[K] = \frac{AE}{L} \begin{bmatrix} l^2 & lm & -l^2 & -lm \\ lm & m^2 & -lm & -m^2 \\ -l^2 & -lm & l^2 & lm \\ -lm & -m^2 & lm & m^2 \end{bmatrix} = \frac{200 \times 70,000}{50 \times 25} \begin{bmatrix} 16 & 12 & -16 & -12 \\ 12 & 9 & -12 & -9 \\ -16 & -12 & 16 & 12 \\ -12 & -9 & 12 & 9 \end{bmatrix}$$

Example 4.4

Determine the stiffness matrix, stresses and reactions in the truss structure shown below, assuming points 1 and 3 are fixed. Use E = 200 GPa and A = 1000 mm².

Solution

Stiffness matrix of any truss element is given by

$$[K] = \frac{AE}{L} \begin{bmatrix} l^2 & lm & -l^2 & -lm \\ lm & m^2 & -lm & -m^2 \\ -l^2 & -lm & l^2 & lm \\ -lm & -m^2 & lm & m^2 \end{bmatrix}$$

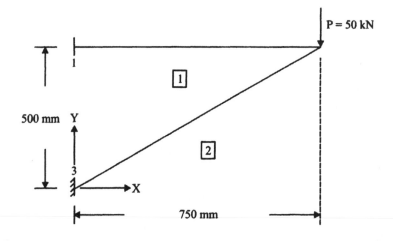

In the given problem, $L_1 = 750$ mm; $L_2 = \sqrt{[750^2 + 500^2]} = 250\sqrt{13}$

For element-1, $l = \dfrac{(x_2 - x_1)}{L_1} = 1$ and $m = \dfrac{y_2 - y_1}{L_1} = 0$

$$[K]_1 = \frac{AE}{750} \begin{bmatrix} 1 & 0 & -1 & 0 \\ 0 & 0 & 0 & 0 \\ -1 & 0 & 1 & 0 \\ 0 & 0 & 0 & 0 \end{bmatrix}$$

$$\frac{AE}{750} = 266.67 \times 10^3$$

For element-2, $l = \dfrac{(x_3 - x_2)}{L_2} = \dfrac{3}{\sqrt{13}}$ and $m = \dfrac{y_3 - y_2}{L_2} = \dfrac{2}{\sqrt{13}}$

$$[K]_2 = \frac{AE}{250 \times 13 \times \sqrt{13}} \begin{bmatrix} 9 & 6 & -9 & -6 \\ 6 & 4 & -6 & -4 \\ -9 & -6 & 9 & 6 \\ -6 & -4 & 6 & 4 \end{bmatrix}$$

$$\frac{AE}{250 \times 13 \sqrt{13}} = 17.07 \times 10^3$$

The assembled stiffness matrix is given by appropriate addition of stiffness coefficients of the two elements,

$$[K] = 10^3 \begin{bmatrix} 266.67 & 0 & -266.67 & 0 & 0 & 0 \\ 0 & 0 & 0 & 0 & 0 & 0 \\ -266.67 & 0 & 266.67-153.63 & 102.42 & -153.63 & -102.42 \\ 0 & 0 & -102.42 & 68.28 & -102.42 & -68.28 \\ 0 & 0 & -153.63 & -102.42 & 153.63 & 102.42 \\ 0 & 0 & -102.42 & -68.28 & 102.42 & 68.28 \end{bmatrix}$$

After applying boundary conditions that $u_1 = v_1 = u_3 = v_3 = 0$, the load-displacement relationships reduce to $\{P\}_R = [K]_R \{u\}_R$

$$\begin{Bmatrix} 0 \\ -50000 \end{Bmatrix} = 10^3 \begin{bmatrix} -266.67 + 153.63 & 102.42 \\ 102.42 & 68.28 \end{bmatrix} \begin{Bmatrix} u_2 \\ v_2 \end{Bmatrix}$$

Solving these two simultaneous equations gives

$$u_2 = 0.2813 \text{ mm} \quad \text{and} \quad v_2 = -1.154 \text{ mm}$$

Displacements of element-1 in local coordinate system are given by

$$\{q_1'\} = \begin{bmatrix} 1 & 0 & 0 & 0 \\ 0 & 0 & 1 & 0 \end{bmatrix} \begin{Bmatrix} 0 \\ 0 \\ 0.2813 \\ -1.154 \end{Bmatrix} = \begin{Bmatrix} 0 \\ 0.2813 \end{Bmatrix}$$

Stress in element-1, $\sigma_1 = E \, \varepsilon_1 = E[-1/L \quad 1/L]\{q_1\}$

$$= 200 \times 10^3 \times 0.2813 / 750 = 75 \text{ N/mm}^2$$

Displacements of element-2 in local coordinate system are given by

$$\{q_2'\} = \begin{bmatrix} 3/\sqrt{13} & 2/\sqrt{13} & 0 & 0 \\ 0 & 0 & 3\sqrt{13} & 2\sqrt{13} \end{bmatrix} \begin{Bmatrix} 0.28313 \\ -1.154 \\ 0 \\ 0 \end{Bmatrix} = \begin{Bmatrix} -0.406 \\ 0 \end{Bmatrix}$$

Stress in element-2, $\sigma_2 = E \, \varepsilon_2 = E\left[\dfrac{-1}{L} \quad \dfrac{1}{L}\right]\{q_2'\}$

$$= 200 \times 10^3 \times \frac{(-0.406)}{250\sqrt{13}} = 90.08 \text{ N/mm}^2$$

Reactions at the two fixed ends are obtained from the equations of the assembled stiffness matrix corresponding to the specified zero displacements

$$\begin{Bmatrix} R_{1-X} \\ R_{1-Y} \\ R_{1-X} \\ R_{1-Y} \end{Bmatrix} = 10^3 \begin{bmatrix} 266.67 & 0 & -266.67 & 0 & 0 & 0 \\ 0 & 0 & 0 & 0 & 0 & 0 \\ 0 & 0 & -153.63 & -102.42 & 153.63 & 102.42 \\ 0 & 0 & -102.42 & -68.28 & 102.42 & 68.28 \end{bmatrix} \begin{Bmatrix} 0 \\ 0 \\ 0.2813 \\ -1.154 \\ 0 \\ 0 \end{Bmatrix}$$

$$= \begin{Bmatrix} -75014.3 \\ 0 \\ 74976.6 \\ 49984.4 \end{Bmatrix} \text{N}$$

The exact solution can be obtained from the equilibrium conditions as follows -

The force in element-2 is such that its vertical component is equal to the applied load P. Horizontal component of this force is given by

P x (750/500) = 75000 N

$$R_{3-Y} + P = 0 \qquad \text{or} \qquad R_{3-Y} = 50,000 \text{ N}$$

$$R_{3-X} + R_{1-X} = 0 \qquad \text{or} \qquad R_{1-X} = -R_{3-X} = 75,000 \text{ N}$$

It can be seen that the approximate solution obtained by FEM is in close agreement with the exact solution obtained from equilibrium consideration.

Example 4.5

Determine the nodal displacements, element stresses and support reactions in the truss structure shown below, assuming points 1 and 3 are fixed. Use E = 70 GPa and A = 200 mm^2.

Solution

Stiffness matrix of any truss element is given by

$$[K]\frac{AE}{L}\begin{bmatrix} l^2 & lm & -l^2 & -lm \\ lm & m^2 & -lm & -m^2 \\ -l^2 & -lm & l^2 & lm \\ -lm & -m^2 & lm & m^2 \end{bmatrix}$$

In the given problem, $L_1 = 500$ mm ; $\quad L_2 = \sqrt{[400^2 + 300^2]} = 500$ mm

For element-1, $\quad l = \dfrac{(x_2 - x_1)}{L_1} = 1 \quad$ and $\quad m = \dfrac{(y_2 - y_1)}{L_1} = 0$

$$[K]_1 = \frac{AE}{500}\begin{bmatrix} 1 & 0 & -1 & 0 \\ 0 & 0 & 0 & 0 \\ -1 & 0 & 1 & 0 \\ 0 & 0 & 0 & 0 \end{bmatrix}$$

$$\frac{AE}{500} = 28 \times 10^3$$

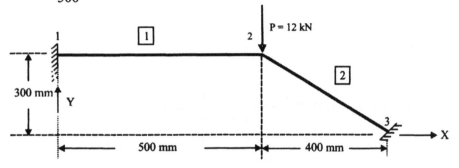

For element-2, $l = \dfrac{(x_3 - x_2)}{L_2} = \dfrac{4}{5}$ and $m = \dfrac{(y_3 - y_2)}{L_2} = \dfrac{-3}{5}$

$$[K]_2 = \dfrac{AE}{500 \times 25} \begin{bmatrix} 16 & -12 & -16 & 12 \\ -12 & 9 & 12 & -9 \\ -16 & 12 & 16 & -12 \\ 12 & -9 & -12 & 9 \end{bmatrix}$$

$$\dfrac{AE}{25 \times 500} = \dfrac{28 \times 10^3}{25}$$

The assembled stiffness matrix is given by appropriate addition of stiffness coefficients of the two elements,

$$[K] = 28 \times 10^3 \begin{bmatrix} 1 & 0 & -1 & 0 & 0 & 0 \\ 0 & 0 & 0 & 0 & 0 & 0 \\ -1 & 0 & 1+(16/25) & -12/25 & -16/25 & 12/125 \\ 0 & 0 & -12/25 & 9/25 & 12/25 & -9/25 \\ 0 & 0 & -16/25 & 12/25 & 16/25 & -12/25 \\ 0 & 0 & 12/25 & -9/25 & -12/25 & 9/25 \end{bmatrix}$$

After applying boundary conditions that $u_1 = v_1 = u_3 = v_3 = 0$, the load-displacement relationships reduce to $\{P\}_R = [K]_R \{u\}_R$

$$\begin{Bmatrix} 0 \\ -12000 \end{Bmatrix} = 28 \times 10^3 \begin{bmatrix} 1+(16/25) & -12/25 \\ -12/25 & 9/25 \end{bmatrix} \begin{Bmatrix} u_2 \\ v_2 \end{Bmatrix}$$

Solving these two simultaneous equations gives

$u_2 = -4/7$ mm and $v_2 = -41/21$ mm

Displacements of element-1 in local coordinate system are given by

$$\{q_1\} = \begin{bmatrix} 1 & 0 & 0 & 0 \\ 0 & 0 & 1 & 0 \end{bmatrix} \begin{Bmatrix} 0 \\ 0 \\ -4/7 \\ -41/21 \end{Bmatrix} = \begin{Bmatrix} 0 \\ -4/7 \end{Bmatrix}$$

Stress in element-1, $\sigma_1 = E\,\varepsilon_1 = E \begin{bmatrix} \dfrac{-1}{L} & \dfrac{1}{L} \end{bmatrix} \{q_1\}$

$$= \dfrac{70 \times 10^3}{500} \times \left(\dfrac{-4}{7} \right) = -80 \text{ N/mm}^2$$

Displacements of element-2 in local coordinate system are given by

$$\{q_2\} = \begin{bmatrix} 4/5 & -3/5 & 0 & 0 \\ 0 & 0 & 4/5 & -3/5 \end{bmatrix} \begin{Bmatrix} -4/7 \\ -41/21 \\ 0 \\ 0 \end{Bmatrix} = \begin{Bmatrix} 15/21 \\ 0 \end{Bmatrix}$$

Stress in element-2, $\sigma_2 = E\,\varepsilon_2 = E\left[\dfrac{-1}{L} \quad \dfrac{1}{L}\right]\{q_2\}$

$$= -\frac{70 \times 10^3}{500} \times \left(\frac{15}{21}\right) = -100 \text{ N/mm}^2$$

Reactions at the two fixed ends are obtained from the equations of the assembled stiffness matrix corresponding to the specified zero displacements

$$\begin{Bmatrix} R_{1-X} \\ R_{1-Y} \\ R_{3-X} \\ R_{3-Y} \end{Bmatrix} = 28 \times 10^3 \begin{bmatrix} 1 & 0 & -1 & 0 & 0 & 0 \\ 0 & 0 & 0 & 0 & 0 & 0 \\ 0 & 0 & -16/25 & 12/25 & 16/25 & -12/25 \\ 0 & 0 & 12/25 & -9/25 & -12/25 & 9/25 \end{bmatrix} \begin{Bmatrix} 0 \\ 0 \\ -4/7 \\ -41/21 \\ 0 \\ 0 \end{Bmatrix}$$

$$= \begin{Bmatrix} 16000 \\ 0 \\ -16000 \\ 12000 \end{Bmatrix} \text{N}$$

Example 4.6

Estimate the displacement vector, stresses and reactions for the truss structure as shown below.

Solution

Stiffness matrix of any truss element is given by

$$[K] = \frac{AE}{L} \begin{bmatrix} l^2 & lm & -l^2 & -lm \\ lm & m^2 & -lm & -m^2 \\ -l^2 & -lm & l^2 & lm \\ -lm & -m^2 & lm & m^2 \end{bmatrix}$$

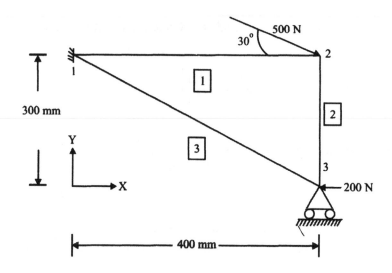

In the given problem,

$$L_1 = 400 \text{ mm} ;$$

$$L_2 = 300 \text{ mm} ;$$

and $\quad L_3 = \sqrt{400^2 + 300^2} = 500 \text{ mm}$

For element-1, $\quad l = \dfrac{(x_2 - x_1)}{L_1} = 1$ and $\quad m = \dfrac{(y_2 - y_1)}{L_1} = 0$

For element-2, $\quad l = \dfrac{(x_3 - x_2)}{L_2} = 0$ and $\quad m = \dfrac{(y_3 - y_2)}{L_2} = -1$

$$[K]_1 = \frac{AE}{400}\begin{bmatrix} 1 & 0 & -1 & 0 \\ 0 & 0 & 0 & 0 \\ -1 & 0 & 1 & 0 \\ 0 & 0 & 0 & 0 \end{bmatrix} ; \quad [K]_2 = \frac{AE}{300}\begin{bmatrix} 0 & 0 & 0 & 0 \\ 0 & 1 & 0 & -1 \\ 0 & 0 & 0 & 0 \\ 0 & -1 & 0 & 1 \end{bmatrix}$$

For element-3, $\quad l = \dfrac{(x_3 - x_1)}{L_3} = \dfrac{4}{5}$ and $\quad m = \dfrac{(y_3 - y_1)}{L_3} = \dfrac{-3}{5}$

$$[K]_3 = \frac{AE}{12500}\begin{bmatrix} 16 & -12 & -16 & 12 \\ -12 & 9 & 12 & -9 \\ -16 & 12 & 16 & -12 \\ 12 & -9 & -12 & 9 \end{bmatrix}$$

The assembled stiffness matrix is given by appropriate addition of stiffness coefficients of the two elements,

$$[K] = \frac{AE}{100} \begin{bmatrix} 1/4+16/25 & -12/125 & -1/4 & 0 & -16/125 & 12/125 \\ -12/125 & 9/125 & 0 & 0 & 12/125 & -9/125 \\ -1/4 & 0 & 1/4 & 0 & 0 & 0 \\ 0 & 0 & 0 & 1/3 & 0 & -1/3 \\ -16/125 & 12/125 & 0 & 0 & 16/25 & -12/125 \\ 12/125 & -9/125 & 0 & -1/3 & -12/125 & 1/3+9/125 \end{bmatrix}$$

After applying boundary conditions that $u_1 = v_1 = v_2 = 0$, the load-displacement relationships reduce to $\{P\}_R = [K]_R \{u\}_R$

$$\begin{Bmatrix} 500\,Cos\,30 \\ -500\,Sin\,30 \\ -200 \end{Bmatrix} = \frac{AE}{100} \begin{bmatrix} 1/4 & 0 & 0 \\ 0 & 1/3 & 0 \\ 0 & 0 & 16/125 \end{bmatrix} \begin{Bmatrix} u_2 \\ v_2 \\ u_3 \end{Bmatrix}$$

These three equations give $u_2 = 400 \times 500 \dfrac{\sqrt{3}}{2\,AE} = 100000 \dfrac{\sqrt{3}}{AE}\,mm$

$$v_2 = \frac{-500 \times 300}{2\,AE} = -\frac{75000}{AE}\,mm \quad u_3 = \frac{-200 \times 12500}{16\,AE} = \frac{-250000}{16\,AE}\,mm$$

Displacements of element-3 in local coordinate system are given by

$$\{q_3'\} = \begin{bmatrix} 4/5 & -3/5 & 0 & 0 \\ 0 & 0 & 4/5 & -3/5 \end{bmatrix} \begin{Bmatrix} 0 \\ 0 \\ -250000/16\,AE \\ 0 \end{Bmatrix} = \begin{Bmatrix} 0 \\ 200000/16\,AE \end{Bmatrix}$$

Stress in element-1,

$$\sigma_1 = E\,\varepsilon_1 = \frac{E\,u_2}{L_1} = E \times 100000 \frac{\sqrt{3}}{(400\,AE)} = \frac{250\sqrt{3}}{A}\,N/mm^2$$

Stress in element-2,

$$\sigma_2 = E\,\varepsilon_2 = \frac{E\,v_2}{L_2} = \frac{E \times 75000}{(300\,AE)} = \frac{250}{A}\,N/mm^2$$

Stress in element-3,

$$\sigma_3 = E\,\varepsilon_3 = \frac{E\,q_2}{L_3} = \frac{E \times 200000}{(500\,AE)} = \frac{400}{A}\,N/mm^2$$

Reactions are obtained from the equations corresponding to the fixed DOFs in the assembled stiffness matrix as given below:

$$R_{1-x} = \left(\frac{AE}{100}\right)\left[\left(\frac{1}{4}+\frac{16}{125}\right) \quad -\frac{12}{125} \quad \frac{-1}{4} \quad 0 \quad \frac{-16}{125} \quad \frac{12}{125}\right]\begin{bmatrix} u_1 \\ v_1 \\ u_2 \\ v_2 \\ u_3 \\ v_3 \end{bmatrix}$$

$$= \frac{AE}{100}\left[\left(\frac{-1}{4}\right)\left(\frac{100000\sqrt{3}}{AE}\right) + \left(\frac{-16}{125}\right)\left(\frac{-250000}{16\,AE}\right)\right]$$

$$= -250\sqrt{3} + 200 \text{ N}$$

$$R_{1-Y} = \left(\frac{AE}{100}\right)\left[\frac{-12}{125} \quad \frac{9}{125} \quad 0 \quad 0 \quad \frac{12}{125} \quad \frac{-9}{125}\right]\begin{Bmatrix} u_1 \\ v_1 \\ u_2 \\ v_2 \\ u_3 \\ v_3 \end{Bmatrix}$$

$$= \frac{AE}{100}\left[(12/125)\,(-250000/16\,AE)\right] = -150 \text{ N}$$

$$R_{3-Y} = \left(\frac{AE}{100}\right)\left[\frac{12}{125} \quad \frac{-9}{125} \quad 0 \quad \frac{-1}{3} \quad \frac{-12}{125} \quad \left(\frac{1}{3}+\frac{9}{125}\right)\right]\begin{Bmatrix} u_1 \\ v_1 \\ u_2 \\ v_2 \\ u_3 \\ v_3 \end{Bmatrix}$$

$$= \frac{AE}{100}\left[\left(\frac{-1}{3}\right)\times\left(\frac{-75000}{AE}\right) + \left(\frac{-12}{125}\right)\left(\frac{-250000}{16\,AE}\right)\right] = 400 \text{ N}$$

Check : For equilibrium of the truss, from basic strength of materials,

$$R_{1-X} = -500 \cos 30 + 200 \text{ N}$$

and $R_{1-Y} + R_{3-Y} = 500 \sin 30 = 250 \text{ N}$

These two equations are satisfied by the results obtained

Example 4.7

For the three bar truss shown in figure below, determine the displacements of node 'A' and the stress in element 3.

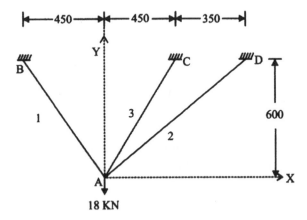

$$A = 250 \text{ mm}^2 ; \quad E = 200 \text{ GPa}$$

For a truss element with direction cosines l and m, w.r.t. global X-axis,

$$\left[K_e' \right] = \frac{AE}{L} \begin{bmatrix} l^2 & lm & -l^2 & -lm \\ lm & m^2 & -lm & -m^2 \\ -l^2 & -lm & l^2 & lm \\ -lm & -m^2 & lm & m^2 \end{bmatrix}$$

where,	Elem.No.	Nodes	L (mm)	$l = X_{ji}/L$	$m = (Y_{ji}/L)$
	1	A B	750	–0.6	0.8
	2	A C	1000	0.8	0.6
	3	A D	750	0.6	0.8

After assembling and applying boundary conditions, u = v = 0   at b, c and d we get

$$\begin{Bmatrix} 0 \\ -18000 \end{Bmatrix} = \left(250 \times 200 \times \frac{10^3}{250} \right) \begin{bmatrix} 0.36/3 + 0.64/4 + 0.36/3 & -0.48/3 + 0.48/4 + 0.48/3 \\ -0.48/3 + 0.48/4 + 0.48/3 & 0.64/3 + 0.36/4 + 0.64/3 \end{bmatrix} \begin{Bmatrix} u_1 \\ v_1 \end{Bmatrix}$$

Solving these two equations, we get

$u_1 = 0.06036$ mm; $v_1 = -0.2012$ mm

Stress in element 3,

$$\sigma_3 = E\,\varepsilon_3 = E\left[B_3\right]\{q'\}_{A-C} = E\left[\dfrac{-1}{L} \quad \dfrac{1}{L}\right]\{q'\}_{A-C} = \left(\dfrac{E}{L}\right)\left[-l \quad -m \quad l \quad m\right]\{q\}_{A-C}$$

$$= \dfrac{200 \times 10^3}{750}\left[-0.6 \quad -0.8 \quad 0.6 \quad 0.8\right]\begin{Bmatrix} 0.06036 \\ -0.2012 \\ 0.0 \\ 0.0 \end{Bmatrix}$$

$$= 33.264 \text{ N/mm}^2$$

Example 4.8

A concentrated load P = 50 kN is applied at the center of a fixed beam of length 3m, depth 200 mm and width 120 mm. Calculate the deflection and slope at the mid point. Assume E = 2 × 10⁵ N/mm².

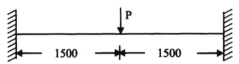

Solution

The finite element model consists of 2 beam elements, as shown here, with nodes 1 and 3 at the two fixed supports and node 2 at the location where load P is applied.

$$1 \rule{1.5cm}{0.4pt} 2 \rule{1.5cm}{0.4pt} 3$$
$$\boxed{1} \qquad \boxed{2}$$

Stiffness matrices of elements 1 and 2 (connected by nodes 1 and 2 ; 2 and 3 respectively, each with L = 1500 mm) are given by,

$$[K] = \dfrac{E\,I_z}{L^3}\begin{bmatrix} 12 & 6L & -12 & 6L \\ 6L & 4L^2 & -6L & 2L^2 \\ -12 & -6L & 12 & -6L \\ 6L & 2L^2 & -6L & 4L^2 \end{bmatrix} = \dfrac{2\times10^5 \times \dfrac{(120\times200^3)}{12}}{L^3}\begin{bmatrix} 12 & 6L & -12 & 6L \\ 6L & 4L^2 & -6L & 2L^2 \\ -12 & -6L & 12 & -6L \\ 6L & 2L^2 & -6L & 4L^2 \end{bmatrix}$$

Assembling the element stiffness matrices, we get

$$\begin{Bmatrix} P_1 \\ M_1 \\ P_2 \\ M_2 \\ P_3 \\ M_3 \end{Bmatrix} = \dfrac{2\times10^5 \times \dfrac{120\times200^3}{12}}{1500^3}\begin{bmatrix} 12 & 6L & -12 & 6L & 0 & 0 \\ 6L & 4L^2 & -6L & 2L^2 & 0 & 0 \\ -12 & -6L & 12+12 & -6L+6L & -12 & 6L \\ 6L & 2L^2 & -6L+6L & 4L^2+4L^2 & -6L & 2L^2 \\ 0 & 0 & -12 & -6L & 12 & -16L \\ 0 & 0 & 6L & 2L^2 & -6L & 4L^2 \end{bmatrix}\begin{Bmatrix} w_1 \\ \theta_1 \\ w_2 \\ \theta_2 \\ w_3 \\ \theta_3 \end{Bmatrix}$$

After applying boundary conditions $v_1 = v_3 = 0$ and $(\theta_z)_1 = (\theta_z)_3 = 0$, the equations reduce to

$$\begin{Bmatrix} P_2 \\ M_2 \end{Bmatrix} = \frac{2 \times 10^5 \times \dfrac{\left(120 \times 200^3\right)}{12}}{1500^3} \begin{bmatrix} 12 + 12 & -6L + 6L \\ -6L + 6L & 4L^2 + 4L^2 \end{bmatrix} \begin{Bmatrix} v_2 \\ (\theta_z)_2 \end{Bmatrix}$$

The applied loads are $P_2 = -50000$ N and $M_2 = 0$

Therefore, $v_2 = \dfrac{-50000 \times 1500^3}{\left[2 \times 10^5 \times \dfrac{\left(120 \times 200^3\right)}{12} \times 24 \right]} = -0.4395$ mm

and $(\theta_z)_2 = 0$

Check : From strength of materials approach, $v_3 = \dfrac{-P\,L^3}{24\,EI}$ or $\dfrac{P\left(2\,L\right)^3}{192\,EI}$

$$= -0.4395 \text{ mm}$$

and the deflection being symmetric, slope at the center $(\theta_z)_2 = 0$.

CHAPTER 5

CONTINUUM (2-D & 3-D) ELEMENTS

5.1   2-D ELEMENTS SUBJECTED TO IN-PLANE LOADS

When one of the cross sectional dimensions, width is significant compared to the length of the member while the thickness is very small, it is considered as a 2-D element. Displacement variation is therefore neglected across the thickness. Let us consider the element in the X-Y plane while dimension in the Z-direction represents the thickness of the element. The load is assumed to be acting in the plane of the element, along X-direction and/or Y-direction. Such a plane element has two degrees of freedom per node, displacements along X and Y directions.

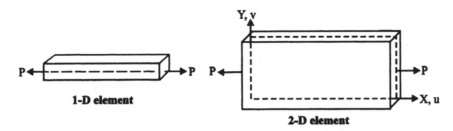

1-D element

2-D element

If a concentrated load is applied at a point on the width of the plate, load can not be considered as uniformly distributed over the width and hence the displacement 'u' at any point is a function of its x and y coordinates. Load along X-direction produces lateral strain and, hence, a displacement 'v' in the y-direction (because of Poisson's effect). Thus, displacements u and v are functions of x and y coordinates of the point.

In the case of discrete structures, with each member treated as a 1-D element, nodes are chosen at junctions of two discrete members, junctions of two different materials, at points of change of cross section or at points of load application. However, in the case of continuum, which is modeled by 2-D or 3-D elements, there is no unique finite element model for analysis. *Each engineer may use a particular number of nodes and a particular orientation of elements. Hence, the results obtained by different engineers may vary.* Mandatory safety codes for the design of pressure vessels are not therefore based on the results of FEM. *The results obtained by FEM have to be suitably modified for compliance with mandatory safety codes.* Varying number or type of elements, but at a higher computational cost, may improve accuracy. A judicial compromise has to be made between better accuracy of results and computational cost. This aspect is further discussed under 'modelling techniques'

5.2 SIMPLEX, COMPLEX AND MULTIPLEX ELEMENTS

Finite elements are classified into three categories.

- **Simplex** elements are those obtained by joining n + 1 nodes in n-dimensional space. Displacement functions of such elements consist of only constant terms and linear terms, if nodal DOFs include only translational modes.

  Ex :   2-noded truss (1-D) element, displacement represented by $u = a_1 + a_2x$

  3-noded triangular (2-D) element, displacement represented by

  $u = a_1 + a_2x + a_3y$

  4-noded tetrahedron (3-D) element, displacement represented by
  $u = a_1 + a_2x + a_3y + a_4z$

- **Complex** elements are those elements whose displacement function consists of quadratic or higher order terms. Such elements naturally need additional boundary nodes and, sometimes, internal nodes.

  Ex :   Quadratic models like 6-noded triangular element and

  10-noded tetrahedron element

  Cubic models like 10-noded triangular element and

  20-noded tetrahedron element

- **Multiplex** elements are those elements whose boundaries are parallel to the coordinate axes and whose displacement function consists of higher order terms.

  Ex :   4-noded rectangle (2-D) element

  8-noded hexahedron (3-D brick) element

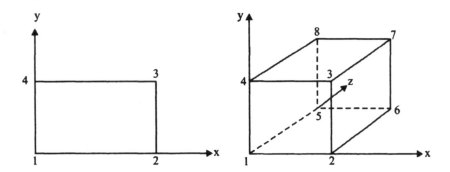

Calculation of stiffness matrix for a triangular element is first considered since triangular elements are the simplest and can be used to define arbitrary boundaries of a component more conveniently, by approximating curved boundary with a large number of elements having straight edges.

5.3 STIFFNESS MATRIX OF A CST ELEMENT

Let $u(x, y) = a_1 + a_2.x + a_3.y$ and $v(x,y) = a_4 + a_5.x + a_6.y$ be the displacements in the element. Displacement function $f(x,y)$, representing u or v, can be graphically represented by the following figure. In general, 1-1', 2-2' and 3-3' are not equal.

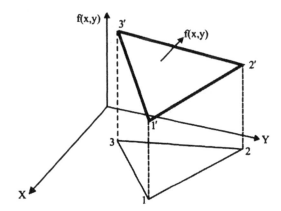

FIGURE 5.1 Graphical representation of displacement function on a triangle

Substituting nodal coordinates, while the element 1-2-3 displaces to 1'-2'-3' on application of load, we get nodal displacement vector as

$$\begin{Bmatrix} u_1 \\ u_2 \\ u_3 \\ v_1 \\ v_2 \\ v_3 \end{Bmatrix} = \begin{bmatrix} 1 & x_1 & y_1 & 0 & 0 & 0 \\ 1 & x_2 & y_2 & 0 & 0 & 0 \\ 1 & x_3 & y_3 & 0 & 0 & 0 \\ 0 & 0 & 0 & 1 & x_1 & y_1 \\ 0 & 0 & 0 & 1 & x_2 & y_2 \\ 0 & 0 & 0 & 1 & x_3 & y_3 \end{bmatrix} \begin{Bmatrix} a_1 \\ a_2 \\ a_3 \\ a_4 \\ a_5 \\ a_6 \end{Bmatrix}$$

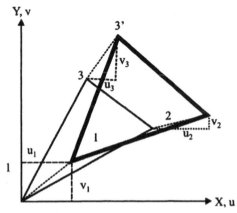

$$\{u_e\} = [G]\{a\} = \begin{bmatrix} [A] & [O] \\ [O] & [A] \end{bmatrix}\{a\}$$

where, $[A] = \begin{bmatrix} 1 & x_1 & y_1 \\ 1 & x_2 & y_2 \\ 1 & x_3 & y_3 \end{bmatrix}$ and $[O] = \begin{bmatrix} 0 & 0 & 0 \\ 0 & 0 & 0 \\ 0 & 0 & 0 \end{bmatrix}$

or $\{a\} = [G]^{-1}\{u_e\} = \begin{bmatrix} [A]^{-1} & [O] \\ [O] & [A]^{-1} \end{bmatrix}\{u_e\}$

$$\begin{Bmatrix} u \\ v \end{Bmatrix} = [f(x,y)]\{a\} = [f(x,y)][G]^{-1}\{u_e\}$$

$$\begin{Bmatrix} \varepsilon_x \\ \varepsilon_y \\ \gamma_{xy} \end{Bmatrix} = \begin{Bmatrix} \partial u/\partial x \\ \partial v/\partial y \\ \partial u/\partial y + \partial v/\partial x \end{Bmatrix} = [f'(x,y)][G]^{-1}$$

$$\{\varepsilon\} = \begin{bmatrix} 0 & 1 & 0 & 0 & 0 & 0 \\ 0 & 0 & 0 & 0 & 0 & 1 \\ 0 & 0 & 1 & 0 & 1 & 0 \end{bmatrix}[G]^{-1}\{u_e\} = [B]\{u_e\}$$

where, strain-displacement matrix,

$$[B] = \frac{1}{Det\ J} \begin{bmatrix} y_{23} & 0 & y_{31} & 0 & y_{12} & 0 \\ 0 & x_{32} & 0 & x_{13} & 0 & x_{21} \\ x_{32} & y_{23} & x_{13} & y_{31} & x_{21} & y_{12} \end{bmatrix} \text{where [J] is Jacobian}$$

and

$$Det\ J = \begin{vmatrix} 1 & x_1 & y_1 \\ 1 & x_2 & y_2 \\ 1 & x_3 & y_3 \end{vmatrix} = \begin{vmatrix} 0 & x_1 - x_3 & y_1 - y_3 \\ 0 & x_2 - x_3 & y_2 - y_3 \\ 1 & x_3 & y_3 \end{vmatrix} = \begin{vmatrix} 0 & x_{13} & y_{13} \\ 0 & x_{23} & y_{23} \\ 1 & x_3 & y_3 \end{vmatrix}$$

$$= x_{13}\ y_{23} - y_{13}\ x_{23}$$

It can be seen that area of the triangle, $A = \frac{1}{2} Det\ J = \left(\frac{1}{2}\right) \begin{vmatrix} 0 & x_{13} & y_{13} \\ 0 & x_{23} & y_{23} \\ 1 & x_3 & y_3 \end{vmatrix}$

If nodes are numbered counter-clockwise, in right-handed coordinate system Det J is +ve.

Stiffness matrix of the 3-node triangular element can now be obtained from

$$[K]_v = \int [B]^T [D][B] dV = t \iint [B]^T [D][B] dx\ dy$$

where $t = \int dz$ is the thickness of the element and

[B] is a function of X and Y only.

Here, matrix [D] corresponding to the particular deformation pattern of the component (plane stress, plane strain or axi-symmetric) is to be used. An explicit evaluation of the stiffness matrix is not generally feasible (except for a few special cases).

This triangular element, with 3 nodes and 2 DOF per node chooses linear displacement functions for u and v and hence gives constant strain terms over the entire element as seen from $[f'(x,y)]$ or [B] and hence is popularly known as 'Constant Strain Triangle (CST)' element.

5.3.1 Stiffness Matrix of Right Angled Triangle

For the particular case of a right angled triangle with coordinates 1(0,0), 2(a, b) and 3(0, b), let $u = a_1 + a_2\ x + a_3\ y$ and $v = a_4 + a_5\ x + a_6\ y$ represent the displacement field. Substituting nodal coordinates, vector of nodal displacements can be written as

$$\begin{Bmatrix} u_1 \\ u_2 \\ u_3 \\ v_1 \\ v_2 \\ v_3 \end{Bmatrix} = \begin{bmatrix} 1 & 0 & 0 & 0 & 0 & 0 \\ 1 & a & b & 0 & 0 & 0 \\ 1 & 0 & b & 0 & 0 & 0 \\ 0 & 0 & 0 & 1 & 0 & 0 \\ 0 & 0 & 0 & 1 & a & b \\ 0 & 0 & 0 & 1 & 0 & b \end{bmatrix} \begin{Bmatrix} a_1 \\ a_2 \\ a_3 \\ a_4 \\ a_5 \\ a_6 \end{Bmatrix} \quad \text{or } \{u_e\} = \begin{bmatrix} [A] & [O] \\ [O] & [A] \end{bmatrix} \{a\} = [G]\,\{a\}$$

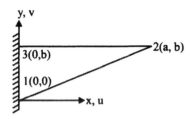

Evaluating coefficients a_1 to a_6 in terms of nodal displacements,

$$\begin{Bmatrix} a_1 \\ a_2 \\ a_3 \\ a_4 \\ a_5 \\ a_6 \end{Bmatrix} = \begin{bmatrix} 1 & 0 & 0 & 0 & 0 & 0 \\ 0 & 1/a & -1/a & 0 & 0 & 0 \\ -1/b & 0 & 1/b & 0 & 0 & 0 \\ 0 & 0 & 0 & 1 & 0 & 0 \\ 0 & 0 & 0 & 0 & 1/a & -1/a \\ 0 & 0 & 0 & -1/b & 0 & 1/b \end{bmatrix} \begin{Bmatrix} u_1 \\ u_2 \\ u_3 \\ v_1 \\ v_2 \\ v_3 \end{Bmatrix}$$

or $\{a\} = \{G\}^{-1}\{u_e\} = \{a\} = \begin{bmatrix} [A]^{-1} & [O] \\ [O] & [A]^{-1} \end{bmatrix}\{u_e\}$

Strain, $\{\varepsilon\} = \begin{Bmatrix} \partial u/\partial x \\ \partial u/\partial y \\ \partial u/\partial y + \partial v/\partial x \end{Bmatrix} = \begin{bmatrix} 0 & 1 & 0 & 0 & 0 & 0 \\ 0 & 0 & 0 & 0 & 0 & 1 \\ 0 & 0 & 1 & 0 & 1 & 0 \end{bmatrix} \{a\} = [B]\,\{u_e\}$

where, $[B] = \left(\dfrac{1}{ab}\right) \begin{bmatrix} 0 & b & -b & 0 & 0 & 0 \\ 0 & 0 & 0 & -a & 0 & a \\ -a & 0 & a & 0 & b & -b \end{bmatrix}$

$$[K] = \int_V [B]^T\,[D][B]\,dv = t \int_0^a \int_0^b [B]^T\,[D][B]\,dx\,dy = tA\,[B]^T\,[D][B]$$

since elements of matrices [B] and [D] are not functions of x or y

For a plane stress case, stress-strain relationship is given by

$$[D] = \frac{E}{(1-v^2)}\begin{bmatrix} 1 & v & 0 \\ v & 1 & 0 \\ 0 & 0 & (1-v)/2 \end{bmatrix} \quad \text{Let} \quad \beta = \frac{(1-v)}{2}$$

Then, with element DOFs arranged in the sequence of $[u_1 \ u_2 \ u_3 \ v_1 \ v_2 \ v_3]^T$

$$[K] = \frac{EtA}{a^2b^2(1-v^2)}\begin{bmatrix} \beta a^2 & & & & & \\ 0 & b^2 & & & & \\ -\beta a^2 & -b^2 & b^2+\beta a^2 & & \text{Symmetric} & \\ 0 & -v\,ab & v\,ab & a^2 & & \\ -\beta\,ab & 0 & \beta\,ab & 0 & \beta b^2 & \\ \beta\,ab & v\,ab & -ab(v+\beta) & -a^2 & -\beta b^2 & a^2+\beta b^2 \end{bmatrix}$$

If the element DOFs are arranged in the sequence of $[u_1 \ v_1 \ u_2 \ v_2 \ u_3 \ v_3]^T$, the elements of stiffness matrix are rearranged as

$$[K] = \frac{EtA}{a^2b^2(1-v^2)}\begin{bmatrix} \beta a^2 & & & & & \\ 0 & a^2 & & & & \\ 0 & -v\,ab & b^2 & & \text{Symmetric} & \\ -\beta\,ab & 0 & 0 & \beta b^2 & & \\ -\beta a^2 & v\,ab & -b^2 & \beta\,ab & b^2+\beta a^2 & \\ \beta\,ab & -a^2 & v\,ab & -\beta b^2 & -ab(v+\beta) & a^2+\beta b^2 \end{bmatrix}$$

5.4 CONVERGENCE CONDITIONS
(TO BE SATISFIED BY THE DISPLACEMENT FUNCTION)

While choosing the function to represent u and v displacements at any point in the element, care should be taken to ensure that the following conditions are satisfied.

(i)   The function should be continuous and differentiable (to obtain strains) within the element. This is automatically satisfied with polynomial functions.

(ii) The displacement polynomial should include constant term, representing rigid body displacement, which any unrestrained portion of a component should experience when subjected to external loads.

(iii) The polynomial should include linear terms, which on differentiation give constant strain terms. Constant strain is the logical condition as the element size reduces to a point in the limit.

(iv) Compatibility of displacement and its derivatives, up to the required order, must be satisfied across inter-element boundaries. Otherwise the displacement solution may result in separated or overlapped inter-element boundaries when the displacement patterns of deformed elements with a common boundary are plotted separately (explained in more detail in section 5.7).

(v) The polynomial shall satisfy geometric isotropy (terms symmetric in terms of coordinate axes x, y and z). Otherwise, different users analysing the same component may get different results by following different node number sequence to define the elements (different local coordinate systems).

Terms used in the polynomial, satisfying all the above conditions, are represented by **Pascal triangle** given below, for a 2-D element.

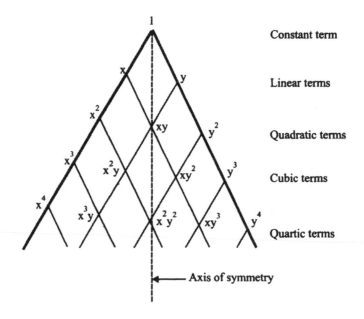

Similarly, the polynomial for a 3-D element is represented by the terms of **Pascal tetrahedron,** as given below.

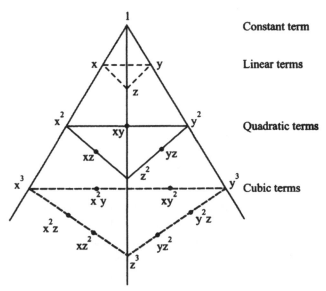

This can also be represented by the nodes of a **hypercube,** as given below. Here, terms with other combinations are on the invisible sides of the cube.

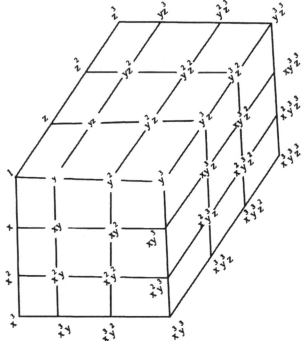

5.5 GEOMETRIC ISOTROPY

If all the terms of a particular order are included in the polynomial, it is called the complete set. If the terms are symmetric w.r.t. x and y, it is called geometric isotropy.

Based on the terms included in the polynomial, the function may be termed as

        complete   and isotropic

        complete   and non-isotropic

        incomplete and isotropic

or   incomplete and non-isotropic

As the number of terms in the polynomial depend on the number of DOF and the number of nodes, use of complete set of terms of a particular order may not be possible in all cases. But, isotropy can be maintained in all the cases and is preferable so that user of a general-purpose program can start with any particular node of his choice for defining the nodal sequences (which decide the local coordinate systems) of different elements of the structure.

For each element, local coordinate system is usually defined with node 1 as the origin; X-axis along 1-2 and Y-axis perpendicular to X-axis in the plane of nodes 1-2-3.

The displacement function $u(x) = a_1 + a_2.x + a_3.y$ of a triangular element is complete and isotropic while $u(x) = a_1 + a_2.x + a_3.y + a_4.xy$ of a quadrilateral element is incomplete but isotropic.

Higher order elements are broadly classified as -

• **Serendipity elements** – These are the elements having no internal nodes

    Ex : 8-noded quadrilateral, 12-noded quadrilateral, etc.

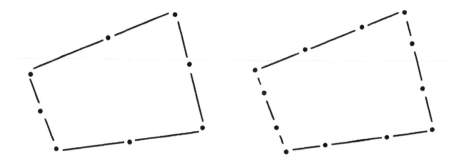

- **Lagrange elements** – These are the elements having internal nodes which can be condensed out at the element level before assembling.

Ex : 9-noded quadrilateral, 16-noded quadrilateral, etc.

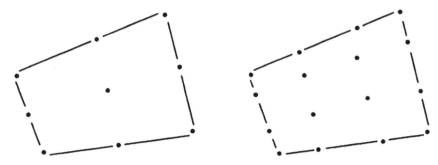

Polynomials used for some 2-D elements (subjected to in-plane loads), satisfying the convergence and isotropy conditions, are given below.

Element	Order of Displacement	No.of nodes	Terms included	Polynomial type
Triangle (Fig.5.2 a)	Linear	3	$a_1 + a_2.x + a_3.y$	Complete & Isotropic
Triangle (Fig.5.2 b)	Quadratic	6	$a_1 + a_2.x + a_3.y + a_4.x^2 + a_5.xy + a_6.y^2$	Complete & Isotropic
Triangle (Fig.5.2 c)	Cubic	9	$a_1 + a_2.x + a_3.y + a_4.x^2 + a_5.xy + a_6.y^2 +$ $a_7.x^2y + a_8.xy^2 + a_9.x^2y^2$	Incomplete, Isotropic
			$a_1 + a_2x + a_3y + a_4x^2 + a_5xy + a_6y^2$ $+ a_7x^2y + a_8xy^2 + a_9x^3$ (Not preferred)	Incomplete Non-Isotropic
Triangle (Fig.5.2 d)	Cubic	10	$a_1 + a_2.x + a_3.y + a_4.x^2 + a_5.xy + a_6.y^2 +$ $a_7.x^3 + a_8.x^2y + a_9.xy^2 + a_{10}.y^3$	Complete, Isotropic
Quadrilateral (Fig.5.2 e)	Linear	4	$a_1 + a_2.x + a_3.y + a_4.xy$	Incomplete, Isotropic
			$a_1 + a_2x + a_3y + a_4x^2$ (Not preferred)	Incomplete, Non-Isotropic
Quadrilateral (Fig.5.2 f)	Quadratic	8	$a_1 + a_2.x + a_3.y + a_4.x^2 + a_5.xy + a_6.y^2 +$ $a_7.x^2y + a_8.xy^2$	Incomplete, Isotropic
Quadrilateral (Fig.5.2 g)	Cubic	12	$a_1 + a_2.x + a_3.y + a_4.x^2 + a_5.xy + a_6.y^2 +$ $a_7.x^3 + a_8.x^2y + a_9.xy^2 + a_{10}.y^3 + a_{11}.x^3y$ $+ a_{12}.xy^3$	Incomplete, Isotropic

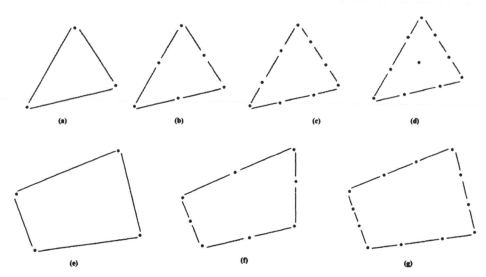

FIGURE 5.2 Some 2-D elements for in-plane loads

5.6 ASPECT RATIO

In 2-D and 3-D elements, the displacement function is symmetric in x, y and z, whether it is complete or not in terms of coefficients of a particular order as given by Pascal triangle or Pascal tetrahedron. Hence, the shape of the finite element in the idealised structure should also be oriented equally to all the relevant axes. For this purpose, certain conditions are generally specified in the standard packages on the sizes and included angles for various elements. Aspect ratio is defined for this purpose as the ratio of the longest side to the

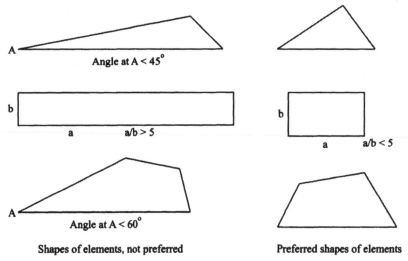

Shapes of elements, not preferred   Preferred shapes of elements

shortest side. It is usually limited to 5, while the included angle is usually limited to 45^0 to 135^0 for a triangular element and to 60^0 to 120^0 for a quadrilateral or 3-D element. A few 2-D elements with valid and invalid shapes are shown here.

5.7 INTER-ELEMENT COMPATIBILITY

The polynomial used to represent variation of displacement over the element should ensure compatibility of displacement along inter-element boundary. If this condition is not satisfied, inter-element boundary of two adjacent elements may overlap or show void on application of external loads, when the displacement pattern of different elements with a common boundary are

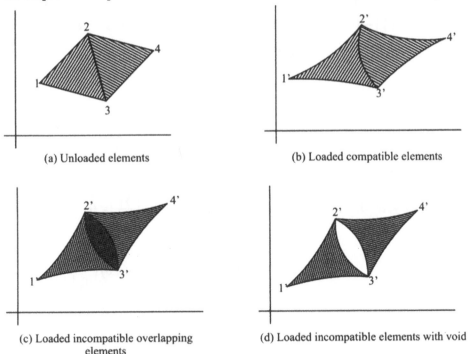

(a) Unloaded elements

(b) Loaded compatible elements

(c) Loaded incompatible overlapping elements

(d) Loaded incompatible elements with void

plotted separately. The inter-element compatibility condition is satisfied when displacement at any point along a common edge, of all elements joining along that edge, is a function of displacements of nodes on that edge only. This concept is demonstrated here for displacement along x axis of a right angled triangular element, with the two sides of lengths a and b coinciding with the coordinate axes.

$$u = \begin{bmatrix} 1 & x & y \end{bmatrix} \begin{Bmatrix} a_1 \\ a_2 \\ a_3 \end{Bmatrix} \quad \therefore \begin{Bmatrix} u_1 \\ u_2 \\ u_3 \end{Bmatrix} = \begin{bmatrix} 1 & 0 & 0 \\ 1 & a & 0 \\ 1 & 0 & b \end{bmatrix} \begin{Bmatrix} a_1 \\ a_2 \\ a_3 \end{Bmatrix} \text{ by substituting nodal coordinates}$$

or

$$\begin{Bmatrix} a_1 \\ a_2 \\ a_3 \end{Bmatrix} = \begin{bmatrix} 1 & 0 & 0 \\ -1/a & 1/a & 0 \\ -1/b & 0 & 1/b \end{bmatrix} \begin{Bmatrix} u_1 \\ u_2 \\ u_3 \end{Bmatrix}$$

Then,

$$u = \begin{bmatrix} 1 & x & y \end{bmatrix} \begin{bmatrix} 1 & 0 & 0 \\ -1/a & 1/a & 0 \\ -1/b & 0 & 1/b \end{bmatrix} \begin{Bmatrix} u_1 \\ u_2 \\ u_3 \end{Bmatrix} = \begin{bmatrix} 1-x/a-y/b & x/a & y/b \end{bmatrix} \begin{Bmatrix} u_1 \\ u_2 \\ u_3 \end{Bmatrix}$$

At $R(x_1,0)$, $u = \begin{bmatrix} 1-x_1/a & x_1/a & 0 \end{bmatrix} \begin{Bmatrix} u_1 \\ u_2 \\ u_3 \end{Bmatrix} = \begin{bmatrix} 1-x_1/a & x_1/a \end{bmatrix} \begin{Bmatrix} u_1 \\ u_2 \end{Bmatrix}$

or u at $R(x_1, 0)$ is a function of u_1 and u_2, displacements of the two end nodes of that edge only.

Similarly, at $S(x_1, y_1)$, $\dfrac{y_1}{b} = \dfrac{(a-x_1)}{a}$ or $1 - \left(\dfrac{x_1}{a}\right) - \left(\dfrac{y_1}{b}\right) = 0$

Then, $u = \begin{bmatrix} 0 & x_1/a & y_1/b \end{bmatrix} \begin{Bmatrix} u_1 \\ u_2 \\ u_3 \end{Bmatrix} = \begin{bmatrix} x_1/a & y_1/b \end{bmatrix} \begin{Bmatrix} u_2 \\ u_3 \end{Bmatrix}$

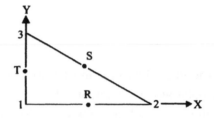

or u at $S(x_1, y_1)$ is a function of u_2 and u_3, displacements of the two end nodes of that edge only.

Similarly,

at $T(0, y_1)$, $u = \begin{bmatrix} 1 - y_1/b & 0 & y_1/b \end{bmatrix} \begin{Bmatrix} u_1 \\ u_2 \\ u_3 \end{Bmatrix} = \begin{bmatrix} 1 - y_1/b & y_1/b \end{bmatrix} \begin{Bmatrix} u_1 \\ u_3 \end{Bmatrix}$

or u at $T(0, y_1)$ is a function of u_1 and u_3, displacements of the two end nodes of that edge only.

The same logic holds good for v, displacement at any point of the element in y direction.

To adequately represent stress concentration in some local regions, it is a common practice to either increase number of elements or increase the order of the polynomial of the displacement function. The first method ensures inter-element displacement compatibility but at a higher computational cost. The second method may not always ensure inter-element displacement compatibility. Transition elements are commonly used in such situations. These are covered in more detail in section 7.10 of this book.

5.8   2-D ELEMENTS SUBJECTED TO BENDING LOADS

Plate bending element : It is a plate element in X-Y plane subjected to bending load P_Z and/or bending moments M_x , M_y. A thin plate (span > 10 x thickness) with small deflection (< thickness/10) follows Kirchhoff's theory and is an extension of 1-D beam element into two dimensions. It will have three degrees of freedom at each node, displacement normal to the plate (w) and rotations about the two major axes of the element represented by derivatives of w about x and y (θ_x and θ_y).

For a triangular Plate bending element, normal deflection is assumed by the polynomial.

$$w = a_1 + a_2.x + a_3.y + a_4.x^2 + a_5.y^2 + a_6.x^2y + a_7.xy^2 + a_8.x^3 + a_9.y^3$$

$$\theta_x = \frac{\partial w}{\partial x} = a_2 + 2a_4.x + 2a_6.xy + a_7.y^2 + 3a_8.x^2$$

$$\theta_y = \frac{\partial w}{\partial y} = a_3 + 2a_5.y + a_6.x^2 + 2a_7.xy + 3a_9.y^2$$

Displacement function of triangular plate bending element is incomplete but isotropic.

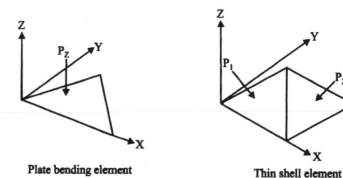

Plate bending element   **Thin shell element**

Thin plate with large deflection is characterised by large tensile or compressive stresses in the middle plane. The corresponding u and v displacements are given by

$$u = -z\left(\frac{\partial w}{\partial x}\right) \quad ; \quad v = -z\left(\frac{\partial w}{\partial y}\right)$$

$$\varepsilon_x = \left(\frac{\partial u}{\partial x}\right) = -z\left[\frac{\partial^2 w}{\partial x^2}\right]$$

$$\varepsilon_y = \left(\frac{\partial v}{\partial y}\right) = -z\left[\frac{\partial^2 w}{\partial y^2}\right]$$

$$\gamma_{xy} = \left(\frac{\partial u}{\partial y}\right) + \left(\frac{\partial y}{\partial x}\right) = -2z\left(\frac{\partial^2 w}{\partial x \, \partial y}\right)$$

The stress-strain matrix is given by $\{\sigma\} = [D]\{\varepsilon\}$

$$\text{or} \quad \begin{Bmatrix} \sigma_x \\ \sigma_y \\ \tau_{xy} \end{Bmatrix} = \frac{E}{1-v^2}\begin{bmatrix} 1 & v & 0 \\ v & 1 & 0 \\ 0 & 0 & (1-v)/2 \end{bmatrix}\begin{Bmatrix} \varepsilon_x \\ \varepsilon_y \\ \gamma_{xy} \end{Bmatrix}$$

This is more commonly expressed in terms of moments per unit width (b=1), also called stress resultants, using

$$M_x = \left(\frac{I}{z}\right)\!.\sigma_x = \left(\frac{bh^3/12}{z}\right)\!.\sigma_x = \left(\frac{h^3/12}{z}\right)\!.(E.\varepsilon_x) = -\left(\frac{Eh^3}{12}\right)\!\left(\frac{\partial^2 w}{\partial x^2}\right)$$

$$\text{Thus,} \quad \begin{Bmatrix} M_x \\ M_y \\ M_{xy} \end{Bmatrix} = \frac{-Eh^3}{12(1-v^2)}\begin{bmatrix} 1 & v & 0 \\ v & 1 & 0 \\ 0 & 0 & (1-v)/2 \end{bmatrix}\begin{Bmatrix} \partial^2 w/\partial x^2 \\ \partial^2 w/\partial y^2 \\ \partial^2 w/\partial x \, \partial y \end{Bmatrix}$$

Thin shell element : It is a 2-D element subjected to in-plane loads as well as bending loads. As in the case of a general beam, these two behaviours represent uncoupled degrees of freedom. It can therefore be considered as a combination of plane stress element and plate bending element. In the local coordinate system, this element will have five degrees of freedom since moment about normal to the plate is not included. However, if different elements are inclined to each other, transformation of the combined stiffness matrix of each element with five degrees of freedom per node in local coordinate system results in six degrees of freedom per node in the global coordinate system.

5.9   3-D ELEMENTS

Similar to the elements described above, 3-D elements are of two types based on the type of function used – *solid element* with three translational displacement degrees of freedom per node without significant bending behaviour and *thick shell element* with six degrees of freedom. The displacement functions normally used are given below.

For a **4-noded 3-D solid tetrahedron element**

$$u = a_1 + a_2.x + a_3.y + a_4.z$$

$$v = a_5 + a_6.x + a_7.y + a_8.z$$

$$w = a_9 + a_{10}.x + a_{11}.y + a_{12}.z$$

These functions are complete and isotropic.

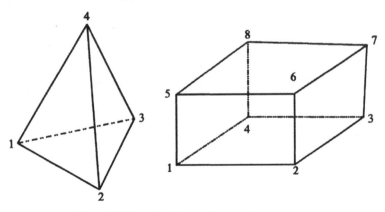

For a **8-noded 3-D solid hexahedron element**

$$u = a_1 + a_2.x + a_3.y + a_4.z + a_5.x^2 + a_6.y^2 + a_7.z^2 + a_8.xyz$$

$$v = a_9 + a_{10}.x + a_{11}.y + a_{12}.z + a_{13}.x^2 + a_{14}.y^2 + a_{15}.z^2 + a_{16}.xyz$$

$$w = a_{17} + a_{18}.x + a_{19}.y + a_{20}.z + a_{21}.x^2 + a_{22}.y^2 + a_{23}.z^2 + a_{24}.xyz$$

The function used is therefore incomplete but isotropic.

Isotropy condition in these elements involves symmetry w.r.t. X, Y and Z axes. The displacement field can also be chosen by a different polynomial

$$u = a_1 + a_2.x + a_3.y + a_4.z + a_5.x^2 + a_6.y^2 + a_7.z^2 + a_8.xyz \; ; \; \ldots$$

Thus, *there is no unique way of choosing an incomplete polynomial*

5.10 AXI-SYMMETRIC ELEMENTS

These are special cases of 3-D components where 2-D analysis can be carried out for evaluating displacements and stresses, saving lot of time and effort. There are many components such as turbine casings, compressor casings, pressure vessels, cylindrical heat exchangers etc., which are 3-D components by the relative dimensions of the component in the three coordinate directions. However, each is symmetric about its axis of rotation and thus deflection and stress along any 2-D radial plane (imaginary section planes A, B,.. in the figure), will be identical. It is often more convenient to represent such components in cylindrical coordinate system, consisting of axial (usually represented by z-axis), radial (or r-axis) and hoop (or circumferential or θ) direction. A section through r-z plane is considered for analysis. Unlike in the case of plane stress or plane strain analysis of 2-D components, it need not be constrained in the radial direction at any point in the component to suppress rigid body motion. This is automatically taken care of by the closed geometry in the hoop direction, thus providing a natural boundary condition.

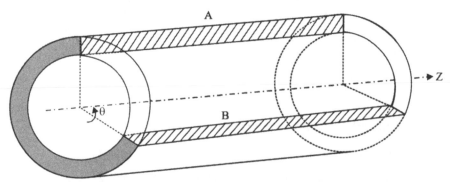

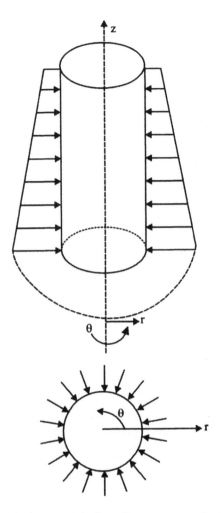

Also, displacement in the radial direction at any point in the component by 'dr' gives rise to a corresponding change in circumferential length by $2\pi(r + dr) - 2\pi r$ or $2\pi(dr)$. It amounts to a hoop strain of $\dfrac{2\pi(dr)}{2\pi r}$ or $\dfrac{dr}{r}$.

Thus the stress-strain relation is a 4×4 matrix relating σ_r, σ_θ, σ_z and $\tau_{r\theta}$ with ε_r, ε_θ, ε_z and $\gamma_{r\theta}$ given by

$$\begin{Bmatrix} \sigma_r \\ \sigma_\theta \\ \sigma_z \\ \tau_{r\theta} \end{Bmatrix} = \frac{E}{(1+v)(1-2v)} \begin{bmatrix} 1-v & v & v & 0 \\ v & 1-v & v & 0 \\ v & v & 1-v & 0 \\ 0 & 0 & 0 & \dfrac{1-2v}{2} \end{bmatrix} \begin{Bmatrix} \varepsilon_r \\ \varepsilon_\theta \\ \varepsilon_z \\ \gamma_{r\theta} \end{Bmatrix} \quad \text{or} \quad \{\sigma\} = [D]\{\varepsilon\}$$

Sometimes, components like shell nozzle junction are analysed as axisymmetric solids, to save time and effort, even though it is not symmetric about nozzle axis or shell axis. For this purpose, cylindrical shell is replaced by a spherical shell of double the radius, because the larger stress component (hoop stress) in a sphere is half the corresponding stress of a cylinder of the same radius. The same is not true for thermal stresses and needs some modification.

5.11 SUMMARY

- While 1-D elements generally form a discrete structure, 2-D and 3-D elements form part of a continuum. In the 1-D element, where axial dimension is very large compared to the cross section, load is assumed to act uniformly over the entire cross section. When one of the cross sectional dimensions, width, is significant compared to the length of the member, it is considered as a 2-D element.

- Finite elements are classified into three categories - *Simplex* elements obtained by joining n+1 nodes in n-dimensional space; *Complex* elements whose displacement function consists of quadratic or higher order terms; *Multiplex* elements whose boundaries are parallel to the coordinate axes.

- The function chosen to represent u and v displacements at any point in the 2-D element should satisfy the following conditions - The function should be continuous and differentiable within the element; should include constant term and linear terms; should satisfy compatibility of displacement and its derivatives, across inter-element boundaries and should satisfy geometric isotropy.

- A model with less number of higher order elements (with more than two nodes along edges of the element) will give better results than more number of lower order elements and is economical in terms of computer memory and time.

- 2-D Plane stress element and 3-D solid element are similar to 1-D truss element, whose degrees of freedom do not include slopes while 2-D Plate bending element and 3-D thick shell element are similar to 1-D beam element, whose degrees of freedom include slopes.

- *Plane stress element* applies to a thin plate with in-plane loads, having zero stress in the normal direction, while *plane strain element* is a thin slice of a large 3-D solid, where load acts in the plane of slices and analysis of one slice represents solution of the 3-D solid, and has zero strain in the normal direction.

- *Axisymmetric element* is also a 2-D model of a 3-D solid with axisymmetric geometry, loads and boundary conditions. It differs from plane strain element, in having finite strain normal to the plane of analysis as a function of radial displacement.

- *Thin shell element* is a superposition of plane stress element and plate bending element, representing uncoupled degrees of freedom.

OBJECTIVE QUESTIONS

1. Complete polynomial is ___ important, compared to symmetry of displacement polynomial w.r.t. coordinate directions

   (a) equally   (b) more   (c) less   (d) unrelated

2. A triangular element with cubic displacement function requires ____ nodes to represent the complete and symmetric polynomial

   (a) 3   (b) 6   (c) 9   (d) 10

3. A triangular element with quadratic displacement function requires ____ nodes to represent the complete polynomial

   (a) 3   (b) 6   (c) 9   (d) 10

4. A triangular 9-noded element will usually have ____ cubic displacement function

   (a) symmetric & complete   (b) symmetric & incomplete

   (c) unsymmetric & complete   (d) unsymmetric & incomplete

5. A constant term in the displacement function ensures

   (a) rigid body mode   (b) constant strain mode

   (c) zero stress   (d) zero deformation

6. Number of terms in the displacement function in relation to the number of nodes in that element is

   (a) more   (b) equal   (c) less   (d) unrelated

7. A linear term in the displacement function ensures

   (a) rigid body mode   (b) constant strain mode

   (c) strain varying in the element   (d) stress varying in the element

8. All stiffness coefficients of a plate bending element have ___ units

   (a) same   (b) different   (c) any set of   (d)   depend on other data

9. All stiffness coefficients of an axisymmetric element have ___ units
    (a) same   (b) different   (c) any set of   (d) depend on other data

10. Displacement method can NOT be used with _____ boundary conditions
    (a) pressure   (b) temperature (c) stress   (d) displacement

11. A triangular plane stress element has ___ D.O.F
    (a) 6   (b) 9   (c) 12   (d) 15

12. A thin shell element has ___ no. of DOF, compared to a plate bending element
    (a) same   (b) more   (c) less   (d) unrelated

13. A plane stress element has ___ no. of DOF, compared to a plate bending element
    (a) same   (b) more   (c) less   (d) unrelated

14. An axisymmetric element has ___ no. of DOF, compared to a plate bending element
    (a) same   (b) more   (c) less   (d) unrelated

15. A structural thin shell triangular element has ___ DOF
    (a) 3   (b) 6   (c) 9   (d) 18

16. A triangular plane strain element has ___ DOF
    (a) 3   (b) 6   (c) 9   (d) 15

17. Number of displacement polynomials used for an element depends on
    (a) No. of nodes/element   (b) No. of DOF/node
    (c) No. of DOF/element   (d) type of element

18. For a plate bending element, number of displacement polynomials and number of D.O.F/node are
    (a) 1,2   (b) 1,3   (c) 2,3   (d) 2,4

19. Accuracy of solution in a 2-D component depends on
    (a) included angle of elements   (b) size of the component
    (c) no. of DOF/ node   (d) type of load

20. Displacement of any point on a side is related to displacements of nodes on that side only, ensures
    (a) equilibrium   (b) compatibility
    (c) energy balance   (d) continuity along inter-element boundary

21. Continuum analysis covers

    (a) all 2-D trusses & frames   (b) all 3-D trusses & frames

    (c) all 2-D and 3-D plates, solids   (d) only 3-D solids

22. Optimum number of elements in finite element model depends on assessment of ___ distribution in the component

    (a) displacement   (b) stress   (c) strain   (d) potential energy

23. Displacement function which matches function value at the specified nodes is classified as

    (a) Lagrange interpolation function   (b) Serendipity function

    (c) Hermite interpolation function   (d) Pascal function

24. Displacement function which matches function value as well as its derivatives (slopes) at the specified nodes is classified as

    (a) Lagrange interpolation function   (b) Serendipity function

    (c) Hermite interpolation function   (d) Pascal function

25. Continuum analysis includes

    (a) trusses   (b) beams   (c) plates   (d) plates & solids

26. Continuum elements and discrete members can be included in a single model for analysis

    (a) always true   (b) sometimes true

    (c) never true   (d) depends on matching DOF

27. Continuum elements in different analysis may vary in

    (a) size   (b) shape   (c) size or shape   (d) size & shape

28. Element formed with edges parallel to coordinate axes is called

    (a) simplex element   (b) complex element

    (c) multiplex element   (d) compound element

29. An element with no internal nodes is classified as

    (a) serendipity element   (b) Lagrange element

    (c) Hermite element   (d) Laplace element

30. An element with internal nodes is classified as

    (a) serendipity element   (b) Lagrange element

    (c) Hermite element   (d) Laplace element

31. A concrete pedestal is represented by
    - (a) plane stress elements
    - (b) plane strain elements
    - (c) 3-D solid elements
    - (d) 3-D shell elements

32. Combination of plane stress element behaviour and plate bending behaviour forms
    - (a) 3-D solid element
    - (b) 3-D shell element
    - (c) Thin shell element
    - (d) thick shell element

33. A 3-D dam is usually modeled with
    - (a) 2-D plane stress elements
    - (b) 2-D plane strain elements
    - (c) 3-D solid elements
    - (d) 3-D shell elements

34. Element formed by joining n+1 nodes in n-dimensional space is called
    - (a) simplex element
    - (b) complex element
    - (c) multiplex element
    - (d) compound element

35. Element formed with quadratic or higher order displacement polynomial is a
    - (a) simplex element
    - (b) complex element
    - (c) multiplex element
    - (d) compound element

36. Elements connecting lower order elements and higher order elements in a mesh are called
    - (a) transition elements
    - (b) sub-parametric elements
    - (c) iso-parametric elements
    - (d) super-parametric elements

37. Elements having mid-side nodes only on some sides are called
    - (a) transition elements
    - (b) sub-parametric elements
    - (c) iso-parametric elements
    - (d) super-parametric elements

38. Stress-strain matrix for plane stress element, if strain is represented by s_{ij} and stress is represented by st_{ij}, is obtained from the condition
    - (a) $s_{zz} = 0$
    - (b) $s_{zx} = 0$
    - (c) $st_{zx} = 0$
    - (d) $st_{zz} = 0$

39. Stress-strain matrix for plane strain element, if strain is represented by s_{ij}, is obtained from the condition
    - (a) $s_{zz} = 0$
    - (b) $s_{zx} = 0$
    - (c) $st_{zx} = 0$
    - (d) $st_{zz} = 0$

40. Stress-strain matrix for axisymmetric element is of order

    (a) 3 × 3   (b) 4 × 4   (c) 6 × 6   (d) 9 × 9

41. Stress-strain matrix for plate bending element is of order

    (a) 3 × 3   (b) 4 × 4   (c) 6 × 6   (d) 9 × 9

42. Elasticity matrix for ___ behaviour is similar to 3-D elasticity matrix

    (a) plane stress   (b) plane strain

    (c) plate bending   (d) axisymmetric

43. Plane stress element is an extension of

    (a) truss element   (b) beam element

    (c) pipe element   (d) spring element

44. Plate bending element is an extension of

    (a) truss element   (b) beam element

    (c) pipe element   (d) spring element

45. Wrong sequencing of nodal connectivity in 2-D & 3-D problems leads to

    (a) +ve Jacobian   (b) zero Jacobian

    (c) −ve Jacobian   (d) No relation with Jacobian

46. Axisymmetric structures are usually modeled in

    (a) element local coordinates   (b) global cartesian coordinates

    (c) global cylindrical coordinates (d) user specified system

47. A plate of 1cm thickness with in-plane loads is modeled by

    (a) plane stress element   (b) plane strain element

    (c) plate bending element   (d) any one of them

48. Actual thickness of plane strain element is

    (a) very small   (b) very large

    (c) any specified value   (d) assumed by software

49. Order of stiffness matrix for a plane stress model with 20 nodes is

    (a) 10   (b) 20   (c) 40   (d) 60

50. Order of stiffness matrix for an axisymmetric model with 20 nodes is

    (a) 10   (b) 20   (c) 40   (d) 60

51. Number of stress components per node calculated for a plane stress quadrilateral element is

    (a) 2   (b) 3   (c) 4   (d) 5

52. Number of stress components per node calculated for a triangular axisymmetric element is

(a) 2 (b) 3 (c) 4 (d) 5

53. A general plate element is a superposition of _____ elements

(a) plane stress & plane strain (b) plane strain & plate bending

(c) plane stress & plate bending (d) plate bending only

54. An element with in-plane loads having 3 nodes along each side is a

(a) constant strain element (b) linear strain element

(c) quadratic strain element (d) constant displacement method

SOLVED PROBLEMS

Example 5.1

Calculate displacements and stress in a triangular plate, fixed along one edge and subjected to concentrated load at its free end. Assume E = 70,000 MPa, t = 10 mm and ν = 0.3.

Solution

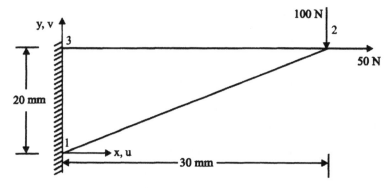

If the element DOFs are arranged in the sequence of $[u_1 \quad v_1 \quad u_2 \quad v_2 \quad u_3 \quad v_3]^T$, the stiffness matrix from 5.3.1 is

$$[K] = \frac{EtA}{a^2 b^2 (1-v^2)} \begin{bmatrix} \beta a^2 & & & & & \\ 0 & a^2 & & & & \\ 0 & -v\,ab & b^2 & & \text{Symmetric} & \\ -\beta\,ab & 0 & 0 & \beta b^2 & & \\ -\beta a^2 & v\,ab & -b^2 & \beta\,ab & b^2 + \beta a^2 & \\ \beta\,ab & -a^2 & v\,ab & -\beta b^2 & -ab(v+\beta) & a^2 + \beta b^2 \end{bmatrix}$$

Substituting the given dimensions and material properties,

$$[K] = \frac{70000 \times 10 \times \left(\dfrac{30 \times 20}{2}\right)}{30^2 \times 20^2 \times \left(1 - 0.3^2\right)} \begin{bmatrix} 315 & & & & & \\ 0 & 900 & & & & \\ 0 & -180 & 400 & & \text{Symmetric} & \\ -210 & 0 & 0 & 140 & & \\ -315 & 180 & -400 & 210 & 715 & \\ 210 & -900 & 180 & -140 & -390 & 1040 \end{bmatrix}$$

After applying boundary conditions, $u_1 = v_1 = u_3 = v_3 = 0$, these equations reduce to

$$\begin{Bmatrix} P_{X2} \\ P_{Y2} \end{Bmatrix} = \begin{Bmatrix} 50 \\ -100 \end{Bmatrix} = 641.026 \begin{bmatrix} 400 & 0 \\ 0 & 140 \end{bmatrix} \begin{Bmatrix} u_2 \\ v_2 \end{Bmatrix}$$

Therefore, $u_2 = 0.000195$ mm and $v_2 = -0.001114$ mm

Note : The given *thick* plate, from university question paper, should not be analysed as a 2-D problem. It can not be solved as a 3-D problem *manually*.

Example 5.2

Compute the plane strain stiffness matrix in terms of the ratio $r = a/b$ for the rectangular element of sides a and b, using $v = 0.2$, $r = 1$ and displacement model

$$u = a_1 + a_2\, x + a_3\, y + a_4\, xy \quad \text{and} \quad v = a_5 + a_6\, x + a_7\, y + a_8\, xy$$

Solution

By substituting coordinates of the four corner nodes in the displacement model,

With origin (0,0) at node 1,

$$\begin{Bmatrix} u_1 \\ u_2 \\ u_3 \\ u_4 \\ v_1 \\ v_2 \\ v_3 \\ v_4 \end{Bmatrix} = \begin{bmatrix} 1 & 0 & 0 & 0 & 0 & 0 & 0 & 0 \\ 1 & a & 0 & 0 & 0 & 0 & 0 & 0 \\ 1 & a & b & ab & 0 & 0 & 0 & 0 \\ 1 & 0 & b & 0 & 0 & 0 & 0 & 0 \\ 0 & 0 & 0 & 0 & 1 & 0 & 0 & 0 \\ 0 & 0 & 0 & 0 & 1 & a & 0 & 0 \\ 0 & 0 & 0 & 0 & 1 & a & b & ab \\ 0 & 0 & 0 & 0 & 1 & 0 & b & 0 \end{bmatrix} \begin{Bmatrix} a_1 \\ a_2 \\ a_3 \\ a_4 \\ a_5 \\ a_6 \\ a_7 \\ a_8 \end{Bmatrix} \quad \text{or} \quad \{u_e\} = \begin{bmatrix} [A] & [O] \\ [O] & [A] \end{bmatrix} \{a\}$$

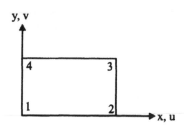

For the particular case of $r = \dfrac{a}{b} = 1$, evaluating coefficients a_1 to a_8 in terms of nodal displacements,

$$
\begin{Bmatrix} a_1 \\ a_2 \\ a_3 \\ a_4 \\ a_5 \\ a_6 \\ a_7 \\ a_8 \end{Bmatrix} = \begin{bmatrix}
1 & 0 & 0 & 0 & 0 & 0 & 0 & 0 \\
-1/a & 1/a & 0 & 0 & 0 & 0 & 0 & 0 \\
-1/a & 0 & 0 & 1/a & 0 & 0 & 0 & 0 \\
1/a^2 & -1/a^2 & 1/a^2 & -1/a^2 & 0 & 0 & 0 & 0 \\
0 & 0 & 0 & 0 & 1 & 0 & 0 & 0 \\
0 & 0 & 0 & 0 & -1/a & 1/a & 0 & 0 \\
0 & 0 & 0 & 0 & -1/a & 0 & 0 & 1/a \\
0 & 0 & 0 & 0 & 1/a^2 & -1/a^2 & 1/a^2 & -1/a^2
\end{bmatrix} \begin{Bmatrix} u_1 \\ u_2 \\ u_3 \\ u_4 \\ v_1 \\ v_2 \\ v_3 \\ v_4 \end{Bmatrix}
$$

or $\quad \{a\} = \begin{bmatrix} [A]^{-1} & [O] \\ [O] & [A]^{-1} \end{bmatrix} \{u_e\}$

Strain, $\{\varepsilon\} = \begin{Bmatrix} \partial u / \partial x \\ \partial v / \partial y \\ \partial u / \partial y + \partial y / \partial x \end{Bmatrix} = \begin{bmatrix} 0 & 1 & 0 & y & 0 & 0 & 0 & 0 \\ 0 & 0 & 0 & 0 & 0 & 0 & 1 & x \\ 0 & 0 & 1 & x & 0 & 1 & 0 & y \end{bmatrix} \{a\} = [B]\{u_e\}$

where,

$$
[B] = \left(\frac{1}{a^2}\right) \begin{bmatrix}
y-a & -(y-a) & y & -y & 0 & 0 & 0 & 0 \\
0 & 0 & 0 & 0 & x-a & -x & x & -(x-a) \\
x-a & -x & x & -(x-a) & y-a & -(y-a) & y & -y
\end{bmatrix}
$$

For a plane strain case, stress-strain relationship is given by

$$
[D] = \frac{E}{(1+v)(1-2v)} \begin{bmatrix}
1-v & v & 0 \\
v & 1-v & 0 \\
0 & 0 & (1-2v)/2
\end{bmatrix}
$$

$$= \frac{E}{7.2} \begin{bmatrix} 8 & 2 & 0 \\ 2 & 8 & 0 \\ 0 & 0 & 3 \end{bmatrix} \text{ for } v = 0.2$$

$$[K] = \int_v [B]^T [D][B] \, dv = t \int_0^a \int_0^b [B]^T [D][B] \, dx \, dy$$

$$= \int_0^a \int_0^b [B]^T [D][B] \, dx \, dy \qquad \text{since } t = 1 \quad \text{for plane strain elements}$$

$$= \frac{E}{86.4} \begin{bmatrix} 44 & & & & & & & \\ -26 & 44 & & & \text{Symmetric} & & & \\ -22 & 4 & 44 & & & & & \\ 4 & -22 & -4 & 44 & & & & \\ 15 & 3 & -15 & -3 & 44 & & & \\ -3 & -15 & 3 & 15 & 4 & 44 & & \\ -15 & -3 & -3 & 3 & -22 & -26 & 44 & \\ 3 & 15 & 15 & -15 & -4 & -22 & 4 & 44 \end{bmatrix}$$

when written corresponding to the displacement vector $[u_1 \ v_1 \ u_2 \ v_2 \ u_3 \ v_3 \ u_4 \ v_4]^T$

Example 5.3

Compute the plane strain stiffness matrix of a square, treating this as an assembly of two triangular elements with the displacement field in these elements expressed as $u = a_1 + a_2 x + a_3 y$ and $v = a_4 + a_5 x + a_6 y$. Assume $v = 0.2$

Solution

Let the square of side 'a' be represented by two triangular elements identified by nodes 1, 2, 3 and 1, 3, 4 respectively. Choosing node 1 as the origin of the X-Y coordinate system,

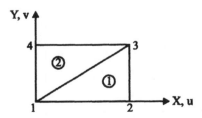

For element-1,

$$\begin{Bmatrix} u_1 \\ u_2 \\ u_3 \end{Bmatrix} = \begin{bmatrix} 1 & 0 & 0 \\ 1 & a & 0 \\ 1 & a & a \end{bmatrix} \begin{Bmatrix} \alpha_1 \\ \alpha_2 \\ \alpha_3 \end{Bmatrix}$$

Then

$$\begin{Bmatrix} \alpha_1 \\ \alpha_2 \\ \alpha_3 \end{Bmatrix} = \begin{bmatrix} 1 & 0 & 0 \\ -1/a & 1/a & 0 \\ 1 & -1/a & 1/a \end{bmatrix} \begin{Bmatrix} u_1 \\ u_2 \\ u_3 \end{Bmatrix}$$

Strain,
$$\{\varepsilon\} = \begin{Bmatrix} \partial u/\partial x \\ \partial v/\partial y \\ \partial u/\partial y + \partial v/\partial x \end{Bmatrix} = \begin{bmatrix} 0 & 1 & 0 & 0 & 0 & 0 \\ 0 & 0 & 0 & 0 & 0 & 1 \\ 0 & 0 & 1 & 0 & 1 & 0 \end{bmatrix} \{a\} = [B]\{u_e\}$$

where,
$$[B] = \left(\frac{1}{a}\right) \begin{bmatrix} -1 & 1 & 0 & 0 & 0 & 0 \\ 0 & 0 & 0 & 0 & -1 & 1 \\ 0 & -1 & 1 & -1 & 1 & 0 \end{bmatrix}$$

For a plane strain case, stress-strain relationship is given by

$$[D] = \frac{E}{(1+v)(1-2v)} \begin{bmatrix} 1-v & v & 0 \\ v & 1-v & 0 \\ 0 & 0 & (1-2v)/2 \end{bmatrix}$$

$$= \frac{E}{7.2} \begin{bmatrix} 8 & 2 & 0 \\ 2 & 8 & 0 \\ 0 & 0 & 3 \end{bmatrix} \text{ for } v = 0.2$$

$$[K] = \int_v [B]^T [D] [B] \, dv = t \int_0^a \int_0^b [B]^T [D] [B] \, dx \, dy$$

$$= \int_0^a \int_0^b [B]^T [D] [B] \, dx \, dy \text{ since } t = 1 \text{ for plane strain elements}$$

$$= \frac{E}{14.4} \begin{bmatrix} 8 & & & & & \\ 0 & 4 & & \text{Symmetric} & & \\ -2 & 4 & 6 & & & \\ 2 & -4 & -6 & 12 & & \\ 0 & -4 & -4 & 4 & 4 & \\ -2 & 0 & 2 & -8 & 0 & 8 \end{bmatrix}$$

when written corresponding to the displacement vector $[u_1 \ v_1 \ u_2 \ v_2 \ u_3 \ v_3]^T$

For element-2,

$$\begin{Bmatrix} u_1 \\ u_3 \\ u_4 \end{Bmatrix} = \begin{bmatrix} 1 & 0 & 0 \\ 1 & a & a \\ 1 & 0 & a \end{bmatrix} \begin{Bmatrix} \alpha_1 \\ \alpha_2 \\ \alpha_3 \end{Bmatrix}$$

Then, $\begin{Bmatrix} \alpha_1 \\ \alpha_2 \\ \alpha_3 \end{Bmatrix} = \begin{bmatrix} 1 & 0 & 0 \\ 0 & 1/a & -1/a \\ -1/a & 0 & 1/a \end{bmatrix} \begin{Bmatrix} u_1 \\ u_3 \\ u_4 \end{Bmatrix}$

$$[B] = \left(\frac{1}{a}\right) \begin{bmatrix} 0 & 1 & -1 & 0 & 0 & 0 \\ 0 & 0 & 0 & -1 & 0 & 1 \\ -1 & 0 & 1 & 0 & 1 & -1 \end{bmatrix}$$

$$[K] = \frac{E}{14.4} \begin{bmatrix} 3 & & & & & \\ 0 & 8 & & \text{Symmetric} & & \\ 0 & -2 & 8 & & & \\ -3 & 0 & 0 & 3 & & \\ -3 & 2 & -8 & 3 & 12 & \\ 3 & -8 & 2 & -3 & -5 & 11 \end{bmatrix}$$

when written corresponding to the displacement vector $[u_1 \ v_1 \ u_3 \ v_3 \ u_4 \ v_4]^T$

The *stiffness matrix of the square plate* is obtained by adding relevant coefficients of the stiffness matrices of elements 1 and 2, as

$$[K] = \frac{E}{14.4} \begin{bmatrix} 3+8 & & & & & & & \\ 0+0 & 8+4 & & & & & & \\ -2 & 4 & 6 & & \text{Symmetric} & & & \\ 2 & -4 & -6 & 12 & & & & \\ 0+0 & -2-4 & -4 & 4 & 8+4 & & & \\ -3-2 & 0+0 & 2 & -8 & 0+0 & 3+8 & & \\ -3 & 2 & 0 & 0 \text{`} & -8 & 3 & 12 & \\ 3 & -8 & 0 & 0 & 2 & -3 & -5 & 11 \end{bmatrix}$$

corresponding to the displacement vector $[u_1 \ v_1 \ u_2 \ v_2 \ u_3 \ v_3 \ u_4 \ v_4]^T$

Note :

1. The stiffness matrix obtained for the combination of two triangular elements in Ex. 5.3 is different from the one obtained for the square element in Ex. 5.2.

2. In the first case, node 2 and node 4 are not linked and hence terms k_{37}, k_{38}, k_{47}, k_{48}, k_{73}, k_{74}, k_{83} and k_{84} are equal to zero.

3. Even other terms are different, indicating that the *results obtained for a given problem will depend on the model* used for analysis and *differ marginally*.

CHAPTER 6

HIGHER ORDER AND ISOPARAMETRIC ELEMENTS

6.1 HIGHER ORDER ELEMENTS

When geometry is modeled with CST elements, large number of small-size elements need to be used, in order to accommodate variation of strains over the entire geometry. In view of the constraints on computer memory and time for solving large size problems, an alternative method of using a small number of higher order (refined) elements can also be considered. In these elements, a higher order polynomial is used to include variation of strain over the element by choosing additional nodes.

In most cases, axially loaded spars or truss elements have uniform cross section and hence stress/strain in the element is constant along the length of the member. A few special cases may involve stress/strain varying along the length of a truss element, such as a vertical column with distributed self weight. For such applications, a 3-noded truss element is used. However, in the case of beam elements, elementary beam equation predicts deflection of a beam varying parabolically along the length. Basic beam element in FEM caters to this variation. *Higher order beam element is not of any practical use* and is, therefore, not discussed further.

(a)   **3-noded truss element :** This element has 3 nodes including one extra node in the middle of the element. 2nd order displacement function is used. Stiffness matrix for this element is derived here for an understanding of the subject.

Let $u(x) = a_1 + a_2.x + a_3.x^2 = [1 \quad x \quad x^2] \begin{Bmatrix} a_1 \\ a_2 \\ a_3 \end{Bmatrix} = [f(x)]^T \{ a \}$

Substituting the values of x for the two end points and middle point of the truss element, (x = 0 at node i, x = L at node j and x = L/2 at node k), nodal displacement vector is written as

$$\begin{Bmatrix} u_i \\ u_j \\ u_k \end{Bmatrix} = \begin{bmatrix} 1 & 0 & 0 \\ 1 & L & L^2 \\ 1 & L/2 & L^2/4 \end{bmatrix} \begin{Bmatrix} a_1 \\ a_2 \\ a_3 \end{Bmatrix} \quad \text{or} \quad \{u_e\} = [G] \{a\}$$

Solving for the coefficients a and substituting in the earlier equation,

$$u(x) = [f(x)]^T [G]^{-1} \{u_e\}$$

$$= [1 \quad x \quad x^2] \begin{bmatrix} 1 & 0 & 0 \\ -3/L & -1/L & 4/L \\ 2/L^2 & 2/L^2 & -4/L^2 \end{bmatrix} \begin{Bmatrix} u_i \\ u_j \\ u_k \end{Bmatrix} = [N]^T \{u_e\}$$

Strain, $\varepsilon = \dfrac{du}{dx} = [f'(x)]^T [G]^{-1} \{u_e\} = [B] \{u_e\}$

where, $[B] = [0 \quad 1 \quad 2x] \begin{bmatrix} 1 & 0 & 0 \\ -3/L & -1/L & 4/L \\ 2/L^2 & 2/L^2 & -4/L^2 \end{bmatrix}$

$$= \left(\dfrac{1}{L^2} \right) [(-3L + 4x) \quad (-L + 4x) \quad (4L - 8x)]$$

and $\{\delta\varepsilon\} = [B] \{\delta u_e\}$; $\{\delta\varepsilon\}^T = \{\delta u_e\}^T [B]^T$

Stress, $\{\sigma\} = [D] \{\varepsilon\} = E \{\varepsilon\} = E [B] \{u_e\}$

$$\therefore \{P_e\} = [K_e] \{u_e\}$$

where, $[K_e] = \int_v [B]^T [D][B] \, dv = \iiint [B]^T E [B] \, dx \, dy \, dz$

$$= AE \int_L [B]^T [B] \, dx$$

since [B] is independent of y and z and $\iint dy \, dz = A$

$$= \frac{AE}{3L} \begin{bmatrix} 7 & 1 & -8 \\ 1 & 7 & -8 \\ -8 & -8 & 16 \end{bmatrix}$$

A higher order truss element, which accounts for variable stress/strain, has very few applications. Some of them are shown here.

Example 6.1

Calculate the displacement at the free end of a 50cm long tapered bar of area of cross section 1000 mm^2 at its fixed end and 600 mm^2 at the free end, subjected to an axial tensile load of 1kN at the free end. Assume E = 200GPa.

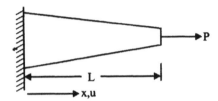

Solution

In order to explain the advantage of a higher order element, the problem is solved first by using basic truss element and then by using higher order element.

(a) The bar is identified by two 2-noded elements and mean area of cross section is considered for both elements.

At mid-point of the bar (node 2),

$$A = \frac{(1000 + 600)}{2} = 800 \text{mm}^2$$

For element 1, connecting nodes 1 and 2,

$$A_1 = \frac{(1000 + 800)}{2} = 900 \text{ mm}^2 = 9 \text{ cm}^2$$

Stiffness matrix for element-1 is,

$$[K]_1 = \frac{A_1 E}{L} \begin{bmatrix} 1 & -1 \\ -1 & 1 \end{bmatrix} = \left(\frac{E}{L}\right) \begin{bmatrix} 9 & -9 \\ -9 & 9 \end{bmatrix}$$

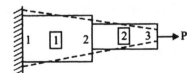

For element 2, connecting nodes 2 and 3,

$$A_2 = \frac{(800 + 600)}{2} = 700 \text{ mm}^2 = 7 \text{ cm}^2$$

Stiffness matrix for element-1 is,

$$[K]_2 = \frac{A_2 E}{L}\begin{bmatrix} 1 & -1 \\ -1 & 1 \end{bmatrix} = \frac{E}{L}\begin{bmatrix} 7 & -7 \\ -7 & 7 \end{bmatrix}$$

Assembled stiffness matrix is then obtained as

$$[K] = \frac{E}{L}\begin{bmatrix} 9 & -9 & 0 \\ -9 & 9+7 & -7 \\ 0 & -7 & 7 \end{bmatrix} \quad \text{or} \quad \begin{Bmatrix} P_1 \\ P_2 \\ P_3 \end{Bmatrix} = \frac{E}{L}\begin{bmatrix} 9 & -9 & 0 \\ -9 & 16 & -7 \\ 0 & -7 & 7 \end{bmatrix}\begin{Bmatrix} u_1 \\ u_2 \\ u_3 \end{Bmatrix}$$

Applying boundary condition $u_1 = 0$, reduced stiffness matrix is obtained as

$$\begin{Bmatrix} P_2 \\ P_3 \end{Bmatrix} = \begin{Bmatrix} 0 \\ 1000 \end{Bmatrix} = \frac{E}{L}\begin{bmatrix} 16 & -7 \\ -7 & 7 \end{bmatrix}\begin{Bmatrix} u_2 \\ u_3 \end{Bmatrix}$$

Solving these two simultaneous equations, we get

$$\begin{Bmatrix} u_2 \\ u_3 \end{Bmatrix} = \frac{1}{\text{Det}}\begin{bmatrix} 7 & 7 \\ 7 & 16 \end{bmatrix}\begin{Bmatrix} 0 \\ 1000 \end{Bmatrix}$$

where, $\text{Det} = \dfrac{E}{L}(7 \times 16 - 7 \times 7) = \dfrac{63E}{L}$

$$u_2 = \frac{7 \times 1000}{\left(\dfrac{63E}{L}\right)} = 111\left(\frac{L}{E}\right)$$

and $u_3 = \dfrac{16 \times 1000}{\left(\dfrac{63E}{L}\right)} = 254\left(\dfrac{L}{E}\right)$

Corresponding stresses are

$$\sigma_1 = \frac{E(u_2 - u_1)}{L} = 111 - 0 = 111 \text{ N/cm}^2 \text{ constant along element-1}$$

and $\sigma_2 = \dfrac{E(u_3 - u_2)}{L} = 254 - 111 = 143 \text{ N/cm}^2 \text{ constant along element-2}$

(b) The bar is modeled by one 3-noded element with uniform area of cross section (mean of max and min areas). Thus, with $A_2 = 800$ mm^2, element stiffness matrix is derived as

$$[K] = \left(\frac{A_2 E}{3L}\right) \begin{bmatrix} 7 & 1 & -8 \\ 1 & 7 & -8 \\ -8 & -8 & 16 \end{bmatrix} \begin{Bmatrix} u_1 \\ u_3 \\ u_2 \end{Bmatrix}$$

Since the chosen model has only one element, assembled stiffness matrix is also the same. Therefore,

$$\begin{Bmatrix} P_1 \\ P_2 \\ P_3 \end{Bmatrix} = \frac{A_2 E}{3L} \begin{bmatrix} 7 & 1 & -8 \\ 1 & 7 & -8 \\ -8 & -8 & 16 \end{bmatrix} \begin{Bmatrix} u_1 \\ u_2 \\ u_3 \end{Bmatrix}$$

Note : Care should be taken while numbering the load and displacement components, since the 3 rows of the stiffness matrix correspond to the end nodes at $x = 0$, $x = L$ and the mid point (at $x = L/2$) respectively.

Applying the boundary condition, $u_1 = 0$, the above equations reduce to

$$\begin{Bmatrix} P_3 \\ P_2 \end{Bmatrix} = \begin{Bmatrix} 1000 \\ 0 \end{Bmatrix} = \frac{A_2 E}{3L} \begin{bmatrix} 7 & -8 \\ -8 & 16 \end{bmatrix} \begin{Bmatrix} u_3 \\ u_2 \end{Bmatrix}$$

Solving these equations, we get

$$\begin{Bmatrix} u_3 \\ u_2 \end{Bmatrix} = \frac{1}{\text{Det}} \begin{bmatrix} 16 & 8 \\ 8 & 7 \end{bmatrix} \begin{Bmatrix} 1000 \\ 0 \end{Bmatrix}$$

where, $\quad \text{Det} = \dfrac{A_2 E}{3L} (16 \times 7 - 8 \times 8) = \dfrac{128E}{L}$

Therefore, $\quad u_2 = \dfrac{8 \times 1000}{\left(\dfrac{128E}{L}\right)} = 62.5 \left(\dfrac{L}{E}\right)$

and $u_3 = \dfrac{16 \times 1000}{\left(\dfrac{128E}{L}\right)} = 125\left(\dfrac{L}{E}\right)$

Stress in the bar is given by

$$\sigma = E\,[B]\,\{u_e\} = E\left(\dfrac{1}{L^2}\right)[(-3L + 4x) \quad (-L + 4x) \quad (4L - 8x)]\begin{Bmatrix} u_1 \\ u_3 \\ u_2 \end{Bmatrix}$$

$$= E\left(\dfrac{1}{L^2}\right)[(-3L + 4x)\,u_1 + (-L + 4x)\,u_3 + (4L - 8x)\,u_2]$$

$$= E\left(\dfrac{1}{L^2}\right)[(-3L + 4x).0 + (-L + 4x).(125\ L/E) + (4L - 8x).(62.5\ L/E)]$$

$$= \left(\dfrac{1}{L}\right)[(-125L + 500x) + (250L - 500x)]$$

$$= \left(\dfrac{1}{L}\right)(125L) = 125\ \text{N/cm}^2$$

Since a single element model with constant area of cross section is used, stress along the element remains constant and is not realistic.

(c) The bar is modeled by one 3-noded element with area of cross section varying along the length of the bar. The element stiffness matrix is obtained from

$$[K_e] = \int_v [B]^T[D][B]\,dv = \int_L [B]^T E\,[B]\,A(x)\,dx = E\int_L [B]^T[B]\,A(x)\,dx$$

where $[B] = \left(\dfrac{1}{L^2}\right)[(-3L + 4x) \quad (-L + 4x) \quad (4L - 8x)\,]$ as derived earlier.

On integration of the above with $A(x) = A_1 - (A_1 - A_3).x/L = a - b\,x$,

where, $a = A_1$ and $b = (A_1 - A_3)/L$, we get

$$[K] = \int \dfrac{E(a - bx)}{L^2}\begin{bmatrix} 9L^2 - 24Lx + 16x^2 & 3L^2 - 16Lx + 16x^2 & -12L^2 + 40Lx - 32x^2 \\ 3L^2 - 16Lx + 16x^2 & L^2 - 8Lx + 16x^2 & -4L^2 + 24Lx - 32x^2 \\ -12L^2 + 40Lx - 32x^2 & -4L^2 + 24Lx - 32x^2 & 16L^2 - 64Lx + 64x^2 \end{bmatrix}$$

$$= \dfrac{E}{6L}\begin{bmatrix} 14a - 3b\,L & 2a - b\,L & -16a + 4b\,L \\ 2a - b\,L & 14a - 11b\,L & -16a + 12b\,L \\ -16a + 4b\,L & -16a + 12b\,L & 32a - 16b\,L \end{bmatrix}$$

$$= \frac{E}{6L} \begin{bmatrix} 11A_1 + 3A_3 & A_1 + A_3 & -12A_1 - 4A_3 \\ A_1 + A_3 & 3A_1 + 11A_3 & -4A_1 - 12A_3 \\ -12A_1 - 4A_3 & -4A_1 - 12A_3 & 16A_1 + 16A_3 \end{bmatrix} = \frac{E}{6L} \begin{bmatrix} 128 & 16 & -144 \\ 16 & 96 & -112 \\ -144 & -112 & 256 \end{bmatrix}$$

Since the chosen model has only one element, assembled stiffness matrix is also the same. Therefore,

$$\begin{Bmatrix} P_1 \\ P_3 \\ P_2 \end{Bmatrix} = \frac{8E}{3L} \begin{bmatrix} 8 & 1 & -9 \\ 1 & 6 & -7 \\ -9 & -7 & 16 \end{bmatrix} \begin{Bmatrix} u_1 \\ u_3 \\ u_2 \end{Bmatrix}$$

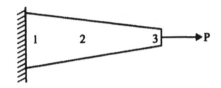

Note : Care should be taken while numbering the load and displacement components, since the 3 rows of the stiffness matrix correspond to the end nodes at x = 0, x = L and the mid point (at x = L/2) respectively.

Applying the boundary condition, $u_1 = 0$, the above equations reduce to

$$\begin{Bmatrix} P_3 \\ P_2 \end{Bmatrix} = \begin{Bmatrix} 1000 \\ 0 \end{Bmatrix} = \frac{8E}{3L} \begin{bmatrix} 6 & -7 \\ -7 & 16 \end{bmatrix} \begin{Bmatrix} u_3 \\ u_2 \end{Bmatrix}$$

Solving these equations, we get

$$\begin{Bmatrix} u_3 \\ u_2 \end{Bmatrix} = \frac{1}{\text{Det}} \begin{bmatrix} 16 & 7 \\ 7 & 6 \end{bmatrix} \begin{Bmatrix} 1000 \\ 0 \end{Bmatrix}$$

where, $\quad \text{Det} = \left(\frac{8E}{3L}\right)(6 \times 16 - 7 \times 7) \simeq 125\left(\frac{E}{L}\right)$

Therefore, $\quad u_2 = \dfrac{7 \times 1000}{\left(125\dfrac{E}{L}\right)} = \dfrac{56L}{E}$

and $\quad u_3 = \dfrac{16 \times 1000}{\left(125\dfrac{E}{L}\right)} = \dfrac{128L}{E}$

Stress in the bar is given by

$$\sigma = E\,[B]\,\{u_e\} = E\left(\frac{1}{L}\right)^2 [(-3L + 4x) \quad (-L + 4x) \quad (4L - 8x)] \begin{Bmatrix} u_1 \\ u_3 \\ u_2 \end{Bmatrix}$$

$$= E\left(\frac{1}{L^2}\right)(-3L + 4x)\,u_1 + (-L + 4x)\,u_3 + (4L - 8x)\,u_2]$$

$$= E\left(\frac{1}{L^2}\right)\left[(-3L + 4x).0 + (-L + 4x).\left(\frac{128L}{E}\right) + (4L - 8x).\left(\frac{56L}{E}\right)\right]$$

$$= \left(\frac{1}{L}\right)[(-128L + 512x) + (224L - 448x)] = \left(\frac{16}{L}\right)(6L + 4x)$$

Stress at node 1, for x = 0, is $\sigma_1 = \left(\frac{16}{L}\right)(6L + 4 \times 0)$ = 96 N/cm^2

Stress at node 2, for $x = \dfrac{L}{2}$, is $\sigma_2 = \left(\frac{16}{L}\right)\left(6L + 4 \times \dfrac{L}{2}\right)$ = 128 N/cm^2

Stress at node 3, for x = L, is $\sigma_3 = \left(\frac{16}{L}\right)(6L + 4 \times L) = 160$ N/cm^2

Analysis of results

All the three models have 3-nodes, but gave different solutions. Exact solution depends on how closely the assumed displacement field or stress distribution matches with the actual displacement field. In this example, it is clear from basic equilibrium condition that stress is linearly increasing from P/A_1 at node-1 to P/A_3 at node-3 and hence, displacement must increase in a parabolic form. The results are tabulated below.

Model	Displacements at		Stresses (N/mm^2) at		
	Node-2	Node-3	Node-1	Node-2	Node-3
Two 2-noded	111 L/E	254 L/E	111	111 / 143 Ave 127	143
One 3-noded, Constant area	62.5 L/E	125 L/E	125	125	125
One 3-noded, varying area	56 L/E	128 L/E	96	128	160
Exact			P/A_1= 100	P/A_2=125	P/A_3=167

(i) The 1st model assumes linear displacement field and hence constant stress over each of the two elements. At the common node between elements, stress has a step function and the average value can be taken to represent stress at that node. Results can be improved by taking more such elements in the model.

(ii) The 2nd model uses a single 3-noded element, assuming constant (mean) area of cross section along the element. Hence, even though parabolic displacement is considered, linear displacement field and constant stress are obtained.

(iii) The 3rd model consisting of a single 3-noded element, whose stiffness matrix is derived for varying cross section area along the element, gives parabolic displacement and linear stress representing the true situation. Hence, this model gives best results.

(b) Higher order Continuum elements

Higher order elements are more commonly used for analysing 2-D and 3-D structures. **Linear strain triangle (LST)** will have six unknown coefficients to include all terms upto second order, as shown in Pascal's triangle. For evaluation of these six coefficients, six nodal values are required in each element. Thus, the 6-noded element is formed by including midpoints of the three sides as the additional nodes. The functions for u and v displacements are

$$u(x) = a_1 + a_2.x + a_3.y + a_4.x^2 + a_5.\ xy + a_6.\ y^2$$

and $v(x) = a_7 + a_8.x + a_9.y + a_{10}.x^2 + a_{11}.\ xy + a_{12}.\ y^2$

The function for LST element, as in the case of CST element, is also complete and isotropic.

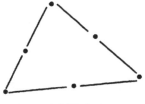

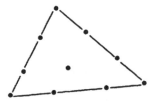

| 6-noded CST element | 10-noded QST element |

Quadratic strain triangle (QST) will have ten unknown coefficients to include all terms upto third order, as shown in Pascal's triangle. For evaluation of these ten coefficients, ten nodal values are required in each element. Thus, the 10-noded element is formed with 6 additional mid-side nodes and 1 internal node. The functions for u and v displacements (complete and isotropic) are

$$u(x) = a_1 + a_2.x + a_3.y + a_4.x^2 + a_5.\ xy + a_6.\ y^2 + a_7.x^3 + a_8.x^2y + a_9.xy^2 + a_{10}.y^3$$

and

$$v(x) = a_{11} + a_{12}.x + a_{13}.y + a_{14}.x^2 + a_{15}.\ xy + a_{16}.y^2 + a_{17}.x^3 + a_{18}.x^2y + a_{19}.xy^2 + a_{20}.y^3$$

Similarly, the displacement functions for a **10-noded 3-D solid tetrahedron element** with additional mid side nodes along its 6 sides (complete and isotropic function) are

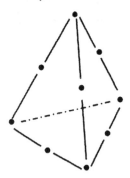

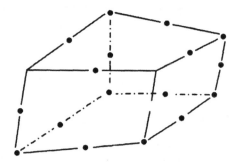

10-noded tetrahedron element   20-noded hexahedron element

$$u = a_1 + a_2.x + a_3.y + a_4.z + a_5.x^2 + a_6.y^2 + a_7.z^2 + a_8.xy + a_9.yz + a_{10}.zx$$

$$v = a_{11} + a_{12}.x + a_{13}.y + a_{14}.z + a_{15}.x^2 + a_{16}.y^2 + a_{17}.z^2 + a_{18}.xy + a_{19}.yz + a_{20}.zx$$

$$w = a_{21} + a_{22}.x + a_{23}.y + a_{24}.z + a_{25}.x^2 + a_{26}.y^2 + a_{27}.z^2 + a_{28}.xy + a_{29}.yz + a_{30}.zx$$

and the displacement functions for a **20-noded 3-D solid hexahedron element** with additional mid side nodes along its 12 sides (incomplete but isotropic) are

$$
\begin{aligned}
u = \ & a_1 \quad + a_2.x \quad + a_3.y \quad + a_4.z \quad + a_5.x^2 \quad + a_6.y^2 + a_7.z^2 \quad + a_8.xy \\
& + a_9.yz \quad + a_{10}.zx \quad + a_{11}.x^3 \quad + a_{12}.x^2y + a_{13}.xy^2 + a_{14}.y^3 + a_{15}.y^2z + a_{16}.yz^2 \\
& + a_{17}.z^3 \quad + a_{18}.x^2z + a_{19}.xz^2 + a_{20}.xyz
\end{aligned}
$$

$$
\begin{aligned}
v = \ & a_{21} \quad + a_{22}x \quad + a_{23}.y \quad + a_{24}.z \quad + a_{25}.x^2 \quad + a_{26}.y^2 + a_{27}.z^2 \quad + a_{28}.xy \\
& + a_{29}.yz \quad + a_{30}.zx \quad + a_{31}.x^3 \quad + a_{32}.x^2y + a_{33}.xy^2 + a_{34}.y^3 + a_{35}.y^2z + a_{36}.yz^2 \\
& + a_{37}.z^3 \quad + a_{38}.x^2z + a_{39}.xz^2 + a_{40}.xyz
\end{aligned}
$$

$$
\begin{aligned}
w = \ & a_{41} \quad + a_{42}.x \quad + a_{43}.y \quad + a_{44}.z \quad + a_{45}.x^2 \quad + a_{46}.y^2 + a_{47}.z^2 \quad + a_{48}.xy \\
& + a_{49}.yz \quad + a_{50}.zx \quad + a_{51}.x^3 \quad + a_{52}.x^2y + a_{53}.xy^2 + a_{54}.y^3 + a_{55}.y^2z + a_{56}.yz^2 \\
& + a_{57}.z^3 \quad + a_{58}.x^2z + a_{59}.xz^2 + a_{60}.xyz
\end{aligned}
$$

Note : Whenever possible, complete and isotropic displacement polynomial is used. If, however, it is not possible, preference is given to isotropy rather than completeness of terms of a particular order. In the above cases of higher order elements with incomplete polynomials, there is no unique combination of the terms and many other combinations of terms in the selected polynomial are possible. Different software may thus use different polynomials.

6.2 ISOPARAMETRIC ELEMENTS

The derivation of stiffness matrix by the method described so far involves integration of the strain energy over the surface or volume. For straight boundaries, this integration can be carried out by numerical techniques. Higher order elements are developed with better displacement functions so that accurate results can be obtained with lesser number of elements. Inherent disadvantage with these elements is that as the size of the element increases, accuracy of boundary representation reduces since edges of element boundary are always assumed as straight lines. Also, as the size of the polynomial increases, inversion of G matrix, linking nodal displacements to the coefficients of the polynomial, takes more time.

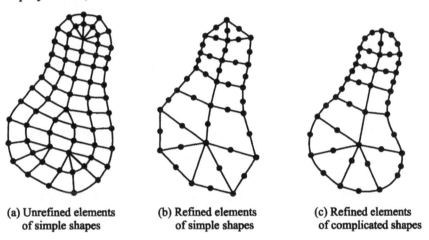

| (a) Unrefined elements of simple shapes | (b) Refined elements of simple shapes | (c) Refined elements of complicated shapes |

FIGURE 6.1 Representation of curved boundaries

A need was therefore felt to improve the method, in order to idealise the given structure with curved boundaries more accurately. In this method, element geometry as well as displacements are interpolated over the element **using shape functions or interpolation functions N_i in terms of natural or intrinsic or non-dimensional coordinates**.

Two types of shape functions are commonly used.

- **Lagrange interpolation function**, which matches the function value (displacement) at specified points or nodes
- **Hermite interpolation function**, which matches function value (displacement) as well as its derivatives (slopes) at the specified nodes.

Curvilinear orthogonal coordinates ξ, η and ζ, whose magnitudes vary from -1 to $+1$ in any element, are used in place of cartesian coordinates x, y and z or cylindrical coordinates R, θ and Z. The shape functions are then defined in terms of the natural coordinates, which link displacement at any point in the element to the nodal displacements thus avoiding the need for inversion of G matrix. It is obvious that there will be as many shape functions as the number of nodes in the element. Each shape function will have a value equal to unity at one node and a value equal to zero at all other nodes

Three types of elements are possible.

- If a higher order function is used to represent displacement and a lower order function is used to represent geometry, it is called a **sub-parametric element**.

- If a lower order function is used to represent displacement and a higher order function is used to represent geometry, it is called a **super-parametric element**.

- If functions of same order are used to represent displacement as well as geometry, then the element is called an **iso-parametric element**.

Of these, iso-parametric elements are most commonly used.

6.3 STIFFNESS MATRICES OF SOME ISO-PARAMETRIC ELEMENTS

(a) 1-D linear interpolation for a truss element

Let the non-dimensional coordinate for the 1-D element be defined by

$$\xi = \frac{2(x - x_1)}{(x_2 - x_1)} - 1$$

so that at node 1, $x = x_1$ and $\xi = -1$ while at node 2, $x = x_2$ and $\xi = 1$

Then, the shape functions N_1 and N_2 are defined in terms of ξ, as shown in Fig.6.2, by

$$N_1(\xi) = \frac{(1 - \xi)}{2}; \quad N_2(\xi) = \frac{(1 + \xi)}{2}$$

It can be seen that the shape function $N_1 = 1$ for $x = x_1$ and $\xi = -1$ at node 1 and $N_1 = 0$ for $x = x_2$ and $\xi = +1$ at node 2, while $N_2 = 0$ at node 1 and $N_2 = 1$ at node 2.

Then, cartesian coordinate 'x' and axial displacement 'u' at any point on the element are given by

$$x = N_1 x_1 + N_2 x_2 = [N] \{x\}$$

and $\quad u = N_1 u_1 + N_2 u_2 = [N] \{q\}$

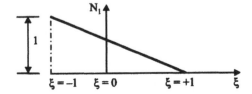 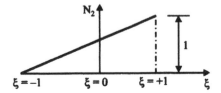

FIGURE 6.2 Shape functions of a 2-noded truss element (Case-1)

The strain $\varepsilon = \dfrac{du}{dx}$ can now be expressed, using chain rule of differentiation, as

$$\varepsilon = \frac{du}{d\xi} \frac{d\xi}{dx} = \left[\frac{dN_1}{d\xi} u_1 + \frac{dN_2}{d\xi} u_2 \right] \frac{d\xi}{dx}$$

$$= \left[\frac{-u_1}{2} + \frac{u_2}{2} \right] \frac{2}{(x_2 - x_1)} = \frac{-u_1 + u_2}{L} = [B]\{q\}$$

where, $\quad [B] = \left[\dfrac{-1}{L} \quad \dfrac{1}{L} \right]$ and $\quad L = x_2 - x_1$

Then, $\quad [K_e] = \displaystyle\int_V [B]^T [D][B] dV = \int_0^L A [B]^T E [B] dx = \frac{AE}{L} \begin{bmatrix} 1 & -1 \\ -1 & 1 \end{bmatrix}$

This stiffness matrix is identical to the one obtained by polynomial method described earlier.

Alternative coordinates – There is no restriction on the choice of the origin of local coordinate system. For example, if the origin is taken at the left end of the truss element, as shown in Fig. 6.3, then the non-dimensional coordinate ξ is given by

$$\xi = (x - x_1) / (x_2 - x_1)$$

At node 1, $x = x_1$ and $\xi = 0$ while at node 2, $x = x_2$ and $\xi = 1$

The shape functions are now defined by $N_1(\xi) = 1-\xi$; $N_2(\xi) = \xi$

which give the values $N_1 = 1$ for $\xi = 0$ at node 1 and $N_1 = 0$ for $\xi = 1$ at node 2 while $N_2 = 0$ at node 1 and $N_2 = 1$ at node 2.

The Cartesian coordinate x and displacement u are again defined by

$$x = N_1 x_1 + N_2 x_2 = [N]\,\{x\}$$

and $$u = N_1 u_1 + N_2 u_2 = [N]\,\{q\}$$

 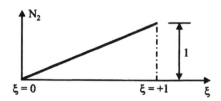

FIGURE 6.3 Shape functions of a 2-noded truss element (Case-2)

$$\varepsilon = \frac{du}{d\xi}\frac{d\xi}{dx} = (-u_1 + u_2)\cdot\frac{1}{x_2 - x_1} = \frac{-u_1 + u_2}{L} = [B]\,\{q\}$$

where, $$[B] = \left(\frac{1}{L}\right)[-1 \quad 1] \; ; \quad L = x_2 - x_1$$

$$[K_e] = \int_v [B]^T [D][B]\,dv = \int_0^L [B]^T E[B]A\,dx = \frac{AE}{L}\begin{bmatrix} 1 & -1 \\ -1 & 1 \end{bmatrix}$$

This stiffness matrix is also identical to the one obtained by polynomial method described earlier.

(b) 1-D Quadratic interpolation for a truss element

The non-dimensional coordinate, with mid-point of the element (node 3) as the origin, is given by $\xi = \dfrac{2(x - x_3)}{(x_2 - x_1)}$

so that with $x_2 - x_1 = L$, $x_1 - x_3 = -L/2$ and $x_2 - x_3 = L/2$ the non-dimensional coordinates at the three nodes are $\xi_1 = -1$; $\xi_2 = 1$; $\xi_3 = 0$

Then, $$\frac{d\xi}{dx} = \frac{2}{(x_2 - x_1)} = \frac{2}{L}$$

Corresponding shape functions N_1, N_2 and N_3 are given by

$$N_1(\xi) = -\xi(1 - \xi)/2; \quad N_2(\xi) = \xi(1 + \xi)/2; \quad N_3(\xi) = (1 + \xi)(1 - \xi)$$

They are graphically represented in Fig. 6.4.

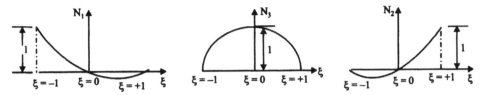

FIGURE 6.4 Shape functions of a 3-noded truss element

The local Cartesian coordinate x and the displacement u are then given by

$$x = N_1\, x_1 + N_2\, x_2 + N_3\, x_3 = [N]\, \{x\}$$

$$u = N_1\, u_1 + N_2\, u_2 + N_3\, u_3 = [N]\, \{q\}$$

Strain, expanded by the chain rule of differentiation, is given by

$$\varepsilon = \frac{du}{d\xi}\frac{d\xi}{dx} = \left[\frac{dN_1}{d\xi}u_1 + \frac{dN_2}{d\xi}u_2 + \frac{dN_3}{d\xi}u_3\right]\frac{d\xi}{dx}$$

$$= \frac{2}{L}\left[-\frac{(1-2\xi)u_1}{2} + \frac{(1+2\xi)u_2}{2} - 2\xi u_3\right] = [B]\{q\}$$

where, $\quad [B] = \dfrac{2}{L}\left[-\dfrac{(1-2\xi)}{2} \quad \dfrac{(1+2\xi)}{2} \quad -2\xi\right]$

$$[K_e] = \int_v [B]^T\,[D][B]\,dV = \int_0^L A\,[B]^T\,E\,[B]\,dx = \frac{AE}{3L}\begin{bmatrix} 7 & 1 & -8 \\ 1 & 7 & -8 \\ -8 & -8 & 16 \end{bmatrix}$$

This stiffness matrix is identical to the matrix obtained earlier for the 3-noded truss element.

(c) 1-D interpolation for a beam element

Shape functions for the beam elements differ from those of truss elements, since derivatives of displacement (slopes) are also involved. Hermite functions, H_i, which satisfy deflection and slope continuity, are used so that deflection at any point is given by

$$w = H_1 w_1 + H_2 \theta_1 + H_3 w_2 + H_4 \theta_2 \text{ or } H_1 w_1 + H_2 (dw/dx)_1 + H_3 w_2 + H_4 (dw/dx)_2$$

$$= H_1 w_1 + (2/L)\, H_2 (dw/d\xi)_1 + H_3 w_2 + (2/L)\, H_4 (dw/d\xi)_2$$

since $\dfrac{dw}{dx} = \left(\dfrac{dw}{d\xi}\right)\left(\dfrac{d\xi}{dx}\right) = \left(\dfrac{2}{L}\right)\left(\dfrac{dw}{d\xi}\right)$

Here, Cartesian coordinate x of any point in the element is related to non-dimensional coordinate ξ by

where, $x = x_1\left(\dfrac{1-\xi}{2}\right) + x_2\left(\dfrac{1+\xi}{2}\right) = \dfrac{(x_1+x_2)}{2} + \dfrac{\xi(x_2-x_1)}{2}$

for $\quad -1 \le \xi \le +1$

and $\quad \dfrac{dx}{d\xi} = \dfrac{(x_2-x_1)}{2} = (L/2) \quad$ or $\quad \dfrac{d\xi}{dx} = \dfrac{2}{L}$

$$H_i = a_i + b_i\xi + c_i\xi^2 + d_i\xi^3 \qquad \forall\ i = 1, 2, 3, 4$$

By imposing the end conditions

$$w_1 = 0 \text{ and } \left(\dfrac{dw}{d\xi}\right)_1 = 0 \text{ at } \xi = -1$$

and $\quad w_2 = 0$ and $\left(\dfrac{dw}{d\xi}\right)_2 = 0$ at $\xi, = 1$

these H_i functions and their derivatives take the following values.

	at $\xi = -1$	at $\xi = +1$		at $\xi = -1$	at $\xi = +1$
H_1	1	0	H_1'	0	0
H_2	0	0	H_2'	1	0
H_3	0	1	H_3'	0	0
H_4	0	0	H_4'	0	1

Coefficients a, b, c, d can then be obtained as

$a_1 = \dfrac{1}{2}$; $b_1 = -\dfrac{3}{4}$; $c_1 = 0$; $d_1 = \dfrac{1}{4}$

$a_2 = \dfrac{1}{4}$; $b_2 = \dfrac{-1}{4}$; $c_2 = \dfrac{-1}{4}$; $d_2 = \dfrac{1}{4}$

$a_3 = \dfrac{1}{2}$; $b_3 = \dfrac{3}{4}$; $c_3 = 0$; $d_3 = \dfrac{-1}{4}$

$a_4 = \dfrac{-1}{4}$; $b_4 = \dfrac{-1}{4}$; $c_4 = \dfrac{1}{4}$; $d_4 = \dfrac{1}{4}$

Thus, $H_1 = \dfrac{(1-\xi)^2(2+\xi)}{4}$ or $\dfrac{(2-3\xi+\xi^3)}{4}$

$H_2 = \dfrac{(1-\xi)^2(\xi+1)}{4}$ or $\dfrac{(1-\xi-\xi^2+\xi^3)}{4}$

$H_3 = \dfrac{(1+\xi)^2(2-\xi)}{4}$ or $\dfrac{(2+3\xi-\xi^3)}{4}$

$H_4 = \dfrac{(1+\xi)^2(\xi-1)}{4}$ or $\dfrac{(-1-\xi+\xi^2+\xi^3)}{4}$

Variation of these functions over the beam element is plotted in Fig. 6.5.

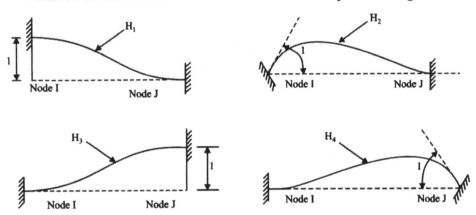

FIGURE 6.5 Hermite Shape functions of a 2-noded beam element

(d) 2-D linear interpolation for a triangular element

The non-dimensional coordinates ξ and η and the shape functions N_1, N_2 and N_3 are also at any point P are given by above the figures.

$$N_1 = \frac{A_1}{A} = \xi ; \quad N_2 = \frac{A_2}{A} = \eta ; \quad N_3 = \frac{A_3}{A} = \zeta$$

Where A is the area of the triangle; A_1 is the area of the triangle formed by points P, 2 and 3; A_2 is the area of the triangle formed by points P, 3 and 1; and A_3 is the area of the triangle formed by points P, 1 and 2, as shown in Fig. 6.6. Hence, the shape functions N_1, N_2 and N_3 are also called **area coordinates**.

It can be seen that for any point P, $A = A_1 + A_2 + A_3$

and so $N_1 + N_2 + N_3 = 1$ or $N_3 = 1 - N_1 - N_2 = 1 - \xi - \eta$

The local Cartesian coordinates x and y and displacements u and v along these Cartesian coordinates are given by

$$x = N_1 x_1 + N_2 x_2 + N_3 x_3 = x_1\xi + x_2\eta + x_3 (1-\xi-\eta)$$

$$= x_{13}\xi + x_{23}\eta + x_3 \quad \text{where,} \quad x_{13} = x_1 - x_3 \text{ and } x_{23} = x_2 - x_3$$

Similarly, $y = y_{13}\xi + y_{23}\eta + y_3$ where, $y_{13} = y_1 - y_3$ and $y_{23} = y_2 - y_3$

.....(6.1)

The Cartesian coordinates as well as corresponding non-dimensional coordinates are shown in Fig. 6.6 for a 3-noded triangular element.

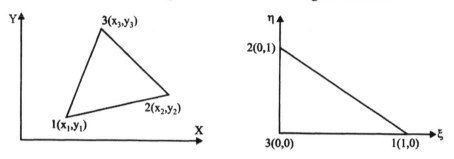

FIGURE 6.6 Mapping of a triangular element in $\xi-\eta$ coordinate system

The displacements u and v can be represented in terms of the same non-dimensional coordinates as

$$u = N_1 u_1 + N_2 u_2 + N_3 u_3$$

$$= (u_1 - u_3)\xi + (u_2 - u_3)\eta + u_3$$

$$v = N_1 v_1 + N_2 v_2 + N_3 v_3$$

$$= (v_1 - v_3)\xi + (v_2 - v_3)\eta + v_3$$

These equations can also be represented in matrix form by

$$\begin{Bmatrix} u \\ v \end{Bmatrix} = \begin{bmatrix} N_1 & 0 & N_2 & 0 & N_3 & 0 \\ 0 & N_1 & 0 & N_2 & 0 & N_3 \end{bmatrix} \begin{Bmatrix} u_1 \\ v_1 \\ u_2 \\ v_2 \\ u_3 \\ v_3 \end{Bmatrix} \quad \text{or } \{u\} = [N]\,\{q\}$$

Shape functions are plotted in Fig. 6.7 with non-dimensional coordinates indicated for each node. The three shape functions have a value of unity at one node and a value of zero at all other nodes. In the figure, A_1, A_2 and A_3 indicate areas used for the calculation of non-dimensional coordinates ξ and η of any point P.

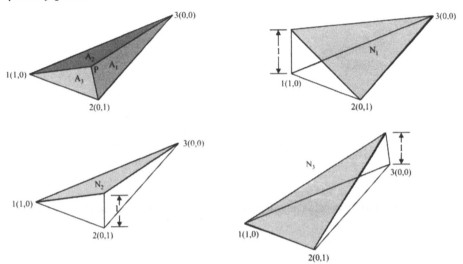

FIGURE 6.7 Shape functions of a 3-noded triangular element

(e) 2-D quadratic interpolation for a triangular element

The non-dimensional coordinates ξ, η and ζ remain same as above but the number of shape functions are increased to six corresponding to the six nodes of the element and are given below

$$N_1 = \xi(2\xi - 1)$$

$$N_2 = \eta(2\eta - 1)$$

$$N_3 = \zeta(2\zeta - 1)$$

$$N_4 = 4\xi\eta$$

$$N_5 = 4\eta\zeta$$

$$N_6 = 4\zeta\xi$$

Lines with constant values of N_1, N_2 and N_3 are shown in Fig. 6.8 while variation of shape functions is plotted in Fig. 6.9.

The cartesian coordinates x and y and displacements u and v in the element are given by

$$x = N_1 x_1 + N_2 x_2 + N_3 x_3 + N_4 x_4 + N_5 x_5 + N_6 x_6$$

$$y = N_1\,y_1 + N_2\,y_2 + N_3\,y_3 + N_4\,y_4 + N_5\,y_5 + N_6\,y_6$$
$$u = N_1\,u_1 + N_2\,u_2 + N_3\,u_3 + N_4\,u_4 + N_5\,u_5 + N_6\,u_6$$
$$v = N_1\,v_1 + N_2\,v_2 + N_3\,v_3 + N_4\,v_4 + N_5\,v_5 + N_6\,v_6$$

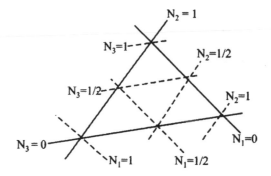

FIGURE 6.8 Lines of constant shape function value over the triangular element

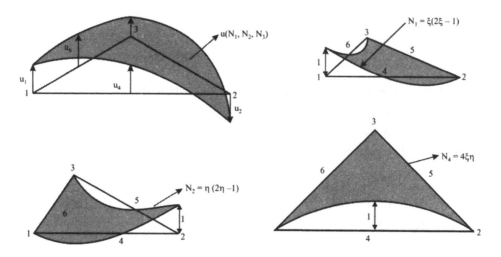

FIGURE 6.9 Shape functions of a 6-noded triangular element

(f) 2-D linear interpolation for a quadrilateral element

Unlike in the case of triangular element, which is identified by three non-dimensional coordinates each having values in the range 0 to 1, a quadrilateral element is identified by two non-dimensional coordinates each having values in the range –1 to +1, as shown in Fig. 6.10.

$$x = N_1\,x_1 + N_2\,x_2 + N_3\,x_3 + N_4\,x_4$$
$$y = N_1\,y_1 + N_2\,y_2 + N_3\,y_3 + N_4\,y_4$$

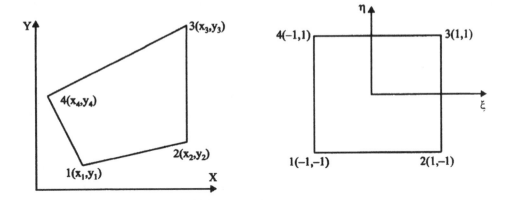

FIGURE 6.10 4-noded quadrilateral element in $\xi - \eta$ coordinate system

The coordinates and displacements at every point in the element are expressed in terms of nodal values, using shape functions N_1, N_2, N_3 and N_4 as

$$u = N_1\, u_1 + N_2\, u_2 + N_3\, u_3 + N_4\, u_4$$

$$v = N_1\, v_1 + N_2\, v_2 + N_3\, v_3 + N_4\, v_4$$

Shape functions at each node of a 2-D element can be derived as the product of shape functions along ξ direction and η direction passing through the particular node.

Thus, $N_1 = N_{1\xi} \cdot N_{1\eta}$

where $N_{1\xi}$ and $N_{1\eta}$ are the shape functions of 1-D elements with

$\xi = -1$ to $+1$ and $\eta = -1$ to $+1$

$\therefore$ $N_1 = \dfrac{(1-\xi)}{2} \cdot \dfrac{(1-\eta)}{2} = \dfrac{(1-\xi)(1-\eta)}{4}$

Similarly, we can get $N_2 = \dfrac{(1+\xi)(1-\eta)}{4}$; $N_3 = \dfrac{(1+\xi)(1+\eta)}{4}$

and $N_4 = \dfrac{(1-\xi)(1+\eta)}{4}$

Their values are graphically represented in Fig. 6.11.

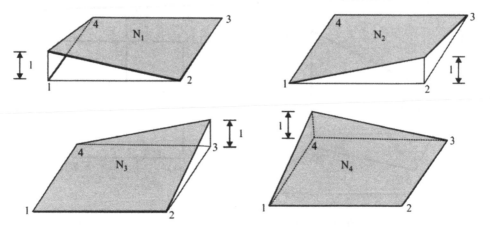

FIGURE 6.11 Shape functions of a 4-noded quadrilateral element

(g) 2-D quadratic interpolation for a quadrilateral element

This quadrilateral element, with 4 corner nodes and 4 mid-side nodes, is also identified by two non-dimensional coordinates each having values in the range -1 to +1. But, the coordinates and displacement of any point in the element are expressed by using 8 shape functions (Ref. Fig. 6.12), as

$$x = N_1 x_1 + N_2 x_2 + N_3 x_3 + N_4 x_4 + N_5 x_5 + N_6 x_6 + N_7 x_7 + N_8 x_8$$

$$y = N_1 y_1 + N_2 y_2 + N_3 y_3 + N_4 y_4 + N_5 y_5 + N_6 y_6 + N_7 y_7 + N_8 y_8$$

$$u = N_1 u_1 + N_2 u_2 + N_3 u_3 + N_4 u_4 + N_5 u_5 + N_6 u_6 + N_7 u_7 + N_8 u_8$$

$$v = N_1 v_1 + N_2 v_2 + N_3 v_3 + N_4 v_4 + N_5 v_5 + N_6 v_6 + N_7 v_7 + N_8 v_8$$

where,

$$N_1 = \frac{-(1-\xi)(1-\eta)(1+\xi+\eta)}{4} \quad ; \quad N_5 = \frac{(1-\xi^2)(1-\eta)}{2}$$

$$N_2 = \frac{-(1+\xi)(1-\eta)(1-\xi+\eta)}{4} \quad ; \quad N_6 = \frac{(1+\xi)(1-\eta^2)}{2}$$

$$N_3 = \frac{-(1+\xi)(1+\eta)(1-\xi-\eta)}{4} \quad ; \quad N_7 = \frac{(1-\xi^2)(1+\eta)}{2}$$

$$N_4 = \frac{-(1-\xi)(1+\eta)(1+\xi-\eta)}{4} \quad ; \quad N_8 = \frac{(1-\xi)(1-\eta^2)}{2}$$

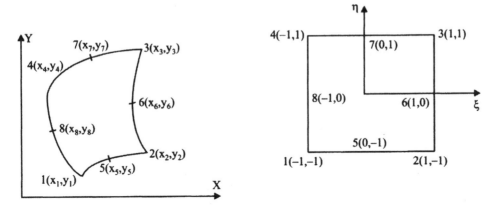

FIGURE 6.12 Mapping of 8-noded quadrilateral in $\xi-\eta$ coordinate system

(h)   3-D linear interpolation

A 8-noded 3-D element is identified by three non-dimensional coordinates each having values in the range −1 to +1, as shown Fig. 6.13.

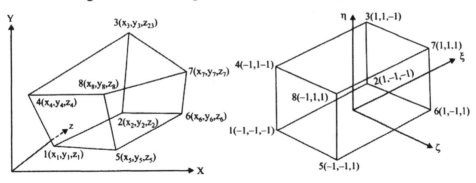

FIGURE 6.13 Mapping of 8-noded 3-D solid element in $\xi-\eta$ coordinate system

The coordinates and displacement functions are given by

$$x = N_1 x_1 + N_2 x_2 + N_3 x_3 + N_4 x_4 + N_5 x_5 + N_6 x_6 + N_7 x_7 + N_8 x_8$$

$$y = N_1 y_1 + N_2 y_2 + N_3 y_3 + N_4 y_4 + N_5 y_5 + N_6 y_6 + N_7 y_7 + N_8 y_8$$

$$z = N_1 z_1 + N_2 z_2 + N_3 z_3 + N_4 z_4 + N_5 z_5 + N_6 z_6 + N_7 z_7 + N_8 z_8$$

and

$$u = N_1 u_1 + N_2 u_2 + N_3 u_3 + N_4 u_4 + N_5 u_5 + N_6 u_6 + N_7 u_7 + N_8 u_8$$

$$v = N_1 v_1 + N_2 v_2 + N_3 v_3 + N_4 v_4 + N_5 v_5 + N_6 v_6 + N_7 v_7 + N_8 v_8$$

$$w = N_1 w_1 + N_2 w_2 + N_3 w_3 + N_4 w_4 + N_5 w_5 + N_6 w_6 + N_7 w_7 + N_8 w_8$$

Shape functions at each node of a 3-D element can be derived as the product of shape functions of 1-D elements along ξ direction, η direction and ζ direction passing through the particular node.

Thus,

$$N_1 = N_{1\xi} \cdot N_{1\eta} \cdot N_{1\zeta}$$

where $N_{1\xi}$, $N_{1\eta}$ and $N_{1\zeta}$ are the shape functions of 1-D elements with

$$\xi = -1 \text{ to } +1, \ \eta = -1 \text{ to } +1 \text{ and } \zeta = -1 \text{ to } +1$$

$\therefore$
$$N_1 = \frac{(1-\xi)}{2} \frac{(1-\eta)}{2} \frac{(1-\zeta)}{2}$$

$$= \frac{(1-\xi)(1-\eta)(1-\zeta)}{8}$$

Similarly, we can get $N_2 = \dfrac{(1+\xi)(1-\eta)(1-\zeta)}{8}$

$$N_3 = \frac{(1+\xi)(1+\eta)(1-\zeta)}{8}$$

$$N_4 = \frac{(1-\xi)(1+\eta)(1-\zeta)}{8}$$

$$N_5 = \frac{(1-\xi)(1-\eta)(1+\zeta)}{8}$$

$$N_6 = \frac{(1+\xi)(1-\eta)(1+\zeta)}{8}$$

$$N_7 = \frac{(1+\xi)(1+\eta)(1+\zeta)}{8}$$

$$N_8 = \frac{(1-\xi)(1+\eta)(1+\zeta)}{8}$$

6.4 JACOBIAN

In order to have unique mapping of elements, there should be only one set of cartesian coordinates for each set of corresponding non-dimensional coordinates.

For a 2-D plate element in x and y coordinates, using chain rule for partial derivatives,

$$\frac{\partial u}{\partial \xi} = \frac{\partial u}{\partial x} \cdot \frac{\partial x}{\partial \xi} + \frac{\partial u}{\partial y} \cdot \frac{\partial y}{\partial \xi} \qquad \text{Similarly for } \frac{\partial u}{\partial \eta}$$

Expressing them in matrix form,

$$\begin{Bmatrix} \partial u / \partial \xi \\ \partial u / \partial \eta \end{Bmatrix} = \begin{bmatrix} \partial x / \partial \xi & \partial y / \partial \xi \\ \partial x / \partial \eta & \partial y / \partial \eta \end{bmatrix} \begin{Bmatrix} \partial u / \partial x \\ \partial u / \partial y \end{Bmatrix} = [J] \begin{Bmatrix} \partial u / \partial x \\ \partial u / \partial y \end{Bmatrix}$$

where [J] is called the Jacobian or a matrix of partial derivatives of Cartesian coordinates of the element w.r.t. non-dimensional local coordinates of the element

From eq. 6.1, $[J] = \begin{bmatrix} x_{13} & y_{13} \\ x_{23} & y_{23} \end{bmatrix}$ or $[J]^{-1} = \dfrac{1}{\text{Det J}} \begin{bmatrix} y_{23} & -y_{23} \\ -x_{23} & x_{13} \end{bmatrix}$

If nodes are numbered counter-clockwise, in a right-handed coordinate system, det J is +ve.

For a 3-noded triangular plate element, from eq. 6.1

$$[J] = \begin{bmatrix} \dfrac{\partial x}{\partial \xi} & \dfrac{\partial y}{\partial \xi} \\ \dfrac{\partial x}{\partial \eta} & \dfrac{\partial y}{\partial \eta} \end{bmatrix} = \begin{bmatrix} X_{13} & Y_{13} \\ X_{23} & Y_{23} \end{bmatrix}$$

or $[J]^{-1} = \dfrac{1}{\text{Det J}} \begin{bmatrix} Y_{23} & -Y_{13} \\ -X_{23} & X_{13} \end{bmatrix}$

where,   Det J = $X_{13} Y_{23} - X_{23} Y_{13}$

It can also be expressed as

$$\text{Det J} = \begin{vmatrix} 1 & x_1 & y_1 \\ 1 & x_2 & y_2 \\ 1 & x_3 & y_3 \end{vmatrix} = \begin{vmatrix} 0 & x_1 - x_3 & y_1 - y_3 \\ 0 & x_2 - x_3 & y_2 - y_3 \\ 1 & x_3 & y_3 \end{vmatrix} = \begin{vmatrix} 0 & x_{13} & y_{13} \\ 0 & x_{23} & y_{23} \\ 1 & x_3 & y_3 \end{vmatrix}$$

$$= x_{13} y_{23} - y_{13} x_{23}$$

Area of the element, $A = \dfrac{1}{2} \text{Det J} = \dfrac{1}{2} \left(x_{13} y_{23} - x_{23} y_{13} \right)$

and $[K] = \int_v [B]^T [D][B] \, dV = \int_{-1}^{1} \int_{-1}^{1} \int_{-1}^{1} [B]^T [D][B] \, \text{Det}[J] \, d\xi \, d\eta \, d\zeta$

For a 3-D element in x, y and z coordinates, these relations can be expressed in matrix form in a similar way as

$$\begin{Bmatrix} \partial u / \partial \xi \\ \partial u / \partial \eta \\ \partial u / \partial \zeta \end{Bmatrix} = \begin{bmatrix} \partial x / \partial \xi & \partial y / \partial \xi & \partial z / \partial \xi \\ \partial x / \partial \eta & \partial y / \partial \eta & \partial z / \partial \eta \\ \partial x / \partial \zeta & \partial y / \partial \zeta & \partial y / \partial \zeta \end{bmatrix} \begin{Bmatrix} \partial u / \partial x \\ \partial u / \partial y \\ \partial u / \partial z \end{Bmatrix} = [J] \begin{Bmatrix} \partial u / \partial x \\ \partial u / \partial y \\ \partial u / \partial z \end{Bmatrix}$$

where,

$$[J] = \begin{bmatrix} \partial x/\partial\xi & \partial y/\partial\xi & \partial z/\partial\xi \\ \partial x/\partial\eta & \partial y/\partial\eta & \partial z/\partial\eta \\ \partial x/\partial\zeta & \partial y/\partial\zeta & \partial y/\partial\zeta \end{bmatrix} = \begin{bmatrix} x_{14} & y_{14} & z_{14} \\ x_{24} & y_{24} & z_{24} \\ x_{34} & y_{34} & z_{34} \end{bmatrix}$$

The same Jacobian relates derivatives of displacement components v and w also.

6.5 STRAIN-DISPLACEMENT RELATIONS

(a) For a 3-noded triangular element

$$\{\varepsilon\} = \begin{Bmatrix} \partial u/\partial x \\ \partial v/\partial y \\ \partial u/\partial y + \partial v/\partial x \end{Bmatrix} = [J]^{-1} \begin{Bmatrix} \partial u/\partial\xi \\ \partial v/\partial\eta \\ \partial u/\partial\eta + \partial v/\partial\xi \end{Bmatrix}$$

$$= \frac{1}{\text{Det J}} \begin{bmatrix} y_{23} & 0 & y_{31} & 0 \\ 0 & x_{32} & 0 & x_{13} \\ x_{32} & y_{23} & x_{13} & y_{31} \end{bmatrix} \begin{Bmatrix} \partial u/\partial\xi \\ \partial v/\partial\eta \\ \partial u/\partial\eta \\ \partial v/\partial\xi \end{Bmatrix}$$

$$= \left(\frac{1}{\text{Det J}}\right) \begin{Bmatrix} y_{23}\,\partial u/\partial\xi + (-y_{13})\,\partial u/\partial\eta \\ (-x_{23})\,\partial v/\partial\xi + x_{13}\,\partial u/\partial\eta \\ (-x_{23})\,\partial u/\partial\xi + x_{13}\,\partial u/\partial\eta + y_{23}\,\partial v/\partial\xi + (-y_{13})\,\partial v/\partial\eta \end{Bmatrix} \quad \text{Since}$$

$$u = N_1 u_1 + N_2 u_2 + N_3 u_3 = (u_1 - u_3)\,\xi + (u_2 - u_3)\,\eta + u_3$$

$$\partial u/\partial\xi = u_1 - u_3 = u_{13} \ ; \qquad \partial u/\partial\eta = u_2 - u_3 = u_{23}$$

Similarly, $\partial v/\partial\xi = v_1 - v_3 = v_{13} \ ; \qquad \partial v/\partial\eta = v_2 - v_3 = v_{23}$

Therefore,

$$\{\varepsilon\} = \frac{1}{\text{Det J}} \begin{bmatrix} y_{23} & 0 & y_{31} & 0 \\ 0 & x_{32} & 0 & x_{13} \\ x_{32} & y_{23} & x_{13} & y_{31} \end{bmatrix} \begin{Bmatrix} u_{13} \\ v_{13} \\ u_{23} \\ v_{23} \end{Bmatrix} = \frac{1}{\text{Det J}} \begin{bmatrix} y_{23}.u_{13} + y_{31}.u_{23} \\ x_{32}.v_{13} + x_{13}.v_{23} \\ x_{32}.u_{13} + y_{23}.v_{13} + x_{13}.u_{23} + y_{31}.v_{23} \end{bmatrix}$$

$$
= \frac{1}{\text{Det } J}
\begin{bmatrix}
y_{23} & 0 & y_{31} & 0 & y_{12} & 0 \\
0 & x_{32} & 0 & x_{13} & 0 & x_{21} \\
x_{32} & y_{23} & x_{13} & y_{31} & x_{21} & y_{12}
\end{bmatrix}
\begin{Bmatrix}
u_1 \\
v_1 \\
u_2 \\
v_2 \\
u_3 \\
v_3
\end{Bmatrix}
= [B]\{q\}
$$

where strain-displacement matrix,

$$
[B] = \frac{1}{\text{Det } J}
\begin{bmatrix}
y_{23} & 0 & y_{31} & 0 & y_{12} & 0 \\
0 & x_{32} & 0 & x_{13} & 0 & x_{21} \\
x_{32} & y_{23} & x_{13} & y_{31} & x_{21} & y_{12}
\end{bmatrix}
$$

The elements of [B] are constants and not functions of coordinates (since shape functions are linear in x and y) and hence, strain in a 3-node triangular element is constant over the entire element.

Element stiffness matrix $[K_e]$ is obtained from $[K_e] = \int [B]^T [D] [B] \, dV$. Here, the integration is carried out on non-dimensional coordinates with the limits 0 to +1 or −1 to +1, depending on the particular case. Element with curved edges also is mapped into an element with straight edges in the non-dimensional coordinate system. Hence, for an element with curved boundaries, integration becomes much simpler in non-dimensional coordinate system compared to the integration in Cartesian coordinate system.

(b) For a 4-noded quadrilateral (2-D) element

Following the same procedure, we can get [B] matrix. The local non-dimensional coordinates ξ and η take values from −1 to +1.

Here, $x = N_1 x_1 + N_2 x_2 + N_3 x_3 + N_4 x_4$

$y = N_1 y_1 + N_2 y_2 + N_3 y_3 + N_4 y_4$

and $u = N_1 u_1 + N_2 u_2 + N_3 u_3 + N_4 u_4$

$v = N_1 v_1 + N_2 v_2 + N_3 v_3 + N_4 v_4$

$N_1 = \dfrac{(1-\xi)(1-\eta)}{4}; \quad N_2 = \dfrac{(1+\xi)(1-\eta)}{4}$

$N_3 = \dfrac{(1+\xi)(1+\eta)}{4}; \quad N_4 = \dfrac{(1-\xi)(1+\eta)}{4}$

Then, $[J] = \begin{bmatrix} \partial x/\partial\xi & \partial y/\partial\xi \\ \partial x/\partial\eta & \partial y/\partial\eta \end{bmatrix} = \begin{bmatrix} J_{11} & J_{12} \\ J_{21} & J_{22} \end{bmatrix}$

where, $J_{11} = \left(\dfrac{1}{4}\right)[-(1-\eta)x_1 + (1-\eta)x_2 + (1+\eta)x_3 - (1+\eta)x_4]$

$J_{12} = \left(\dfrac{1}{4}\right)[-(1-\eta)y_1 + (1-\eta)y_2 + (1+\eta)y_3 - (1+\eta)y_4]$

$J_{21} = \left(\dfrac{1}{4}\right)[-(1-\xi)x_1 - (1+\xi)x_2 + (1+\xi)x_3 + (1-\xi)x_4]$

$J_{22} = \left(\dfrac{1}{4}\right)[-(1-\xi)y_1 - (1+\xi)y_2 + (1+\xi)y_3 + (1-\xi)y_4]$

Now, $\begin{Bmatrix} \partial u/\partial x \\ \partial u/\partial y \end{Bmatrix} = (1/\text{Det } J)\begin{bmatrix} J_{22} & -J_{12} \\ -J_{21} & J_{11} \end{bmatrix}\begin{Bmatrix} \partial u/\partial\xi \\ \partial u/\partial\eta \end{Bmatrix}$

and $\begin{Bmatrix} \partial v/\partial x \\ \partial v/\partial y \end{Bmatrix} = (1/\text{Det } J)\begin{bmatrix} J_{22} & -J_{12} \\ -J_{21} & J_{11} \end{bmatrix}\begin{Bmatrix} \partial v/\partial\xi \\ \partial v/\partial\eta \end{Bmatrix}$

$\{\varepsilon\} = \begin{Bmatrix} \partial u/\partial x \\ \partial v/\partial y \\ \partial u/\partial y + \partial v + \partial x \end{Bmatrix}$

$= \dfrac{1}{\text{Det } J}\begin{bmatrix} J_{22} & -J_{12} & 0 & 0 \\ 0 & 0 & -J_{21} & J_{11} \\ -J_{21} & J_{11} & J_{22} & -J_{12} \end{bmatrix}\begin{Bmatrix} \partial u/\partial\xi \\ \partial u/\partial\eta \\ \partial v/\partial\xi \\ \partial v/\partial\eta \end{Bmatrix} = [B]\{q\}$

where,

$[B] = \dfrac{1}{\text{Det } J}\begin{bmatrix} J_{22} & -J_{12} & 0 & 0 \\ 0 & 0 & -J_{21} & J_{11} \\ -J_{21} & J_{11} & J_{22} & -J_{12} \end{bmatrix} \times$

$\dfrac{1}{4}\begin{bmatrix} -1(1-\eta) & 0 & (1-\eta) & 0 & (1+\eta) & 0 & -(1+\eta) & 0 \\ -(1-\xi) & 0 & -(1+\xi) & 0 & (1+\xi) & 0 & (1-\xi) & 0 \\ 0 & -(1-\eta) & 0 & (1-\eta) & 0 & (1+\eta) & 0 & -(1+\eta) \\ 0 & -(1-\xi) & 0 & -(1+\xi) & 0 & (1+\xi) & 0 & (1-\xi) \end{bmatrix}$

Elements of this matrix are functions of coordinates, since shape functions include higher order terms and hence strain in this element is _not_ constant over the element.

(c) For a 8-noded quadrilateral (2-D) element

Following the same procedure, we can get [B] matrix. The local non-dimensional coordinates ξ and η take values from -1 to $+1$.

$$x = N_1 x_1 + N_2 x_2 + N_3 x_3 + N_4 x_4 + N_5 x_5 + N_6 x_6 + N_7 x_7 + N_8 x_8$$

$$y = N_1 y_1 + N_2 y_2 + N_3 y_3 + N_4 y_4 + N_5 y_5 + N_6 y_6 + N_7 y_7 + N_8 y_8$$

$$u = N_1 u_1 + N_2 u_2 + N_3 u_3 + N_4 u_4 + N_5 u_5 + N_6 u_6 + N_7 u_7 + N_8 u_8$$

$$v = N_1 v_1 + N_2 v_2 + N_3 v_3 + N_4 v_4 + N_5 v_5 + N_6 v_6 + N_7 v_7 + N_8 v_8$$

where,

$$N_1 = \frac{-(1-\xi)(1-\eta)(1+\xi+\eta)}{4} \quad ; \quad N_5 = \frac{(1-\xi^2)(1-\eta)}{2}$$

$$N_2 = \frac{-(1+\xi)(1-\eta)(1-\xi+\eta)}{4} \quad ; \quad N_6 = \frac{(1+\xi)(1-\eta^2)}{2}$$

$$N_3 = \frac{-(1+\xi)(1+\eta)(1-\xi-\eta)}{4} \quad ; \quad N_7 = \frac{(1-\xi^2)(1+\eta)}{2}$$

$$N_4 = \frac{-(1-\xi)(1+\eta)(1+\xi-\eta)}{4} \quad ; \quad N_8 = \frac{(1-\xi)(1-\eta^2)}{2}$$

(d) For a 4-noded tetrahedron (3-D) element

Following the same procedure, we get

$$\{\varepsilon\} = \begin{Bmatrix} \partial u/\partial x \\ \partial u/\partial y \\ \partial w/\partial z \\ \partial u/\partial y + \partial v/\partial x \\ \partial v/\partial z + \partial w/\partial y \\ \partial w/\partial x + \partial u/\partial z \end{Bmatrix} = [J]^{-1} \begin{Bmatrix} \partial u/\partial \xi \\ \partial u/\partial \eta \\ \partial w/\partial \zeta \\ \partial u/\partial \eta + \partial v/\partial \xi \\ \partial v/\partial \zeta + \partial w/\partial \eta \\ \partial w/\partial \xi + \partial u/\partial \zeta \end{Bmatrix} = [B]\{q\}$$

Here, $[J] = \begin{bmatrix} x_{14} & y_{14} & z_{14} \\ x_{24} & y_{24} & z_{24} \\ x_{34} & y_{34} & z_{34} \end{bmatrix}$

If the elements of $[J]^{-1}$ are designated by the constants A_{ij}, then

$$[J]^{-1} = \begin{bmatrix} A_{11} & A_{12} & A_{13} \\ A_{21} & A_{22} & A_{23} \\ A_{31} & A_{32} & A_{33} \end{bmatrix}$$

and $A_1' = A_{11} + A_{12} + A_{13}$; $A_2' = A_{21} + A_{22} + A_{23}$; $A_3' = A_{31} + A_{32} + A_{33}$

then,

$$[B] = \begin{bmatrix} A_{11} & 0 & 0 & A_{12} & 0 & 0 & A_{13} & 0 & 0 & -A_1' & 0 & 0 \\ 0 & A_{21} & 0 & 0 & A_{22} & 0 & 0 & A_{23} & 0 & 0 & -A_2' & 0 \\ 0 & 0 & A_{31} & 0 & 0 & A_{32} & 0 & 0 & A_{33} & 0 & 0 & -A_3' \\ 0 & A_{31} & A_{21} & 0 & A_{32} & A_{22} & 0 & A_{33} & A_{23} & 0 & -A_3' & -A_2' \\ A_{31} & 0 & A_{11} & A_{32} & 0 & A_{12} & A_{33} & 0 & A_{13} & -A_3' & 0 & -A_1' \\ A_{21} & A_{11} & 0 & A_{22} & A_{12} & 0 & A_{23} & A_{13} & 0 & -A_2' & -A_1' & 0 \end{bmatrix}$$

Here again, elements of [B] matrix are constants since shape functions are linear functions and, hence, strain in this element is also constant throughout the element.

6.6 SUMMARY

- Higher order elements are broadly classified as - **Serendipity elements,** having no internal nodes and **Lagrange elements,** having internal nodes which can be condensed out at the element level before assembling.

- Higher order elements use higher degree displacement polynomial and can represent true situation with lesser number of elements than a model with lower order elements.

- Isoparametric elements model displacement as well as boundary with the same polynomial. A small number of higher order isoparametric elements can model curved boundaries of component accurately. Even a large number of lower order and non-isoparametric elements can only approximate curved boundary by a set of straight lines.

- Isoparametric elements use non-dimensional coordinates / shape functions and facilitate programming for numerical integration on a computer.

OBJECTIVE QUESTIONS

1. Curved boundary is better modeled by using
   (a) non-dimensional shape functions   (b) higher order elements
   (c) more number of simple elements   (d) isoparametric elements

2. Sum of shape functions at a point is
   (a) 1   (b) 0   (c) any +ve integer   (d) any –ve integer

3. When fewer nodes are used to define the geometry than are used to define the displacement, the element is called ____ element
   (a) subparametric   (b) isoparametric
   (c) superparametric   (d) complex

4. When same number of nodes are used to define the geometry and displacement, the element is called ____ element
   (a) subparametric   (b) isoparametric
   (c) superparametric   (d) simple

5. When more nodes are used to define the geometry than are used to define the displacement, the element is called ____ element
   (a) subparametric   (b) isoparametric
   (c) superparametric   (d) complex

6. Derivatives of displacement function with respect to element coordinate system and non-dimensional coordinate system is given by
   (a) Lagrangian   (b) Poisson
   (c) Gaussian   (d) Jacobian

7. Number of shape functions for a triangular plane stress element are
   (a) 2   (b) 3   (c) 4   (d) 6

8. Number of shape functions for a quadrilateral plane stress element are
   (a) 2   (b) 3   (c) 4   (d) 8

9. Number of shape functions for a 8-noded quadrilateral plane stress element is
   (a) 2   (b) 3   (c) 4   (d) 8

10. Shape functions for a triangular plane stress element are also called
    (a) r-s coordinates   (b) area coordinates
    (c) volume coordinates   (d) x-y coordinates

SOLVED PROBLEMS

Example 6.1

For a point P located inside the triangle shown in figure, the shape functions N_1 and N_2 are 0.15 and 0.25 respectively. Determine the x and y coordinates of point P.

Solution

A triangular element will have three natural or non-dimensional coordinates N_1, N_2 and N_3 such that $N_1 + N_2 + N_3 = 1$ or $N_3 = 1 - N_1 - N_2$.

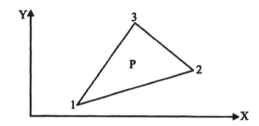

Hence,   coordinates of point P i.e., (x_p, y_p) are given by

$x_P = N_1x_1 + N_2x_2 + N_3x_3 = N_1x_1 + N_2x_2 + (1 - N_1 - N_2)x_3$

$\quad = 0.15 \times 1 + 0.25 \times 4 + (1 - 0.15 - 0.25) \times 3 = 0.15 + 1.0 + 1.8 = 2.95$

$y_P = N_1y_1 + N_2y_2 + N_3y_3 = N_1y_1 + N_2y_2 + (1 - N_1 - N_2)y_3$

$\quad = 0.15 \times 1 + 0.25 \times 2 + (1 - 0.15 - 0.25) \times 5 = 0.15 + 0.5 + 3.0 = 3.65$

Example 6.2

The coordinates and function values at the three nodes of a triangular linear element are given below. Calculate the function value at (20,6).

Node 1 Coordinates (13,1)   Function value 190

Node 2 Coordinates (25,6)   Function value 160

Node 3 Coordinates (13,13)   Function value 185

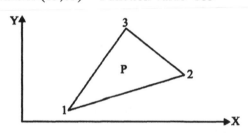

Solution

A triangular element will have three natural or non-dimensional coordinates N_1, N_2 and N_3 such that $N_1 + N_2 + N_3 = 1$ or $N_3 = 1 - N_1 - N_2$

Hence, coordinates of point P i.e., (x_p, y_p) are given by

$$x_P = N_1 x_1 + N_2 x_2 + N_3 x_3 = N_1 (x_1 - x_3) + N_2 (x_2 - x_3) + x_3$$

$$20 = N_1 (13 - 13) + N_2 (25 - 13) + 13$$

$$= 12 N_2 + 13 \quad \text{or} \quad N_2 = 7 / 12$$

$$y_P = N_1 y_1 + N_2 y_2 + N_3 y_3 = N_1 (y_1 - y_3) + N_2 (y_2 - y_3) + y_3$$

$$6 = N_1 (1 - 13) + N_2 (6 - 13) + 13$$

$$= -12 N_1 - 7 \times \left(\frac{7}{12}\right) + 13 \quad \text{or} \quad N_1 = \frac{35}{144}$$

Hence, $N_3 = 1 - N_1 - N_2 = 1 - \left(\frac{35}{144}\right) - \left(\frac{7}{12}\right) = \frac{25}{144}$

Function value at $(x_P, y_P) = N_1 V_1 + N_2 V_2 + N_3 V_3$

$$= \left(\frac{35}{144}\right) \times 190 + \left(\frac{7}{12}\right) \times 160 + \left(\frac{25}{144}\right) \times 185 = 171.632$$

Example 6.3

The nodal coordinates of the triangular element are (1,2), (5,3) and (4,6). At the interior point P, the x-coordinate is 3.3 and the shape function at node 1 is 0.3. Determine the shape functions at nodes 2 and 3 and also the y-coordinate at the point P.

Solution

A triangular element will have three natural or non-dimensional coordinates N_1, N_2 and N_3 such that $N_1 + N_2 + N_3 = 1$ or $N_3 = 1 - N_1 - N_2$.

Hence, coordinates of point P i.e., (x_p, y_p) are given by

$$x_P = N_1 x_1 + N_2 x_2 + N_3 x_3 = N_1 (x_1 - x_3) + N_2 (x_2 - x_3) + x_3$$

$$3.3 = N_1 (1 - 4) + N_2 (5 - 4) + 4 = 0.3 \times (-3) + N_2 + 4$$

$$= N_2 + 3.1 \quad \text{or} \quad N_2 = 0.2$$

Hence, $N_3 = 1 - N_1 - N_2 = 1 - (0.3) - (0.2) = 0.5$

$$y_P = N_1 y_1 + N_2 y_2 + N_3 y_3 = N_1 (y_1 - y_3) + N_2 (y_2 - y_3) + y_3$$

$$= 0.3 \times (2 - 6) + 0.2 \times (3 - 6) + 6 = -1.2 - 0.6 + 6 = 4.2$$

Example 6.4

Triangular elements are used for stress analysis of a plate subjected to in plane load. The components of displacement along x and y axes at the nodes i, j and k of an element are found to be (–0.001, 0.01), (–0.002, 0.01) and (–0.002, 0.02) cm respectively. If the (x, y) coordinates of the nodes i, j and k are (20, 20), (40, 20) and (40, 40) in cm respectively, find (a) the distribution of the two displacement components inside the element and (b) components of displacement of the point $(x_P, y_P) = (30, 25)$ cm.

Solution

(a) Distribution of displacement components u and v inside the element are given by

$$u_P = N_1 u_1 + N_2 u_2 + N_3 u_3 = -0.001 N_1 - 0.002 N_2 - 0.002 N_3$$

and $\quad v_P = N_1 v_1 + N_2 v_2 + N_3 v_3 = 0.01 N_1 + 0.01 N_2 + 0.02 N_3$

(b) A triangular element will have three natural or non-dimensional coordinates N_1, N_2 and N_3 such that $N_1 + N_2 + N_3 = 1$ or $N_3 = 1 - N_1 - N_2$

Hence, coordinates of point P i.e., (x_P, y_P) are given by

$$x_P = N_1 x_1 + N_2 x_2 + N_3 x_3 = N_1 x_1 + N_2 x_2 + (1 - N_1 - N_2) x_3$$

$$30 = 20 N_1 + 40 N_2 + 40 (1 - N_1 - N_2)$$

$$= 40 - 20 N_1$$

Therefore, $20 N_1 = 10$ or $N_1 = 0.5$

Similarly, $y_P = N_1 y_1 + N_2 y_2 + N_3 y_3 = N_1 y_1 + N_2 y_2 + (1 - N_1 - N_2) y_3$

$$25 = 20 N_1 + 20 N_2 + 40 (1 - N_1 - N_2)$$

$$= 40 - 20 N_1 - 20 N_2$$

Therefore, $20 N_2 = 15 - 20 N_1 = 15\text{-}10$ or $N_2 = 0.25$

and $\quad N_3 = 1 - N_1 - N_2 = 1 - 0.5 - 0.25 = 0.25$

The displacements u and v at (x_P, y_P) are given by

$$u_P = N_1 u_1 + N_2 u_2 + N_3 u_3$$

$$= 0.5 \times (-0.001) + 0.25 \times (-0.002) + 0.25 \times (-0.002)$$

$$= -0.0015 \text{ cm}$$

and $\quad v_P = N_1 v_1 + N_2 v_2 + N_3 v_3$

$$= 0.5 \times 0.01 + 0.25 \times 0.01 + 0.25 \times 0.02 = 0.0125 \text{ cm}$$

Example 6.5

The nodal coordinates and the nodal displacements of a triangular element, under a specific load condition are given below.

$X_i = 0$, $Y_i = 0$, $X_j = 1$ mm, $Y_j = 3$ mm, $X_k = 4$ mm, $Y_k = 1$ mm

$U_i = 1$ mm, $V_i = 0.5$ mm, $U_j = -0.05$ mm, $V_j = 1.5$ mm, $U_k = 2$ mm, $V_k = -1$ mm

If $E = 2 \times 10^5$ N/mm^2 and $\mu = 0.3$, find the stresses in the element

Solution

On plotting coordinates of the three nodes of the triangular element, the nodes I, J and K are identified as 1, 3 and 2 respectively to represent the nodes in the counter-clockwise direction around the element.

$Y_{23} = Y_{KJ} = 1 - 3 = -2$; $Y_{31} = Y_{JI} = 3 - 0 = 3$; $Y_{12} = Y_{IK} = 0 - 1 = -1$

$X_{32} = X_{JK} = 1 - 4 = -3$; $X_{13} = X_{IJ} = 0 - 1 = -1$; $X_{21} = X_{KI} = 4 - 0 = 4$

Jacobian of the triangular element, $J = \begin{bmatrix} X_{13} & Y_{13} \\ X_{23} & Y_{23} \end{bmatrix} = \begin{bmatrix} -1 & -3 \\ 3 & -2 \end{bmatrix}$

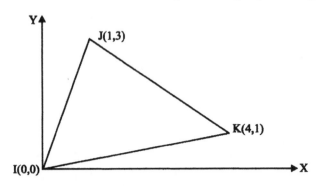

Determinant of Jacobian, $[J] = (-1)(-2) - (-3)(3) = 11$

In a 3-noded triangular element, stresses are constant throughout the element. The three stress components in the element are given by

$\{\sigma\} = [D]\{\varepsilon\} = [D][B]\{u\}$

$= [D]\dfrac{1}{|J|} \begin{bmatrix} Y_{23} & 0 & Y_{31} & 0 & Y_{12} & 0 \\ 0 & X_{32} & 0 & X_{13} & 0 & X_{21} \\ X_{32} & Y_{23} & X_{13} & Y_{31} & X_{21} & Y_{12} \end{bmatrix} \{u\}$

$$= \frac{E}{(1-v^2)} \begin{bmatrix} 1 & v & 0 \\ v & 1 & 0 \\ 0 & 0 & (1-v)/2 \end{bmatrix} \cdot \frac{1}{11} \begin{bmatrix} -2 & 0 & 3 & 0 & -1 & 0 \\ 0 & -3 & 0 & -1 & 0 & 4 \\ -3 & -2 & -1 & 3 & 4 & -1 \end{bmatrix} \begin{Bmatrix} 1.0 \\ 0.5 \\ -0.05 \\ 1.5 \\ 2.0 \\ -1.0 \end{Bmatrix}$$

$$= \frac{2 \times 10^5}{11 \times (1 - 0.3^2)} \begin{bmatrix} 1 & 0.3 & 0 \\ 0.3 & 1 & 0 \\ 0 & 0 & 0.35 \end{bmatrix} \begin{Bmatrix} -4.15 \\ -7.0 \\ 9.55 \end{Bmatrix}$$

$$= \frac{2 \times 10^5}{10.01} \begin{Bmatrix} -6.25 \\ -8.245 \\ 3.3425 \end{Bmatrix} N/mm^2$$

Example 6.6

x, y, z coordinates of nodes of a tetrahedron element are (30,0,0), (0,10,0), (0,0,20) and (20, 20, 10). Formulate strain-displacement matrix [B].

Solution

$$[J] = \begin{bmatrix} x_{14} & y_{14} & z_{14} \\ x_{24} & y_{24} & z_{24} \\ x_{34} & y_{34} & z_{34} \end{bmatrix} = \begin{bmatrix} 10 & -20 & -10 \\ -20 & -10 & -10 \\ -20 & -20 & 10 \end{bmatrix}$$

$$[J]^{-1} = [A] = \begin{bmatrix} A_{11} & A_{12} & A_{13} \\ A_{21} & A_{22} & A_{23} \\ A_{31} & A_{32} & A_{33} \end{bmatrix} = \left(\frac{-1}{13000} \right) \begin{bmatrix} -300 & 400 & 100 \\ 400 & -100 & 300 \\ 200 & 600 & -500 \end{bmatrix}$$

and $A_1' = \dfrac{-(A_{11} + A_{12} + A_{13})}{13000} = \dfrac{-(-300 + 400 + 100)}{13000} = \dfrac{-2}{130}$

$A_2' = \dfrac{-(A_{21} + A_{22} + A_{23})}{13000} = \dfrac{-(400 - 100 + 300)}{13000} = \dfrac{-6}{130}$

$A_3' = \dfrac{-(A_{31} + A_{32} + A_{33})}{13000} = \dfrac{-(200 + 600 - 500)}{13000} = \dfrac{-3}{130}$

Then,

$$[B] = \begin{bmatrix} A_{11} & 0 & 0 & A_{12} & 0 & 0 & A_{13} & 0 & 0 & -A_1' & 0 & 0 \\ 0 & A_{21} & 0 & 0 & A_{22} & 0 & 0 & A_{23} & 0 & 0 & -A_2' & 0 \\ 0 & 0 & A_{31} & 0 & 0 & A_{32} & 0 & 0 & A_{33} & 0 & 0 & -A_3' \\ 0 & A_{31} & A_{21} & 0 & A_{32} & A_{22} & 0 & A_{33} & A_{23} & 0 & -A_3' & -A_2' \\ A_{31} & 0 & A_{11} & A_{32} & 0 & A_{12} & A_{33} & 0 & A_{13} & -A_3' & 0 & -A_1 \\ A_{21} & A_{11} & 0 & A_{22} & A_{12} & 0 & A_{23} & A_{13} & 0 & -A_2' & -A_1' & 0 \end{bmatrix}$$

Thus,

$$[B] = \left(\frac{-1}{130}\right) \begin{bmatrix} -3 & 0 & 0 & 4 & 0 & 0 & 1 & 0 & 0 & -2 & 0 & 0 \\ 0 & 4 & 0 & 0 & -1 & 0 & 0 & 3 & 0 & 0 & -6 & 0 \\ 0 & 0 & 2 & 0 & 0 & 6 & 0 & 0 & -5 & 0 & 0 & -3 \\ 0 & 2 & 4 & 0 & 6 & -1 & 0 & -5 & 3 & 0 & -3 & -6 \\ 2 & 0 & -3 & 6 & 0 & 4 & -5 & 0 & 1 & -3 & 0 & -2 \\ 4 & -3 & 0 & -1 & 4 & 0 & 3 & 1 & 0 & -6 & -2 & 0 \end{bmatrix}$$

Example 6.7

Determine the deflection at the point of load application using a one-element model for the configuration shown in figure.

Solution

Numbering the nodes of the element in counter-clockwise direction as shown,

$$Y_{23} = -20 \; ; \; Y_{31} = 0 \; ; \; Y_{12} = 20$$

$$X_{32} = 30 \; ; \; X_{13} = -30 \; ; \; X_{21} = 0$$

Jacobian of the triangular element, $[J] = \begin{bmatrix} X_{13} & Y_{13} \\ X_{23} & Y_{23} \end{bmatrix} = \begin{bmatrix} -30 & 0 \\ -30 & -20 \end{bmatrix}$

Determinant of Jacobian $|J| = (-30)(-20) = 600$

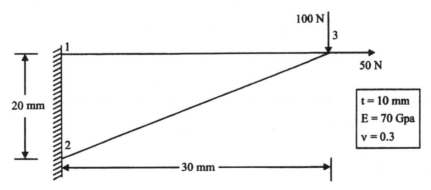

The nodal displacements $\{u\}$ can be calculated from $\{P\} = [K]\{u\}$. So, the stiffness matrix $[K]$ has to be calculated from $[K] = [B]^T[D][B]$

where,

$$[B] = \left(\frac{1}{|J|}\right) \begin{bmatrix} Y_{23} & 0 & Y_{31} & 0 & Y_{12} & 0 \\ 0 & X_{32} & 0 & X_{13} & 0 & X_{21} \\ X_{32} & Y_{23} & X_{13} & Y_{31} & X_{21} & Y_{12} \end{bmatrix}$$

$$= \frac{1}{600} \begin{bmatrix} -20 & 0 & 0 & 0 & 20 & 0 \\ 0 & 30 & 0 & -30 & 0 & 0 \\ 30 & -20 & -30 & 0 & 0 & 20 \end{bmatrix}$$

$$[D][B] = \frac{E}{(1-v)^2} \begin{bmatrix} 1 & v & 0 \\ v & 1 & 0 \\ 0 & 0 & (1-v)/2 \end{bmatrix} \frac{1}{600} \begin{bmatrix} -20 & 0 & 0 & 0 & 20 & 0 \\ 0 & 30 & 0 & -30 & 0 & 0 \\ 30 & -20 & -30 & 0 & 0 & 20 \end{bmatrix}$$

$$= \frac{70 \times 10^3}{600 \times (1-0.3^2)} \begin{bmatrix} -20 & 9 & 0 & -9 & 20 & 0 \\ -6 & 30 & 0 & -30 & 6 & 0 \\ 10.5 & -7 & -10.5 & 0 & 0 & 7 \end{bmatrix}$$

$$[K] = tA[B]^T[D][B] = \frac{10 \times 300 \times 70 \times 10^3}{600 \times 600 \times 0.91} \begin{bmatrix} -20 & 0 & 30 \\ 0 & 30 & -20 \\ 0 & 0 & -30 \\ 0 & -30 & 0 \\ 20 & 0 & 0 \\ 0 & 0 & 20 \end{bmatrix} \begin{bmatrix} -20 & 9 & 0 & -9 & 20 & 0 \\ -6 & 30 & 0 & -30 & 6 & 0 \\ 10.5 & -7 & -10.5 & 0 & 0 & 7 \end{bmatrix}$$

$$= 641 \begin{bmatrix} 715 & -390 & -315 & 180 & -400 & 210 \\ -390 & 1040 & 210 & -900 & 180 & -140 \\ -315 & 210 & 315 & 0 & 0 & -210 \\ 180 & -900 & 0 & 900 & -180 & 0 \\ -400 & 180 & 0 & -180 & 400 & 0 \\ 210 & -140 & -210 & 0 & 0 & 140 \end{bmatrix}$$

After applying boundary conditions, $u_1 = v_1 = u_2 = v_2 = 0$, we get

$$\{P\}_R = [K]_R\{u\}_R$$

or

$$\begin{Bmatrix} 50 \\ -100 \end{Bmatrix} = 641 \begin{bmatrix} 400 & 0 \\ 0 & 140 \end{bmatrix} \begin{Bmatrix} u_3 \\ v_3 \end{Bmatrix}$$

which give the displacements at node 3 as

$$u_3 = \frac{50}{(641 \times 400)} = 0.000195 \text{ mm}$$

and $\quad v_3 = \frac{-100}{(641 \times 140)} = -0.0011143 \text{ mm}$

By the conventional displacement polynomial method

(**Note :** This solution was already given in chapter-5. However, for a quick comparison between these two methods, it has been repeated here.)

$$\{u\} = [1 \quad x \quad y] \{\alpha\} \text{ and } \{v\} = [1 \quad x \quad y] \{\beta\}$$

The coefficients α_1, α_2, α_3, β_1, β_2 and β_3 are evaluated from the above in terms of nodal displacements as

$$\begin{Bmatrix} u_1 \\ u_2 \\ u_3 \end{Bmatrix} = \begin{bmatrix} 1 & 0 & 20 \\ 1 & 0 & 0 \\ 1 & 30 & 20 \end{bmatrix} \begin{Bmatrix} \alpha_1 \\ \alpha_2 \\ \alpha_3 \end{Bmatrix}$$

or $\quad \begin{Bmatrix} \alpha_1 \\ \alpha_2 \\ \alpha_3 \end{Bmatrix} = \frac{1}{600} \begin{bmatrix} 0 & 600 & 0 \\ -20 & 0 & 20 \\ 30 & -30 & 0 \end{bmatrix} \begin{Bmatrix} u_1 \\ u_2 \\ u_3 \end{Bmatrix} = \frac{1}{60} \begin{bmatrix} 0 & 60 & 0 \\ -2 & 0 & 2 \\ 3 & -3 & 0 \end{bmatrix} \begin{Bmatrix} u_1 \\ u_2 \\ u_3 \end{Bmatrix}$

Similarly,

$$\begin{Bmatrix} v_1 \\ v_2 \\ v_3 \end{Bmatrix} = \begin{bmatrix} 1 & 0 & 20 \\ 1 & 0 & 0 \\ 1 & 30 & 20 \end{bmatrix} \begin{Bmatrix} \beta_1 \\ \beta_2 \\ \beta_3 \end{Bmatrix}$$

or $\quad \begin{Bmatrix} \beta_1 \\ \beta_2 \\ \beta_3 \end{Bmatrix} = \frac{1}{600} \begin{bmatrix} 0 & 600 & 0 \\ -20 & 0 & 20 \\ 30 & -30 & 0 \end{bmatrix} \begin{Bmatrix} v_1 \\ v_2 \\ v_3 \end{Bmatrix} = \frac{1}{60} \begin{bmatrix} 0 & 60 & 0 \\ -2 & 0 & 2 \\ 3 & -3 & 0 \end{bmatrix} \begin{Bmatrix} v_1 \\ v_2 \\ v_3 \end{Bmatrix}$

$$\{\varepsilon\} = \begin{Bmatrix} \partial u/\partial x \\ \partial v/\partial y \\ \partial u/\partial y + \partial v/\partial x \end{Bmatrix} = \begin{bmatrix} 0 & 1 & 0 & 0 & 0 & 0 \\ 0 & 0 & 0 & 0 & 0 & 1 \\ 0 & 0 & 1 & 0 & 1 & 0 \end{bmatrix} \begin{Bmatrix} \alpha_1 \\ \alpha_2 \\ \alpha_3 \\ \beta_1 \\ \beta_2 \\ \beta_3 \end{Bmatrix}$$

$$= \frac{1}{60} \begin{bmatrix} 0 & 1 & 0 & 0 & 0 & 0 \\ 0 & 0 & 0 & 0 & 0 & 1 \\ 0 & 0 & 1 & 0 & 1 & 0 \end{bmatrix} \begin{bmatrix} 0 & 60 & 0 & 0 & 0 & 0 \\ -2 & 0 & 2 & 0 & 0 & 0 \\ 3 & -3 & 0 & 0 & 0 & 0 \\ 0 & 0 & 0 & 0 & 60 & 0 \\ 0 & 0 & 0 & -2 & 0 & 2 \\ 0 & 0 & 0 & 3 & -3 & 0 \end{bmatrix} \begin{Bmatrix} u_1 \\ u_2 \\ u_3 \\ v_1 \\ v_2 \\ v_3 \end{Bmatrix} = [B]\{q\}$$

where, $[B] = \dfrac{1}{60} \begin{bmatrix} -2 & 0 & 2 & 0 & 0 & 0 \\ 0 & 0 & 0 & 3 & -3 & 0 \\ 3 & -3 & 0 & -2 & 0 & 2 \end{bmatrix}$

This matrix is same as the one obtained through iso-parametric approach, except that the elements of the stiffness matrix in this method correspond to the displacement vector $[u_1 \ u_2 \ u_3 \ v_1 \ v_2 \ v_3]^T$ whereas in the previous method they correspond to the displacement vector $[u_1 \ v_1 \ u_2 \ v_2 \ u_3 \ v_3]^T$. The sequence of load components in the load vector has to be correspondingly modified. Rewriting the [B] matrix to correspond with the displacement vector $[u_1 \ v_1 \ u_2 \ v_2 \ u_3 \ v_3]$,

$$[K] = t \, A \, [B]^T \, [D] \, [B] = 640 \begin{bmatrix} 715 & -390 & -315 & 180 & -400 & 210 \\ -390 & 1040 & 210 & -900 & 180 & -140 \\ -315 & 210 & 315 & 0 & 0 & -210 \\ 180 & -900 & 0 & 900 & -180 & 0 \\ -400 & 180 & 0 & -180 & 400 & 0 \\ 210 & -140 & -210 & 0 & 0 & 140 \end{bmatrix}$$

After applying boundary conditions, $u_1 = v_1 = u_2 = v_2 = 0$, we get

$$\{P\}_R = [K]_R \, \{u\}_R$$

or $\begin{Bmatrix} 50 \\ -100 \end{Bmatrix} = \left(\dfrac{1}{0.00156} \right) \begin{bmatrix} 400 & 0 \\ 0 & 140 \end{bmatrix} \begin{Bmatrix} u_3 \\ v_3 \end{Bmatrix}$

or $u_3 = 0.000195$ mm

       $v_3 = -0.001114$ mm

Check Reactions are obtained from the assembled stiffness matrix, corresponding to the fixed degrees of freedom, and checked with

$$R_{1X} + R_{2X} + 50 = 0 \quad ; \quad R_{1Y} + R_{2Y} - 100 = 0$$

$$
\begin{Bmatrix} R_{1X} \\ R_{1Y} \\ R_{2X} \\ R_{2Y} \end{Bmatrix} = 641 \begin{bmatrix} 715 & -390 & -315 & 180 & -400 & 210 \\ -390 & 1040 & 210 & -900 & 180 & -140 \\ -315 & 210 & 315 & 0 & 0 & -210 \\ 180 & -900 & 0 & 900 & -180 & 0 \end{bmatrix} \begin{Bmatrix} 0 \\ 0 \\ 0 \\ 0 \\ 0.000195 \\ -0.001114 \end{Bmatrix}
$$

$$
= \begin{Bmatrix} -199.96 \\ 122.47 \\ 149.96 \\ -22.50 \end{Bmatrix} N
$$

Static equilibrium relations using calculated reactions are

$$R_{1X} + R_{2X} + 50 = -199.86 + 149.96 + 50 \simeq 0$$
$$R_{1Y} + R_{2Y} - 100 = 122.47 - 22.50 - 100 \simeq 0$$

Note : The given *thick* plate, from university question paper, should not be analysed as a 2-D problem. It can not be solved as a 3-D problem *manually*.

Finite element analysis gives approximate results for the engineering problems. In this simple example, this was evident from the fact that the first condition was completely satisfied while a small discrepancy was seen in the second condition with the results obtained. The discrepancy increases with the order of the reduced stiffness matrix due to numerical rounding off errors associated with matrix inversion techniques using digital computers.

CHAPTER 7

FACTORS INFLUENCING
SOLUTION

7.1 DISTRIBUTED LOADS

In the discussion so far, analysis has been confined to structures subjected to specified nodal loads. But, many engineering problems include distributed loads like

- Loads along the length of 1-D elements such as wind load on columns, self weight of beams
- Loads along the edges of 2-D elements such as in-plane pressure on edges of plates or axi-symmetric solids; pressure (bending) load normal to the surface of the plate
- Loads on surfaces of 3-D elements such as pressure on one or more surfaces of a thick solid component
- Loads on volumes of 3-D elements such as self weight, centrifugal force on a rotating component.

Such loads are usually represented by *equivalent loads, based on force equilibrium*, in strength of materials.

For example, a uniformly distributed load 'p' on a beam AB of length 'L', as shown in case-1 of Fig. 7.1, is approximated by two equal parts of the beam as shown in case-2. The distributed load on both the parts is transferred to the ends of the beam as point load of pL/2 and moment due to the distributed load represented by the resultant load of pL/2 acting at a distance of L/4 from beam end, as shown in case-3. Thus,

$$P_1 = P_2 = \frac{p\,L}{2} \quad \text{and} \quad M_1 = -M_2 = \left(\frac{p\,L}{2}\right)\left(\frac{L}{4}\right) = \frac{p\,L^2}{8}$$

These statically equivalent loads are shown in case-4.

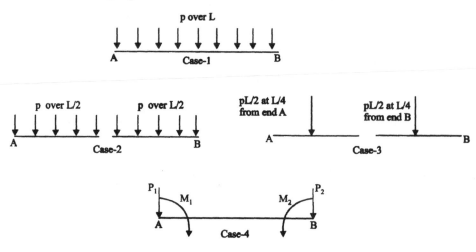

FIGURE 7.1 Statically equivalent loads

If beam AB is simply supported at its two ends, then the reactions based on the static force equilibrium conditions $\Sigma F = 0$ and $\Sigma M = 0$ will be equal and opposite to these equivalent loads.

Finite element method is based on minimum potential energy theory for the calculation of stiffness matrix or load-displacement relations. It will, therefore, be *consistent* if the equivalent *loads* are also *based on energy*. It is noticed that consistent loads used in FEM, give displacements identical with those from closed form solutions.

7.2 STATICALLY EQUIVALENT LOADS VS. CONSISTENT LOADS

Statically equivalent loads, even though satisfy force and moment equilibrium, do not give the same nodal displacements as the actual loads. Consistent loads, based on energy equivalence, give the same displacements as obtained with the actual loads, in addition to satisfying force and moment equilibrium. For this reason, consistent loads are used in FEM. This can be verified by the following example.

Consider a simple cantilever AB of length L, fixed at end A and subjected to uniformly distributed load p, as shown in case-1 of Fig. 7.2. The statically equivalent load system is shown in case-2 of Fig. 7.2 while the consistent load system is shown in case-3 of Fig. 7.2. In these two cases, a point load and a moment at ends A and B can replace the distributed load. Since end A is fixed, the load and moment at A add to the reactions at A without contributing to displacement at any point on the cantilever. Hence, load and moment at end B only are shown in these figures.

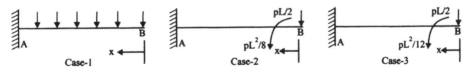

FIGURE 7.2 Distributed and equivalent loads

Case-1 : Actual load

At any section, distance x from B, from simple theory of bending

$$\text{Bending moment } M = p.x\left(\frac{x}{2}\right) = -EI\left(\frac{d^2y}{dx^2}\right)$$

where   v is the displacement normal to the axis of the beam

E is the modulus of elasticity of beam material

and   I is the moment of inertia of the beam cross section

$$\text{Integrating, } -EI\frac{dv}{dx} = \frac{px^3}{6} + C_1$$

The constant C_1 is evaluated from the boundary condition $\frac{dv}{dx} = 0$ at $x = L$

$$\text{Thus, } C_1 = -\frac{pL^3}{6} \text{ and } -EI\frac{dv}{dx} = \frac{px^3}{6} - \frac{pL^3}{6}$$

$$\text{Integrating again, } -EIv = \frac{px^4}{24} - \frac{pL^3x}{6} + C_2$$

The constant C_2 is evaluated from the boundary condition $v = 0$ at $x = L$

$$\text{Thus, } C_2 = -\frac{pL^4}{24} + \frac{pL^4}{6} = \frac{3pL^4}{24} = \frac{pL^4}{8}$$

The maximum displacement is obtained at the free end,   i.e., at $x = 0$

$$-EIv = C_2 \text{ or } v = -\frac{pL^4}{8EI}$$

Case-2 : Statically equivalent loads

As explained in section 7.1, loads $P_B = \frac{pL}{2}$ and $M_B = \frac{pL^2}{8}$, which satisfy static equilibrium conditions, are used as equivalent loads.

At any section, distance x from B, from simple theory of bending

Bending moment $M = \left(\dfrac{pL}{2}\right).x - \dfrac{pL^2}{8} = -EI\left(\dfrac{d^2v}{dx^2}\right)$

Integrating, $-EI\dfrac{dv}{dx} = \dfrac{pLx^2}{4} - \dfrac{pL^2x}{8} + C_1$

The constant C_1 is evaluated from the boundary condition $\dfrac{dv}{dx} = 0$ at $x = L$

Thus, $C_1 = \dfrac{-pL^3}{8}$ and $-EI\dfrac{dv}{dx} = \dfrac{pLx^2}{4} - \dfrac{pL^2x}{8} - \dfrac{pL^3}{8}$

Integrating again, $-EI\,v = \dfrac{pLx^3}{12} - \dfrac{pL^2x^2}{16} - \dfrac{pL^3x}{8} + C_2$

The constant C_2 is evaluated from the boundary condition $v = 0$ at $x = L$

Thus, $C_2 = \left(\dfrac{pL^4}{48}\right).(-4+3+6) = \dfrac{5pL^4}{48}$

The maximum displacement is obtained at the free end, i.e., at $x = 0$

$-EI\,v = C_2$ or $v = \dfrac{-5pL^4}{48\,EI}$

It can be seen that the displacement v in this case is different from that of case-1

Case-3 : Consistent loads

Loads $P_B = \dfrac{pL}{2}$ and $M_B = \dfrac{pL^2}{12}$, based on energy equivalence, as explained later are used to represent distributed load.

At any section, distance x from B, from simple theory of bending

Bending moment $M = \left(\dfrac{pL}{2}\right).x - \dfrac{pL^2}{12} = -EI\dfrac{d^2v}{dx^2}$

Integrating, $-EI\dfrac{dv}{dx} = \dfrac{pLx^2}{4} - \dfrac{pL^2x}{12} + C_1$

The constant C_1 is evaluated from the boundary condition $\dfrac{dv}{dx} = 0$ at $x = L$

Thus, $C_1 = \dfrac{-pL^3}{6}$ and $-EI\dfrac{dv}{dx} = \dfrac{pLx^2}{4} - \dfrac{pL^2x}{12} - \dfrac{pL^3}{6}$

Integrating again, $-EI\,v = \dfrac{pLx^3}{12} - \dfrac{pL^2x^2}{24} - \dfrac{pL^3x}{6} + C_2$

The constant C_2 is evaluated from the boundary condition $v = 0$ at $x = L$

Thus, $C_2 = \left(\dfrac{pL^4}{24}\right)\cdot(-2+1+4) = \dfrac{3pL^4}{24} = \dfrac{pL^4}{8}$

The maximum displacement is obtained at the free end, i.e., at $x = 0$

$$-EI\,v = C_2 \quad \text{or} \quad v = \dfrac{-pL^4}{8EI}$$

It can be seen that the displacement 'v' in this case is identical to the correct value represented by case-1 and hence consistent loads, based on work or energy equivalence, are preferred in FEM.

The shape functions used to define displacement, in natural coordinate system, over a finite element can also be used to calculate nodal loads vector consistent with the loads distributed over an edge or a surface of an element. These consistent loads are calculated for different types of distributed loads (on edge, area or volume), as explained below.

(a)   Consistent nodal loads corresponding to distributed body forces are given by

$$\{P_B\} = \int\limits_v [N]^T \{X\}\,dv$$

where $\{X\}$ indicates distributed load over the volume of the body

For a beam element with uniformly distributed self weight,

$$[N] = [\,L_1^2(3 - 2L_1) \qquad L_1^2L_2 \qquad L_2^2(3 - 2L_2) \qquad -L_1L_2^2\,]$$

$$\{P_B\} = b\,h \int\limits_L [N]^T \{\rho\}\,dL = \left(\dfrac{bh\rho L}{12}\right)\begin{Bmatrix} 6 \\ L \\ 6 \\ -L \end{Bmatrix}$$

(b)   Consistent nodal loads corresponding to distributed surface or traction forces are given by

$$\{P_S\} = \int\limits_s [N]^T \{T\}\,ds$$

where $\{T\}$ indicates distributed load over the surface of the body

For a beam element with uniformly distributed pressure load {p}, in units of force per unit length (w), assuming that the load is uniformly spread across the width 'b' of the cross-section,

$$\int_S p\, ds = b \int_L p\, dL = \int_L w\, dL$$

Thus,

$$\{P_s\} = b \int_L [N]^T \{p\}\, dL$$

$$= \frac{b\,L\,p}{12} \begin{Bmatrix} 6 \\ L \\ 6 \\ -L \end{Bmatrix} = \frac{w\,L}{12} \begin{Bmatrix} 6 \\ L \\ 6 \\ -L \end{Bmatrix}$$

7.3 CONSISTENT LOADS FOR A FEW COMMON CASES

(a) *Load on a beam*

1. Beam of length 'L' with a concentrated load 'P' at distance 'a' from node 1 (r = a/L)

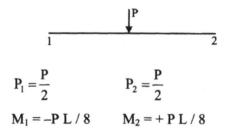

$$P_1 = P\,(1 - 3r^2 + 2r^3); \qquad P_2 = P\,(3r^2 - 2r^3)$$
$$M_1 = -P\,L\,(r - 2r^2 + r^3); \qquad M_2 = -P\,L\,(-r^2 + r^3)$$

For the particular case when the load is at the center of the beam, r = ½ and

$$P_1 = \frac{P}{2} \qquad\qquad P_2 = \frac{P}{2}$$
$$M_1 = -P\,L\,/\,8 \qquad M_2 = +P\,L\,/\,8$$

2. Beam of length 'L' with a uniformly distributed load, w. Let P = w L be the total load

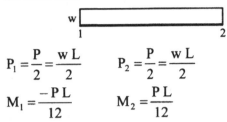

$$P_1 = \frac{P}{2} = \frac{w\,L}{2} \qquad P_2 = \frac{P}{2} = \frac{w\,L}{2}$$

$$M_1 = \frac{-P\,L}{12} \qquad M_2 = \frac{P\,L}{12}$$

3. Beam with a linearly varying load (0 to w). Let P = w L / 2 be the total load

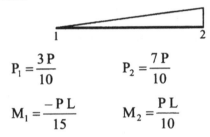

$$P_1 = \frac{3\,P}{10} \qquad P_2 = \frac{7\,P}{10}$$

$$M_1 = \frac{-P\,L}{15} \qquad M_2 = \frac{P\,L}{10}$$

(b) *Load along an edge of a plate*

Moments at nodes are not relevant here, since rotations of nodes are not treated as DOFs.

4. Plate (modeled with CST elements) subjected to uniform pressure 'w' along an edge. Let P = w L be the total load

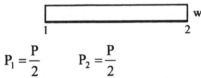

$$P_1 = \frac{P}{2} \qquad P_2 = \frac{P}{2}$$

5. Plate (modeled with CST element) subjected to linearly varying pressure (0 to w) along an edge. Let $P = \dfrac{w\,L}{2}$ be the total load

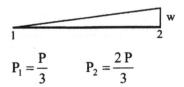

$$P_1 = \frac{P}{3} \qquad P_2 = \frac{2\,P}{3}$$

6. Plate (modeled with LST elements) subjected to uniform pressure 'w' along an edge. Let $P = w\,L$ be the total load

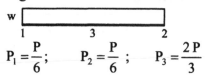

$$P_1 = \frac{P}{6}; \qquad P_2 = \frac{P}{6}; \qquad P_3 = \frac{2\,P}{3}$$

Note : Corresponding statically equivalent nodal loads would be $P_1 = \dfrac{P}{4}$; $P_2 = \dfrac{P}{4}$ and $P_3 = \dfrac{P}{2}$, distributing equally $\dfrac{P}{2}$ over each half of beam (1-3 & 3-2).

7. Plate (modeled with LST element) subjected to linearly varying pressure (0 to w) along an edge. Let $P = \dfrac{w\,L}{2}$ be the total load.

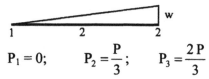

$$P_1 = 0; \qquad P_2 = \frac{P}{3}; \qquad P_3 = \frac{2\,P}{3}$$

(c) *Load normal to the surface of a plate*

8. 3-noded Triangular Plate with uniform pressure normal to the surface, w. Let $P = w\,A$ be the total load.

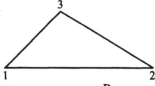

At corner nodes, $P_1 = P_2 = P_3 = \dfrac{P}{3}$

9. 6-noded Triangular Plate with uniform pressure normal to the surface, w. Let $P = w\,A$ be the total load.

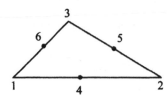

At corner nodes, $P_1 = P_2 = P_3 = 0$

At mid-side nodes, $P_4 = P_5 = P_6 = \dfrac{P}{3}$

10. 4-noded Quadrilateral Plate with uniform pressure normal to the surface, w. Let P = w A be the total load.

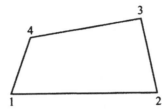

At corner nodes, $P_1 = P_2 = P_3 = P_4 = \dfrac{P}{4}$

11. 8-noded Quadrilateral Plate with uniform pressure normal to the surface, w. Let P = w A be the total load.

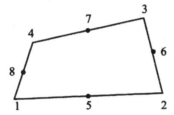

At corner nodes, $P_1 = P_2 = P_3 = P_4 = -\dfrac{P}{12}$

At mid-side nodes, $P_5 = P_6 = P_7 = P_8 = \dfrac{P}{3}$

7.4 ASSEMBLING ELEMENT STIFFNESS MATRICES

Solution of any practical problem by finite element method involves a very large number of simultaneous equations and hence, computer memory needs to be effectively utilised. The following techniques are therefore utilised to effectively use the available computer memory.

Element stiffness matrices are symmetric. Therefore, assembled stiffness matrix of the entire structure is also symmetric. Storing half the matrix is hence adequate. But, for storing even half of n × n stiffness matrix allocation of as many memory locations for the variable K(n, n), no saving of computer memory required.

(a) **Storing a banded matrix** – Stiffness matrix is usually a banded matrix, in addition to being symmetric, depending on the nodal connectivity of the elements, as explained in Fig. 7.3 for the simple case of a rectangular plate modeled with 4-noded quadrilateral elements.

Element-1 is linked to nodes 1, 2, 7 and 6. Hence, contribution of element-1 in the assembled stiffness coefficients will be found in the rows and columns associated with these four nodes only. Likewise, each of the other elements will have contributions in the rows and columns of the assembled stiffness matrix corresponding to their respective four corner nodes only. None of the elements 2-8 are linked to node-1. Hence, assembled stiffness matrix will not have non-zero terms in (i, j) positions, where row 'i' and column 'j', correspond to the degrees of freedom associated with nodes 3, 4, 5, 8, 9, 10, 11, 12, 13, 14 and 15. While assembling such a stiffness matrix in the computer, significant number of memory locations can be saved by avoiding entries corresponding to these zero values.

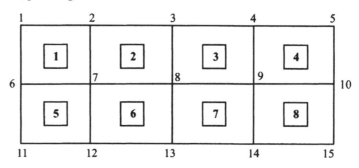

FIGURE 7.3 Numbering of nodes in a model

Half Bandwidth 'b' = (Max. node number difference for any one element + 1) x Number of DOF per node

Storing a symmetric banded matrix requires n × b memory locations as against n × n memory locations required otherwise. It is a rectangular matrix of n rows and b columns, first column representing diagonal elements of the square matrix. The algorithm in the program properly identifies elements of the matrix. For example, $k_{4,4}$ is stored as $k_{4,1}$; $k_{4,5}$ as $k_{4,2}$; $k_{4,6}$ as $k_{4,3}$ and so on till all non-zero elements are covered. In general, $k_{n,m}$ of the assembled matrix is identified by $k_{n,m-n+1}$ in the banded matrix. The banded matrix of order 12 × 7, as stored in this way, is represented below with diagonal element indicated by 'd'. Since the symmetric half of the band matrix has width 'b' in the first few rows and reduces to 1 in the last row, the banded matrix will have some zero values in the last few rows as shown in Fig. 7.4.

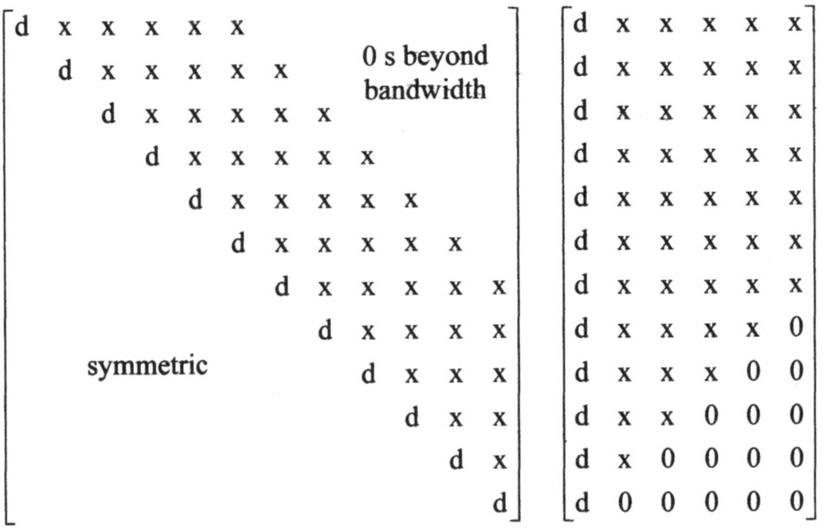

$$
\begin{bmatrix}
d & x & x & x & x & x & & & & & & \\
& d & x & x & x & x & x & & \text{0 s beyond} & & \\
& & d & x & x & x & x & x & \text{bandwidth} & & \\
& & & d & x & x & x & x & x & & \\
& & & & d & x & x & x & x & x & \\
& & & & & d & x & x & x & x & x \\
& & & & & & d & x & x & x & x & x \\
& & & & & & & d & x & x & x & x \\
\text{symmetric} & & & & & & & & d & x & x & x \\
& & & & & & & & & d & x & x \\
& & & & & & & & & & d & x \\
& & & & & & & & & & & d
\end{bmatrix}
\qquad
\begin{bmatrix}
d & x & x & x & x & x \\
d & x & x & x & x & x \\
d & x & x & x & x & x \\
d & x & x & x & x & x \\
d & x & x & x & x & x \\
d & x & x & x & x & x \\
d & x & x & x & x & x \\
d & x & x & x & x & 0 \\
d & x & x & x & 0 & 0 \\
d & x & x & 0 & 0 & 0 \\
d & x & 0 & 0 & 0 & 0 \\
d & 0 & 0 & 0 & 0 & 0
\end{bmatrix}
$$

Half of a symmetric matrix   **Storing in band matrix form**

FIGURE 7.4 Storing half of a banded matrix in different forms

(b) **Minimising bandwidth of stiffness matrix** - Bandwidth of the assembled stiffness matrix can be minimised by renumbering node numbering sequence of a finite element model.

As an example, let us consider a ring modeled by a number of curved beam elements. Case-1 and case-2, shown in Fig. 7.5, follow two different node numbering schemes.

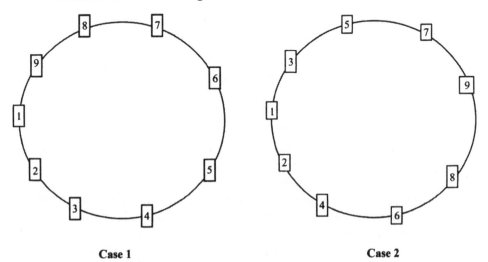

Case 1   Case 2

FIGURE 7.5 Minimising bandwidth by renumbering nodes of a simple ring

In case-1, Maximum node number difference = 9 – 1 = 8
and   half bandwidth b = (8 + 1) × No. DOF/node

while, in case-2, Maximum node number difference = 3 – 1 or 5 – 3,. =2
and   b = (2 + 1) × No. of DOF/node

For all other elements, node number difference is less than or equal to the above values.

Let us consider one more example of a plate with 2 DOF per node. It is shown with two node numbering sequences as case-1 and case-2 in Fig.7.6.

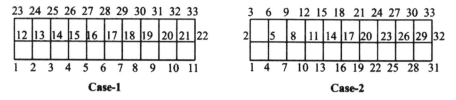

Figure 7.6 Minimising bandwidth by renumbering nodes of a plate

In case-1, Half bandwidth, b = (24– 12 + 1) × 2 = 26 for top left element

In case-2, Half bandwidth, b = (6 – 2 + 1) × 2 = 10 for the same element

The bandwidth of this element happens to be equal to the bandwidth of any other element in the model and, hence, represents the maximum bandwidth for these two cases.

In this simple model,   total number of DOF = 33 × 2 = 66.

If entire stiffness matrix is stored, computer memory required = 66 × 66

If banded stiffness matrix is stored,

   computer memory required = 66 × 26 in case-1 and 66 × 10 in case-2.

Several algorithms are developed for selecting appropriate node numbering sequence in an actual problem so as to minimise bandwidth. Some of them are used, even without the knowledge of the end user, in many general purpose commercial software.

(c)   **Skyline method of assembly** - Bandwidth may not be the same for all the elements in a practical problem with irregular geometry. Then within the maximum bandwidth, b, used for computer memory requirement starting zeroes of all columns can be avoided by following a different method of assembly. Thus, requirement of computer memory can be minimised. This method also avoids the need to store the zero values of the last few rows of the banded matrix. Here, the banded matrix K is stored as a column vector A, with columns of banded matrix stored one after the other. Each column is of variable length and starts with the first non-zero value in that column. Intermediate zeroes need to be stored. Another array variable (ID) stores the sequential address of the last term of each column (diagonal elements) so that the program can identify location of each stiffness coefficient in the column vector.

The elements are identified as shown in the following example, for a simple 7×7 stiffness matrix. In actual practice, the saving of computer memory will be very large, since the stiffness matrix will be of a very large size. The square and banded stiffness matrix

$$[K] = \begin{bmatrix} k_{11} & k_{12} & 0 & k_{14} & 0 & 0 & 0 \\ & k_{22} & k_{23} & k_{24} & k_{25} & 0 & 0 \\ & & k_{33} & k_{34} & 0 & k_{36} & 0 \\ & & & k_{44} & 0 & k_{46} & 0 \\ & \text{symmetric} & & & k_{55} & k_{56} & k_{57} \\ & & & & & k_{66} & k_{67} \\ & & & & & & k_{77} \end{bmatrix}$$

is stored in a single column matrix as

$$\{K\} = [\, k_{11} \ k_{12} \ k_{22} \ k_{23} \ k_{33} \ k_{14} \ k_{24} \ k_{34} \ k_{44} \ k_{25}$$
$$0 \ 0 \ k_{55} \ k_{36} \ k_{46} \ k_{56} \ k_{66} \ k_{57} \ k_{67} \ k_{77} \,]^{T}$$

Note that starting elements k_{13}, k_{15}, k_{16}, k_{26}, k_{17}, k_{27}, k_{37} and k_{47} in 3rd, 5th, 6th and 7th columns having value zero, are not stored while elements k_{35} and k_{45}, also having value zero, are stored since they are included in between elements of a particular column with non-zero value.

Another vector $\{ID\}$, i^{th} element of which corresponds to the position of the ith diagonal element in $[K]$, helps in identifying each element of $\{K\}$ with its corresponding position in $[K]$

$$\{ID\} = [\, 1 \ \ 3 \ \ 5 \ \ 9 \ \ 13 \ 17 \ \ 20 \,]^{T}$$

Position of any element in the jth column of the original stiffness matrix is identified by its position w.r.t. the diagonal element in the jth column represented by ID(j). Position of element in i^{th} row of this column ($i \leq j$) is given by $[\, ID(j) - (j-i) \,]$.

Ex: Position of k_{33} in the array $\{K\} = ID(3) = 5$

Position of k_{23} in the array $\{K\} = ID(3) - (3-2) = 5 - 1 = 4$

Position of k_{46} in the array $\{K\} = ID(6) - (6-4) = 17 - 2 = 15$

7.5   AUTOMATIC MESH GENERATION

One of the difficult tasks in the analysis of a component by finite element method is the need to discretise the component into a large number of elements connected by nodes and specifying the nodal coordinates as well as element connectivity. In most commercial software, this is done by the program from the description of the component geometry as a solid model either as a combination of some primitive shapes or through key points, lines connected by key points, areas connected by lines and volumes specified by the enclosing areas. Type of element (1-D truss, 1-D beam, Plane stress element, plate bending element, thick shell element,...) that the component closely matches in its behaviour, also needs to be specified by the user.

An automatic mesh generation program generates the locations of the node points and elements, labels the nodes and elements and provides the element node connectivity relationships. A set of nodes is identified to represent the component, based on the relative dimensions of the component, choosing a minimum number of nodes across the smallest dimension, which varies with the software. In many programs, the user can also specify the minimum size of the elements to be generated and the ratio of sizes of adjacent elements, for generating mesh which varies from coarse to fine, in the areas of stress concentration.

Two methods are explained here for forming elements, from the given set of nodes.

In the *Tessellation method*, program starts connecting user defined nodes starting with an arbitrary point on the boundary. It creates a simplex element using the neighbouring nodes which give the least distorted element shape. Then, it proceeds to form the next element. An example of Tesselation method is given here. Fig. 7.7 (a) gives nodal set of a component while Fig. 7.7 (b), Fig. 7.7 (c) and Fig. 7.7 (d) show one simplex element formed by joining three nodes. The element in Fig. 7.7 (d) is the least distorted element and is finally selected before generating other elements (shown by dotted lines) in a similar way.

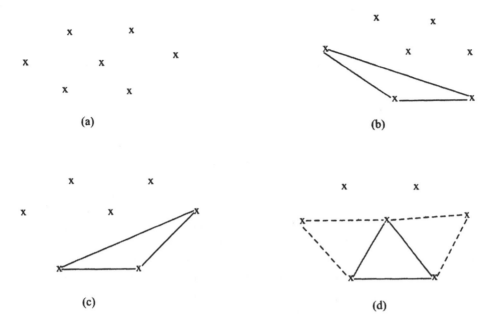

FIGURE 7.7 Mesh generation by Tessellation method

In the **Octree method**, a three dimensional cube is assumed around the object. If the object is partially occupied by the cube, it is subdivided into small cubes and each cube is checked. If any cube is full (completely occupied by the object) or empty, then the cube is not subdivided further. It gives elements of different sizes, irrespective of the stress distribution in the object and hence, is not very much popular.

7.6 OPTIMUM MESH MODEL

Best possible mesh has to be used to obtain solutions as accurately as possible, while minimising the requirement of computer resources. In many cases, it can not be decided before the analysis is completed. In time-dependent iterative problems, mesh refinement between different steps becomes very important to ensure convergence of the solution. There are many mesh refinement methods available.

(a) **Mesh refinement method or h-method** – Refines the element size based on solution gradients.

(b) **Mesh movement method or r-method** – Grid points are moved around (mesh redistribution) to provide clustering in certain regions, based on error indicators.

(c) **Mesh enrichment method or p-method** – Refines degree of polynomial (interpolation function) based on user-specified error tolerance. It is particularly important when singularities are encountered, such as near crack tips in fracture analysis. Side or interior nodes are not installed physically, but higher order modes of the polynomial corresponding to these nodes are combined with corner nodes by means of static condensation (similar to elimination of rows associated with DOFs corresponding to the specified nodal displacements), so that compatibility with adjacent elements is not affected.

7.7 GAUSSIAN POINTS & NUMERICAL INTEGRATION

Closed form solutions for integration, associated with many FEM problems, are not possible using digital computers. Hence, numerical integration is generally used in FEM. Integration of a simple function f(x) over the range (r_1, r_2) amounts to calculating area under the curve. In numerical integration, this is approximated by the sum of areas of a few rectangles (products of functions values and the local range or weight) at a few sampling points as shown. This sum naturally approaches true value of the function area as the number of sampling points increase. Gauss quadrature method of numerical integration is proved to be the most useful in finite element applications.

$$I = \int_{-1}^{1} f(r)\,dr = \sum_{i=1}^{n} w_i\, f(r_i) \text{ with } \Sigma w_i = 1 - (-1) = 2$$

where w_i is the "weight" or range associated with the ith sampling point and n is the number of sampling points within the element.

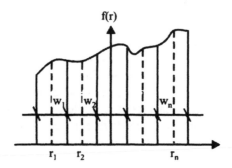

For 1-point integration, $r_1 = 0.0$ $w_1 = 2.0$

2-point integration, $r_1, r_2 = \pm\,0.5774$ (or $\pm\,\dfrac{1}{\sqrt{3}}$) $w_1, w_2 = 1.0$

3-point integration, $r_1, r_3 = \pm\,0.7746$ $w_1, w_3 = 0.5556$

$r_2 = 0.0$ $w_2 = 0.8889$

4-point integration, $r_1, r_4 = \pm 0.8613$ $w_1, w_4 = 0.3479$

$r_2, r_3 = \pm 0.34$ $w_2, w_3 = 0.6521$

etc..

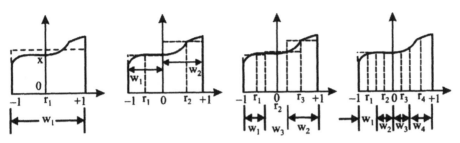

In a similar way, for double integration involving two shape functions r and s,

$$I = \int_{-1}^{1}\int_{-1}^{1} f(r,s)\,dr\,ds \cong \sum_{i=1}^{m}\sum_{j=1}^{n} w_i\, w_j\, f(r_i, s_j)$$

and for triple integration involving three shape functions r, s and t,

$$I = \int_{-1}^{1}\int_{-1}^{1}\int_{-1}^{1} f(r,s,t)\,dr\,ds\,dt \cong \sum_{i=1}^{l}\sum_{j=1}^{m}\sum_{k=1}^{n} w_i\, w_j\, w_k\, f(r_i, s_j, t_k)$$

Since an arbitrary 2-D quadrilateral element is mapped into a square element and an arbitrary 3-D solid element is mapped into a cube in natural (non-dimensional) coordinate system, m = n and l = m = n are commonly used.

(a)   Integration in Natural coordinate system

In terms of non-dimensional shape functions (area coordinates),

$$I = \int_{A}\int f(L_1, L_2, L_3)\,dA \cong \sum_{i=1}^{n} w_i\, f(L_1^i, L_2^i, L_3^i)$$

(i)   For n = 1 (3-noded triangle), Gaussian point of integration is at the center 'O' of the triangle given by the coordinates $L_1^1 = L_2^1 = L_3^1 = 1/3$ with the associated weight $w_1 = 1$

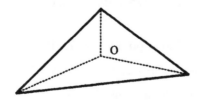

(ii) For $n = 2$ (6-noded triangle), Gaussian points of integration are at the mid-points of the three sides of the triangle given by the coordinates

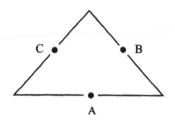

$$A\left(L_1^1 = \frac{1}{2}; L_2^1 = \frac{1}{2}; L_3^1 = 0\right)$$

$$B\left(L_1^2 = 0; L_2^2 = \frac{1}{2}; L_3^2 = \frac{1}{2}\right)$$

and $\quad C\left(L_1^3 = \frac{1}{2}; L_2^3 = 0; L_3^3 = \frac{1}{2}\right)$

The associated weights are $\quad w_1 = w_2 = w_3 = \dfrac{1}{3}$

(c) For a quadrilateral, with two Gaussian points along each coordinate, 2 x 2 points of integration are obtained by extrapolation of 1-D two-point values to 2-D in natural coordinates ξ (± 0.5774) and η (± 0.5774) as :

1 ($-0.5774, -0.5774$);   2 ($+ 0.5774, -0.5774$)

3 ($+0.5774, +0.5774$)   and   4 ($-0.5774, + 0.5774$)

The associated weights with each point are $w_i = w_j = 1.0$

i.e., the function value calculated at the above Gaussian points is associated with a quarter of the square area.

Similarly, with three Gaussian points along each coordinate, 3×3 coordinates and weights are obtained as extensions of 1-D three-point values

$\xi = -0.7746$ at points 1, 4, 7 with weights $w_i = 0.5556$

$= 0.0$ at points 2, 5, 8 with weights $w_i = 0.8889$

and $\quad = + 0.7746$ at points 3, 6, 9 with weights $w_i = 0.5556$

Similarly,

$$\eta = -0.7746 \text{ at points } 1, 2, 3 \text{ with weights } w_i = 0.5556$$

$$= 0.0 \text{ at points } 4, 5, 6 \text{ with weights } w_i = 0.8889$$

and $\quad = +0.7746 \text{ at points } 7, 8, 9 \text{ with weights } w_i = 0.5556$

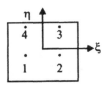

2 × 2 Gaussian integration points

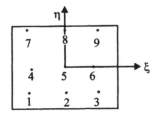

3 × 3 Gaussian integration points

In finite element method, displacement over an element is assumed by an algebraic polynomial and integration of terms of [B] matrix for evaluating element stiffness matrix is carried out term by term, each term being a product of different coordinates in various powers (in general, of the form $L_i^p.L_j^p.L_k^r.L_l^s$. Some mathematicians have derived simple formulae for such integrations.

(i) $$\int_L L_i^p.L_j^q \, dL = L \frac{p!q!}{(p+q+1)!}$$

Ex : $$\int_L L_1 L_2 \, dL = \frac{L}{6} \, ; \quad \int_L L_1^2 \, dL = \frac{L}{3}$$

(ii) $$\iint_A L_i^p.L_j^q.L_k^r \, dA = \frac{2A.p!q!r!}{(p+q+r+2)!} \qquad \text{(Felippa)}$$

Ex : $$\iint_A L_1 L_2 L_3 \, dA = \frac{2A.1!.1!.1!}{5!} = \frac{2A}{120} = \frac{A}{60}$$

$$\iint_A L_1^3 \, dA = \frac{2A.3!.0!.0!}{5!} = \frac{2A \times 6}{120} = \frac{A}{10}$$

$$\iint_A L_1^2 \, dA = \frac{2A.2!.0!.0!}{4!} = \frac{2A \times 2}{24} = \frac{A}{6}$$

(iii) $$\iiint_V L_i^p.L_j^q.L_k^r.L_l^s \, dV = \frac{6V \, p!q!r!s!}{(p+q+r+s+3)!}$$ (Clough)

Ex : $$\iiint_V L_1 L_2 L_3 L_4 \, dV = \frac{6V \, 1!1!1!1!}{7!} = \frac{6V}{5040} = \frac{V}{840}$$

$$\iiint_V L_1^2 \, dV = \frac{6V \, 2!0!0!0!}{5!} = \frac{6V \times 2}{120} = \frac{V}{10}$$

7.8 MODELLING TECHNIQUES

Finite element method does not give a unique solution for any problem. The accuracy of solution depends on many aspects like modelling of the actual component, number and type of elements used, approximation of loads and boundary conditions, solution techniques, etc.. It is for this reason that *design validation of products by the statutory safety codes of many countries is not based on FEM results*. Therefore, the engineer who uses this method (or uses any general purpose software based on this method) should check for the correctness of the results with the expected trend at select locations. In this chapter, important aspects of modelling and boundary conditions are discussed.

An element library is created, in each FEM based software, with each element assumed to have a particular type of deformation. Appropriate *types of elements* are selected to represent the component, based on the dimensions of the component in different directions as well as on the nature of deformation of the component.

For example, if a component has very small dimensions in the transverse directions compared to its length, it can be modeled by 1-D elements. In addition, these 1-D elements may be treated as axial loaded elements (truss elements or torsion elements), laterally loaded elements (beam elements or pipe elements), depending on how close behaviour of the component is to the behaviour assumed for these elements. It may be recalled that most of the trusses analysed do not have pin joints, but their resistance to bending is negligible. A water tank, shown in Fig. 7.8, is an example of one dimensional idealisation when it is analysed for wind loads or seismic loads. In this case, nodal displacements normal to the axis alone are significant. Stresses around the

circumference at any particular section have to be obtained from beam theory
(σ = M y / I), based on the bending moment obtained from this analysis and
geometric properties at that section.

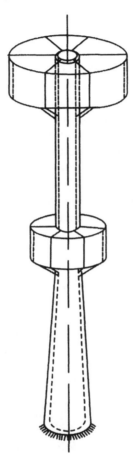

FIGURE 7.8 Modelling of a water tank for dynamic analysis

Similarly, 2-D elements with very small thickness compared to the other two
dimensions may be modeled with plane stress, plane strain, plate bending or
thin shell elements depending on the shape of the component as well as the
loads and boundary conditions applicable to the component.

Component with equally significant dimensions in all the three directions
can be modeled with 3-D solid elements or thick shell elements depending on
whether the component is subjected to significant bending deformation or not. It
may be noted that 3-D solid element will also have bending deformation
included but of a lesser degree (usually first order) whereas shell element has a

more significant bending deformation (displacement modeled by a cubic polynomial) as well as restraining slopes independently (simple supports or fixed supports).

The ***number of nodes and elements*** required for modelling a discrete structure leave very little flexibility or ambiguity to the engineer. But, the number of nodes and elements used in the analysis of a continuum model leaves lot of flexibility to the analyst. The number of elements should be commensurate with the criticality of the component or the desired accuracy of results and is also dependent on the computer time and memory available for this analysis. A lot of work is done to assess whether using a large number of lower order elements is preferable or using a small number of higher order elements is preferable. In addition, aspect ratios and included angles of 2-D and 3-D elements affect the results significantly. Their effect is demonstrated through the following two examples. ***Rigid guidelines applicable to all kinds of problems can NOT be specified.*** These examples are only meant to highlight dependence of results on various parameters.

Example 7.1

A simple beam subjected to bending moment is simulated through a varying load on the ends of a small rectangular plate. It is analysed, with the boundary conditions u = 0 at all nodes along x (neutral axis) and along y (symmetry) and v = 0 at the origin (for suppressing rigid body modes), using six different mesh models. Displacement u at A and v at B obtained using FEM are tabulated below.

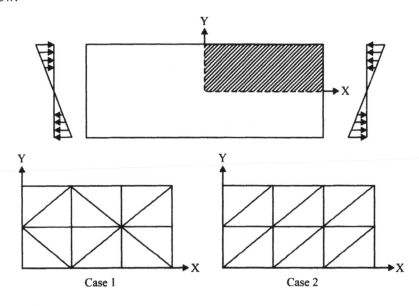

Case 1   Case 2

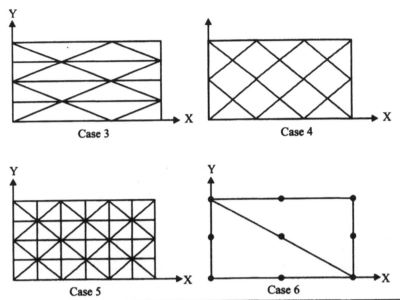

Case	Element type	Number of		FEM Value / Exact Value	
		Nodes	**Elements**	**u at A**	**v at B**
1	CST	12	12	0.84	0.812
2	CST	12	12	0.812	0.916
3	CST	15	16	0.778	0.825
4	CST	18	24	0.940	0.951
5	CST	35	49	0.946	0.960
6	LST	9	2	1.0	1.0

Example 7.2

A square plate in x-y plane simply supported on all four sides and subjected to a uniformly distributed normal load p is analysed using plate bending elements. A quarter plate is modeled taking advantage of symmetry about x and y axes. The results obtained at the center of the plate (C) are tabulated below with quadrilateral elements (cases 1-3) and triangular elements (cases 4-7).

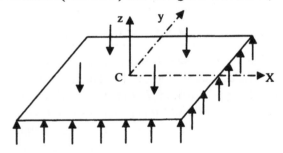

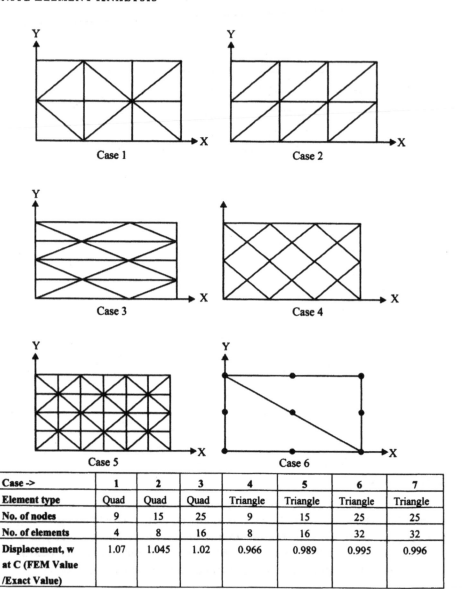

Case ->	1	2	3	4	5	6	7
Element type	Quad	Quad	Quad	Triangle	Triangle	Triangle	Triangle
No. of nodes	9	15	25	9	15	25	25
No. of elements	4	8	16	8	16	32	32
Displacement, w at C (FEM Value /Exact Value)	1.07	1.045	1.02	0.966	0.989	0.995	0.996

7.9 BOUNDARY CONDITIONS FOR CONTINUUM ANALYSIS

Another important choice is to select the region of analysis. If the entire component is not modeled, care should be taken to apply suitable end conditions to simulate the true situation. For example, while analysing a nozzle between a cylindrical shell and a fluid pipe, axial load across the cross section due to fluid pressure need to be applied. Also, reasonable length of the component has to be included in the model around the region of geometric discontinuity to ensure that the results of analysis in the region of discontinuity are not adversely influenced by the imposed end conditions. Some design codes indicate guidelines for the same.

This aspect is explained through the example of a long cylindrical shell, such as a boiler drum, shown in Fig. 7.9. The cylindrical shell is made up of two parts with a circumferential joint and has a mismatch of radius of the two parts by δ at the joint. This joint has been analysed to calculate the stresses generated due to this change of radius. Length of the cylinder considered on either side of the joint influences the stresses at the joint. Analyses carried out with three different models (L) have given three different results. The model in which uniform stresses are observed at the two ends is considered as the most appropriate, since stresses without this discontinuity are uniform along the length and the effect of any local discontinuity should vanish beyond some distance.

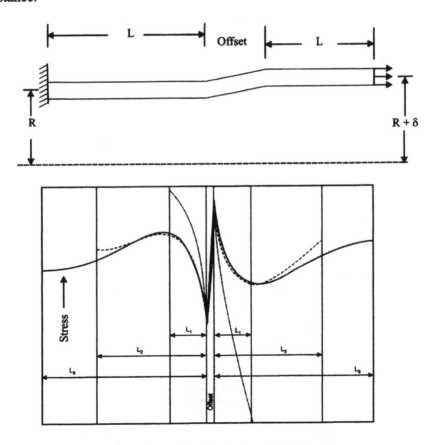

FIGURE 7.9 Effect of size of model around a discontinuity

(a) Symmetric boundary conditions

In some components with one or more lines or planes of symmetry, modelling the entire component for analysis will result in waste of time without any additional information regarding displacements or stresses as compared to the analysis of a symmetric part of the component. Boundaries of this model along these lines/planes of symmetry are represented by a suitable boundary condition to indicate that the model is only a part of the component. Displacements of corresponding nodes on either side of the line of symmetry in the direction normal to the line of symmetry are equal and opposite. Hence, on nodes along the line of symmetry, displacement normal to the line of symmetry is zero. A rectangular plate with a circular hole at the center, shown in Fig. 7.10, is a typical example of symmetry along two lines and hence, only a quarter of the plate can be modeled for analysis, saving time and effort.

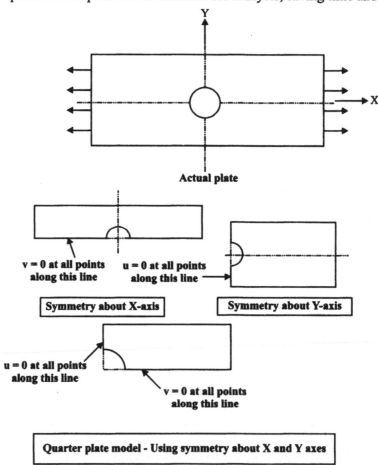

FIGURE 7.10 Symmetric boundary conditions

It should be noted that *symmetry should exist not only in the geometry of the component but also in the loads and boundary conditions of the component,* so that results of one part of the component are applicable to the remaining parts of the component. A few such models, where symmetry can be used in modelling the component for analysis, are shown below.

This aspect is explained through the example of arch of a factory gate, Fig. 7.11. One half of the arch about the vertical line of symmetry can be modeled, if the arch is analysed for a symmetric load such as self weight (Case-a). But the model will not be adequate if it is to be analysed for wind load, say from right to left (Case-b).

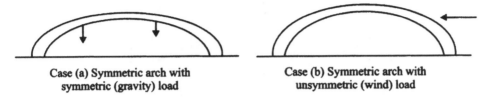

Case (a) Symmetric arch with   Case (b) Symmetric arch with
    symmetric (gravity) load   unsymmetric (wind) load

FIGURE 7.11 Dependence of model on geometry and loads

Symmetry of a component to be analysed need not be limited to the coordinate axes. If an octagonal structure such as a chimney is to be analysed for 2-D heat conduction through its wall, symmetry can be used to model just $1/16^{th}$ of the cross section as shown in Fig. 7.12.

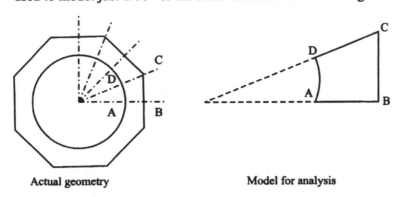

    Actual geometry   Model for analysis

Figure 7.12 Model of a chimney for 2-D heat conduction

(b) Cyclic or Sector symmetry

In rotary components like turbines, fans, compressors, ... one sector covering hub, one blade and rim can be modeled and the boundary condition will be different from that of the symmetry boundary condition. Displacements of nodes along one boundary (A-A) of the sector in cylindrical coordinate system are equal to the displacements of corresponding nodes on the other end (B-B) of the sector. The example of a fan, shown in Fig. 7.13, explains the cyclic symmetry condition.

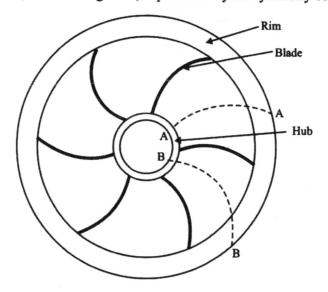

FIGURE 7.13 Cyclic symmetry of a fan

7.10 TRANSITION ELEMENT

The purpose of any analysis is to evaluate stresses at all points in a component when subjected to external or internal loads. Uniform mesh model is adequate when the stress variation in the component is small. However in areas of structural discontinuity or localised loads of high intensity such as thermal stresses during welding, uniform mesh model may not be appropriate. Such situations can be effectively modeled either by choosing fine mesh in the areas of high stress or by using higher degree displacement model for elements in the high stress areas. The latter option will require less computer memory and time. Compatibility conditions are not satisfied on the common edges if elements with linear displacement formulation (with 2 nodes along the common edge) are used on one side and elements with quadratic displacement formulation (with more than 2 nodes along the common edge) are used on the other side.

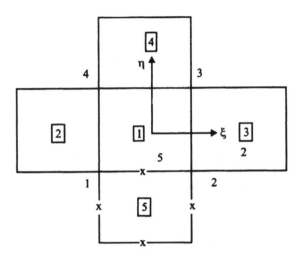

FIGURE **7.14** An example of a transition element

Transition elements, satisfying inter-element compatibility along the common edges, are used to connect different types of elements. These elements will not have same number of nodes on all their edges. For example, a quadrilateral element 1, shown in Fig. 7.14, connecting linear displacement elements 2, 3 and 4 with a quadratic displacement element 5 is one such transition element. Displacement polynomials

$$u = a_1 + a_2\,x + a_3\,y + a_4\,x^2 + a_5\,y^2$$

and $\qquad v = a_6 + a_7\,x + a_8\,y + a_9\,x^2 + a_{10}\,y^2$

are used for that element, ensuring symmetry w.r.t. coordinate axes x and y. In iso-parametric formulation, shape functions of this particular **5-node quadrilateral transition element** are given by

$$N_1 = \frac{\xi(\xi-1)(1-\eta)}{4}$$

$$N_2 = \frac{\xi(\xi+1)(1-\eta)}{4}$$

$$N_3 = \frac{(\xi+1)(1+\eta)}{4}$$

$$N_4 = \frac{(1-\xi)(1+\eta)}{4} \; ; \quad N_5 = \frac{(1-\xi^2)(1-\eta)}{2}$$

We can have, in a similar way, transition quadrilateral elements with 6 or 7 nodes depending on the displacement function used in elements 2, 3 and

4 surrounding the transition element 1. The terms in the displacement polynomial for each type of such elements will be different to ensure symmetry w.r.t. coordinate axes. The same logic can also be extended to other 2-D and 3-D elements. Transition elements are not relevant for discrete structures.

7.11 SUBSTRUCTURING OR SUPER ELEMENT APPROACH

In the finite element analysis of large systems, the number of equations to be solved for an accurate solution will be very large. Computational cost of matrix inversion is proportional to cube of the order of stiffness matrix. Thus, it is cheaper to invert three square matrices of order 500 than inverting one square matrix of order 1500. It also reduces computer memory requirement, since all substructures are not simultaneously processed. Method of substructures can be used to reduce the number of equations. In this method, the structure is divided into a number of parts, called substructures or super elements, each of which can be subdivided into a large number of elements. Each substructure is treated as one large element with many interior and boundary nodes. Assembled stiffness matrix of a substructure is rearranged to group displacements of all boundary nodes (elements with suffix 'b') and displacements of internal nodes (elements with suffix 'i') separately. Using *static condensation* procedure, this stiffness matrix is reduced to include modified contributions of boundary nodes only, depending on the type of elements being used, as explained below.

$$\begin{bmatrix} [K_{ii}] & [K_{ib}] \\ [K_{bi}] & [K_{bb}] \end{bmatrix} \begin{Bmatrix} \{u_i\} \\ \{u_b\} \end{Bmatrix} = \begin{Bmatrix} \{P_i\} \\ \{P_b\} \end{Bmatrix}$$

From the first set of equations $[K_{ii}] \{u_i\} + [K_{ib}] \{u_b\} = \{P_i\}$, we can express $\{u_i\}$ as

$$\{u_i\} = [K_{ii}]^{-1} (\{P_i\} - [K_{ib}] \{u_b\})$$

Substituting these values of $\{u_i\}$ in the 2nd set of equations

$$[K_{bi}]\{u_i\} + [K_{bb}]\{u_b\} = \{P_b\},$$

we get $\quad [K_{bi}][K_{ii}]^{-1} (\{P_i\} - [K_{ib}] \{u_b\}) + [K_{bb}]\{u_b\} = \{P_b\}$

or $\quad ([K_{bb}] - [K_{bi}][K_{ii}]^{-1}[K_{ib}]) \{u_b\} = \{P_b\} - [K_{bi}][K_{ii}]^{-1} \{P_i\}$

This can be written in a different notation as $\quad [K^*] \{u_b\} = \{P^*\}$

where $\quad [K^*]$ is the condensed stiffness matrix of the substructure, including DOFs associated with boundary nodes only

and $\quad \{P^*\}$ is the corresponding modified load vector.

This process of condensation is carried out, in many software, by Gauss elimination procedure. These reduced matrices of different substructures can be assembled together to get the stiffness matrix of the complete structure. The remaining procedure of applying boundary conditions and solving for the unknown displacements and stresses is same. There can be many levels of substructures, the highest level substructure consisting of ordinary finite elements only.

A typical application of this procedure can be seen in the case of an aircraft, where fuselage (central body), main wings, tail wings, etc.. are meshed as independent substructures. They are assembled together, after static condensation of the stiffness matrix of each substructure, to analyse the entire aircraft structure for the specific loads. This method saves considerable time and memory of the computer.

7.12 DEFORMED AND UNDEFORMED PLOTS

One problem with FEM is the generation of a large output consisting of displacement at all nodes and in all active degrees of freedom. It is difficult to scan for useful values from this large output to arrive at any meaningful conclusion. To overcome this difficulty, many general purpose software include options for plotting deformed shape as well as iso-stress and iso-temperature contours. Deformed shape of the component is better appreciated when the same is superimposed on undeformed geometry, with or without node/element numbers. Data errors related to load direction and area of application as well as boundary conditions are often checked with these deformed plots.

Within the elastic limit, the displacements in a component are so small that the deformed and undeformed plots coincide. Hence, nodal displacements are usually multiplied by a factor so that maximum displacement at any point in the element is about 20% of the component size or about 1 cm on A-4 size plot. Thus, deformed plot forms a vital check on the analysis of the component, to assess whether the results are sensible and meaningful or not. *Deformed plot is qualitative and not quantitative*, since physical dimensions of the component and nodal displacements are not plotted to the same scale.

Another **visual anomaly** is in the deformation of frames. Plots are generated by the post processor of the software from the nodal displacements obtained in the solution phase. Thus, the deformed plot of a beam member is a straight line, generated from its two nodal displacements, irrespective of the end conditions (simply supported or fixed ends). Cubic displacement polynomial as well as end conditions on slope are not reflected in these plots. (Ref. Fig. 7.15).

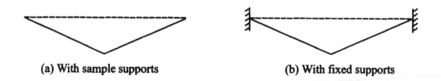

(a) With sample supports   (b) With fixed supports

FIGURE 7.15 Plot of deformation of beam

Note : Plots of deformation of beam between nodes 1 & 3 as well as between 2 and 3 in both the cases of end conditions are linear even through both of them are cubic functions. Even the displacement value at node 3 appears to be of the same magnitude in both cases, though the calculated values are different.

7.13 SUMMARY

- Consistent loads, based on energy equivalence, are used in FEM to represent distributed loads and give better results than equivalent loads

- Nodes, forming elements, are numbered to minimise maximum node number difference in the elements. This affects bandwidth of assembled stiffness matrix

- Many Software, based on FEM, store one half of banded, symmetric stiffness and mass matrices to minimise computer memory requirement

- Modelling a physical problem involves selection of proper types of elements, from the element library of any commercial software, which assume same displacement behaviour as the actual problem. Details, with zero or less influence on the results, can be omitted to simplify the physical model while ensuring a meaningful analysis with minimum computer cost

- Shape and size of elements influence the results significantly

- Use of symmetry in geometry, loads and boundary conditions help in modelling a smaller part of the component, saving computer time and memory

- Substructuring helps in the analysis of very large components by reducing size of matrices

- Deformation plots in many FEM software are qualitative (with magnified displacements), plotting only displacements without indicating correct slopes.

OBJECTIVE QUESTIONS

1. A symmetric structure can be analysed by modelling one symmetric part

    (a)   depending on applied loads

    (b)   depending on boundary conditions

    (c)   always yes

    (d)   depending on applied loads & boundary conditions

2. Anti-symmetric boundary condition along an edge of a 2-D structure implies, applied loads are _____ on either side of the edge

    (a)   opposite   (b)   equal

    (c)   equal and opposite   (d)   unrelated

3. Sector symmetry boundary condition implies _____ along two radial edges of the sector

    (a)   same radial displacements in cartesian coordinate system

    (b)   same circumferential displacements in cylindrical coordinate system

    (c)   equal and opposite radial displacements in cartesian coordinate system

    (d)   equal and opposite circumferential displacements in cylindrical coordinate system

4. Cyclic symmetry boundary condition implies _____ along two edges of the sector

    (a)   same radial displacements in cartesian coordinate system

    (b)   same circumferential displacements in cylindrical coordinate system

    (c)   equal & opposite radial displacements in cartesian coordinate system

    (d)   equal and opposite circumferential displacements in cylindrical coordinate system

5. An octagonal section chimney with hot gases inside can be analysed using ___ model

    (a)   full section   (b)   one half of section

    (c)   one quarter of section   (d) $1/8^{th}$ of section

6. Use of non-dimensional coordinates helps in

   (a)   numerical integration (b)   displacement calculation

   (c)   stress calculation (d)   strain calculation

7. Gaussian points are used for

   (a)   numerical integration (b)   displacement calculation

   (c)   stress calculation (d)   strain calculation

8. Quadrature means

   (a)   calculation of area of element

   (b)   calculation of element stress

   (c)   numerical integration for getting stiffness coefficients

   (d)   calculation of nodal displacements

9. Accuracy of stiffness matrix improves with

   (a)   more number of Gaussian points

   (b)   more number of nodes

   (c)   size of elements

   (d)   shape of elements

10. Sector symmetry and cyclic symmetry differ

    (a)   in the shape of sector edges

    (b)   in the size of sector edges

    (c)   in radial displacements along two sector edges

    (d)   in circumferential displacements along two sector edges

11. Using symmetry condition _____ ; but gives same solution

    (a)   saves computer time

    (b)   saves computer memory

    (c)   saves effort of data preparation

    (d)   all of them

12. Symmetry boundary condition about an edge is applicable when

    (a)   normal loads & normal displacements at nodes along the edge are zero

    (b)   loads & displacements along the edge are zero

    (c)   normal loads & normal displacements at nodes on either side of the edge are equal & opposite

    (d)   loads & displacements along the edge are same

13. A cantilever beam can be analysed as a plate with ____ boundary conditions

    (a) Cartesian symmetric (b) sector symmetry

    (c) cyclic symmetry (d) cartesian anti-symmetric

14. Number of DOF for 2-node cantilever and propped cantilever are

    (a) 1,2 (b) 2,1 (c) 3,4 (d) 2,4

15. Number of DOF for 3-noded simply supported beam and fixed beam are

    (a) 1,2 (b) 2,3 (c) 3,4 (d) 4,2

16. Small region of interest in a big component can be analysed using free body end conditions

    (a) always true (b) sometimes true

    (c) never true (d) depends on other data

17. ___ model of a rectangular plate with a circular hole at the center, and loaded uniformly along the four edges, is adequate for analysis

    (a) full (b) 1/2 (c) 1/4 (d) 1/8

18. ___ model of a square plate with a circular hole at the center, and loaded uniformly along the four edges, is adequate for analysis

    (a) full (b) 1/2 (c) 1/4 (d) 1/8

19. ___ model of a square plate with a rectangular hole at the center (edges parallel to the edges of the plate), and loaded uniformly along the four edges, is adequate for analysis

    (a) full (b) 1/2 (c) ¼ (d) 1/8

20. In statically equivalent loads, free end moment of a cantilever of length 'L' with uniformly distributed load of value 'p' is

    (a) $pL^2/4$ (b) $pL^2/8$ (c) $pL^2/12$ (d) $pL^2/16$

21. In consistent loads, free end moment of a cantilever of length 'L' with uniformly distributed load of value 'p' is

    (a) $pL^2/4$ (b) $pL^2/8$ (c) $pL^2/12$ (d) $pL^2/16$

22. In statically equivalent loads, end moment of a simply supported beam of length 'L' with a concentrated load 'P' at the mid point is

    (a) PL/4 (b) PL/8 (c) PL/12 (d) PL/16

23. In consistent loads, end moment of a simply supported beam of length 'L' with a concentrated load 'P' at the mid point is

     (a)   PL/4   (b)   PL/8   (c)   PL/12   (d)   PL/16

24. In statically equivalent loads, end moment of a simply supported beam of length 'L' with a uniformly distributed load of value 'p' is

     (a) $pL^2/4$ (b) $pL^2/8$ (c) $pL^2/12$ (d) $pL^2/16$

25. In consistent loads, end moment of a simply supported beam of length 'L' with a uniformly distributed load of value 'p' is

     (a) $pL^2/4$ (b) $pL^2/8$ (c) $pL^2/12$ (d) $pL^2//16$

26. Consistent loads for a LST element with uniform pressure 'p' along an edge of length 'L', at the two end nodes and mid-node are

     (a)   pL/2, pL/2, 0   (b)   pL/3, pL/3, pL/3

     (c)   pL/4, pL/4, pL/2   (d)   pL/6, pL/6, 2pL/3

27. The process of reducing number of mid-side or internal nodes before assembling element stiffness matrices is called

     (a)   Gauss reduction   (b)   Jacobi reduction

     (c)   Choleski reduction   (d)   static condensation

28. Lengths of longest side and shortest side of a 2-D or 3-D element decide the

     (a)   aspect ratio

     (b)   shape function

     (c)   order of displacement polynomial

     (d)   included angle

29. Number of nodes along the side of a 2-D or 3-D element decide the

     (a)   aspect ratio

     (b)   shape function

     (c)   order of displacement polynomial

     (d)   nature of deformation

CHAPTER **8**

DYNAMIC ANALYSIS
(UNDAMPED FREE VIBRATION)

Dynamics is a special branch of mechanics where inertia of accelerating masses must be considered in the force-deflection relationships. In order to describe motion of the mass system, a component with distributed mass is approximated by a finite number of mass points. Knowledge of certain principles of dynamics is essential to the formulation of these equations.

Every structure is associated with certain frequencies and mode shapes of free vibration (without continuous application of load), based on the distribution of mass and stiffness in the structure. Any time-dependent external load acting on the structure, whose frequency matches with the natural frequencies of the structure, causes resonance and produces large displacements leading to failure of the structure. Calculation of natural frequencies and mode shapes is therefore very important.

Consider i^{th} mass m_i of a system of connected rigid bodies and the force components F_j ($j = 1,2,..6$) acting upon it in three-dimensional space. If the mass m_i is in equilibrium at rest, then $\Sigma F_i = 0$.

If mass m_j is not in equilibrium, it will accelerate in accordance with Newton's second law i.e., $F_j = m_i \ddot{u}_j$

The force $(-m_i.\ddot{u}_j)$ is called the reversed effective force or inertia force. According to **D'Alembert's principle**, the net external force and the inertia force together keep the body in a state of *'fictitious equilibrium'*. i.e., $\Sigma(F_j - m\ddot{u}_j) = 0$.

If the displacement of the mass m_i is represented by δu_j ($j = 1,2,..6$), then the virtual work done by these force components on the mass m_i in equilibrium is given by

$$\delta W_i = \sum F_j . \delta u_j = 0.$$

D'Alembert's principle rewritten in the form,

$$\delta W_i = \sum F_j \cdot \delta u_j - \sum (m \, \ddot{u}_j) \cdot \delta u_j = 0 \quad \text{is a statement of \textit{virtual work}}$$

for a system in motion.

For a simple spring of stiffness 'k' and a lumped mass 'm' under steady state undamped condition of oscillation without external force, the force equilibrium condition of the system is given by

$$k \, u(t) + m \, \ddot{u}(t) = 0,$$

where, $F_i = - k \, u(t)$ is the reactive elastic force applied to the mass.

Displacement in vibration is a simple harmonic motion and can be represented by a sinusoidal function of time as

$$u(t) = u \sin \omega t$$

where, ω is the frequency of vibration in radians/sec

It is more often expressed in 'f' cycles/sec or Hertz (Hz) where $\omega = 2\pi \, f$

Then, velocity $\dot{u}(t) = -\omega u \cos \omega t$

and acceleration $\ddot{u}(t) = -\omega^2 u \sin \omega t = -\omega^2 u(t)$

$\therefore$ $K.u(t) + m \, \ddot{u}(t) = (k - \omega^2 m) \, u(t) = 0$

In general, for a system with 'n' degrees of freedom, stiffness 'k' and mass 'm' are represented by stiffness matrix [K] and mass matrix [M] respectively.

Then, $([K] - \omega^2 [M]) \{u\} = \{0\}$

or $([M]^{-1}[K] - \omega^2 [I]) \{u\} = \{0\}$

Here, [M] is the mass matrix of the entire structure and is of the same order, say n × n, as the stiffness matrix [K]. This is also obtained by assembling element mass matrices in a manner exactly identical to assembling element stiffness matrices. The mass matrix is obtained by two different approaches, as explained subsequently.

This is a typical eigenvalue problem, with ω^2 as eigenvalues and $\{u\}$ as eigenvectors. A structure with 'n' DOF will therefore have 'n' eigenvalues and 'n' eigenvectors. Some eigenvalues may be repeated and some eigenvalues may be complex, in pairs. The equation can be represented in the standard form, $[A]\{x\}_i = \lambda_i \{x\}_i$. In dynamic analysis, ω_i indicates i[th] natural frequency and $\{x\}_i$ indicates i[th] natural mode of vibration. A natural mode is a *qualitative* plot of nodal displacements. In every natural mode of vibration, all the points on the component will reach their maximum values at the same time and will pass through zero displacements at the same time. Thus, in a particular mode, all the points of a component will vibrate with the same frequency and their relative displacements are indicated by the components of the corresponding eigenvector. These relative (or proportional) displacements at different points on structure remain same at every time instant for undamped free vibration (Ref. Fig. 8.1). Hence, without loss of generality, $\{u(t)\}$ can be written as $\{u\}$.

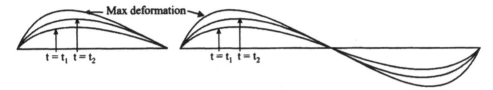

FIGURE 8.1 Mode shape

Since $\{u\} = \{0\}$ forms a trivial solution, the homogeneous system of equations $([A] - \lambda [I]) \{u\} = \{0\}$ gives a non-trivial solution only when

$$([A] - \lambda [I]) = \{0\},$$

which implies   **Det ($[A] - \lambda [I]$) = 0.**

This expression, called **characteristic equation**, results in n^{th} order polynomial in λ and will therefore have n roots. For each λ_i, the corresponding eigenvector $\{u\}_i$ can be obtained from the n homogeneous equations represented by $([K] - \lambda [M]) \{u\} = \{0\}$. The mode shape represented by $\{u(t)\}$ gives relatives values of displacements in various degrees of freedom.

It can also be represented as

$$[A][X] = [X][\Lambda]$$

where, $$[A] = [M]^{-1} [K]$$

[X] is called the **modal matrix**, whose i^{th} column represents i^{th} eigenvector $\{x\}_i$

and [Λ] is called the **spectral matrix** with each diagonal element representing one eigenvalue, corresponding to the eigenvector of that column, and off-diagonal elements equal to zero.

8.1 NORMALISATION OF EIGENVECTORS

The equation of motion of free vibrations $([K] - \omega^2 [M]) \{u\} = \{0\}$ is a system of homogeneous equations (right side vector zero) and hence does not give unique numerical solution. **Mode shape is a set of relative displacements** in various degrees of freedom, while the structure is vibrating in a particular frequency and is usually expressed in normalised form, by following one of the three normalisation methods explained here.

(a) The maximum value of any one component of the eigenvector is equated to '1' and, so, all other components will have a value less than or equal to '1'.

(b) The length of the vector is equated to '1' and values of all components are divided by the length of this vector so that each component will have a value less than or equal to '1'.

(c) The eigenvectors are usually normalised so that

$$\{u\}_i^T [M] \{u\}_i = 1 \quad \text{and} \quad \{u\}_i^T [K] \{u\}_i = \lambda_i$$

For a positive definite symmetric stiffness matrix of size n x n, the eigenvalues are all real and eigenvectors are **orthogonal**

$$\text{i.e.,} \quad \{u\}_i^T [M] \{u\}_j = 0 \quad \text{and} \quad \{u\}_i^T [K] \{u\}_j = 0 \quad \forall \ i \neq j$$

8.2 MODELLING FOR DYNAMIC ANALYSIS

Solution for any dynamic analysis is an iterative process and, hence, is time – consuming. Geometric model of the structure for dynamic analysis can be significantly simplified, giving higher priority for proper representation of distributed mass. An example of a simplified model of a water storage tank is shown in Fig. 8.2, representing the central hollow shaft by long beam elements and water tanks at two levels by a few lumped masses and short beam elements of larger moment of inertia.

8.3 MASS MATRIX

Mass matrix [M] differs from the stiffness matrix in many ways:

(i) The mass of each element is equally distributed at all the nodes of that element

(ii) Mass, being a scalar quantity, has **same** effect along the three translational degrees of freedom (u, v and w) and is **not** shared

(iii) Mass, being a scalar quantity, is not influenced by the local or global coordinate system. Hence, no transformation matrix is used for converting mass matrix from element (or local) coordinate system to structural (or global) coordinate system.

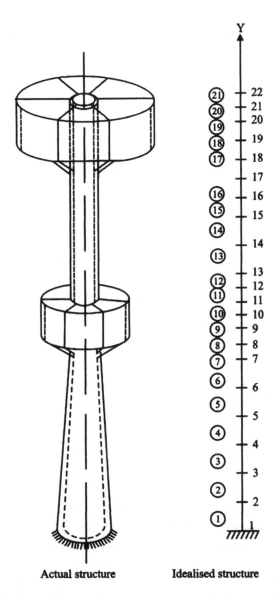

FIGURE 8.2 Finite Element Model of a water tank for dynamic analysis

Two different approaches of evaluating mass matrix [M] are commonly considered.

(a)   Lumped mass matrix

Total mass of the element is assumed equally distributed at all the nodes of the element in each of the translational degrees of freedom. Lumped mass is not used for rotational degrees of freedom. Off-diagonal elements of this matrix are all zero. This assumption *excludes dynamic coupling* that exists between different nodal displacements. Lumped mass matrices [M] of some elements are given here.

Lumped mass matrix of truss element with 1 translational DOF per node along its local X-axis

$$[M] = \frac{\rho AL}{2} \begin{bmatrix} 1 & 0 \\ 0 & 1 \end{bmatrix}$$

Lumped mass matrix of plane truss element in a 2-D plane with 2 translational DOF per node (Displacements along X and Y coordinate axes)

$$[M] = \frac{\rho AL}{2} \begin{bmatrix} 1 & 0 & 0 & 0 \\ 0 & 1 & 0 & 0 \\ 0 & 0 & 1 & 0 \\ 0 & 0 & 0 & 1 \end{bmatrix}$$

Please note that the same lumped mass is considered in each translational degree of freedom (without proportional sharing of mass between them) at each node.

Lumped mass matrix of a beam element in X-Y plane, with its axis along x-axis and with two DOF per node (deflection along Y axis and slope about Z axis) is given below. Lumped mass is not considered in the rotational degrees of freedom.

$$[M] = \frac{\rho AL}{2} \begin{bmatrix} 1 & 0 & 0 & 0 \\ 0 & 0 & 0 & 0 \\ 0 & 0 & 1 & 0 \\ 0 & 0 & 0 & 0 \end{bmatrix}$$

Note that lumped mass terms are not included in 2nd and 4th rows, as well as columns corresponding to rotational degrees of freedom.

Lumped mass matrix of a CST element with 2 DOF per node. In this case, irrespective of the shape of the element, mass is assumed equally distributed at the three nodes. It is distributed equally in all DOF at each node, without any sharing of mass between different DOF

$$[M] = \frac{\rho AL}{3} \begin{bmatrix} 1 & 0 & 0 & 0 & 0 & 0 \\ 0 & 1 & 0 & 0 & 0 & 0 \\ 0 & 0 & 1 & 0 & 0 & 0 \\ 0 & 0 & 0 & 1 & 0 & 0 \\ 0 & 0 & 0 & 0 & 1 & 0 \\ 0 & 0 & 0 & 0 & 0 & 1 \end{bmatrix}$$

(b) Consistent mass matrix

Element mass matrix is calculated here, *consistent* with the assumed displacement field or element stiffness matrix. [M] is a banded matrix of the same order as the stiffness matrix. This is evaluated using the same interpolating functions which are used for approximating displacement field over the element. It yields more accurate results but with more computational cost. Consistent mass matrices of some elements are given here.

Consistent mass matrix of a Truss element along its axis (in local coordinate system)

$$\{u\}^T = [u \quad v]$$

$$[N]^T = [N_1 \quad N_2]$$

where, $N_1 = \dfrac{(1-\xi)}{2}$

and $N_2 = \dfrac{(1+\xi)}{2}$

$$[M] = \int_V [N]\rho[N]^T \, dV = \int_0^L A[N]\rho[N]^T$$

$$dx = \int_{-1}^{+1} A\rho[N][N]^T (\det J)(dx/d\xi)d\xi$$

Here, $x = N_1 x_1 + N_2 x_2 = \dfrac{(x_1 + x_2)}{2} + \dfrac{(x_2 - x_1)\xi}{2}$

and $dx = \dfrac{dx}{d\xi}.d\xi = \det J \, d\xi = \left(\dfrac{L}{2}\right)d\xi$

Using the values of integration in natural coordinate system,

$$[M] = \rho A \left(\frac{L}{2}\right) \int_{-1}^{+1} \begin{bmatrix} (1-\xi)/2 \\ (1+\xi)/2 \end{bmatrix} [(1-\xi)/2 \quad (1+\xi)/2]d\xi$$

$$= \frac{\rho AL}{8} \begin{bmatrix} \int(1-\xi)^2 d\xi & \int(1-\xi^2)d\xi \\ \int(1-\xi^2)d\xi & \int(1+\xi)^2 \, d\xi \end{bmatrix}$$

$$= \frac{\rho AL}{8} \begin{bmatrix} (\xi - \xi^2 + \xi^3/3) & (\xi - \xi^3/3) \\ (\xi - \xi^3/3) & (\xi + \xi^2 + \xi^3/3) \end{bmatrix}$$

$$= \frac{\rho AL}{8} \begin{bmatrix} 8/3 & 4/3 \\ 4/3 & 8/3 \end{bmatrix} = \frac{\rho AL}{6} \begin{bmatrix} 2 & 1 \\ 1 & 2 \end{bmatrix}$$

Consistent mass matrix of a **Plane** *Truss element,* inclined to global X-axis - Same elements of 1-D mass matrix are repeated in two dimensions (along X and Y directions) without sharing mass between them. Mass terms in X and Y directions are uncoupled.

$$[M] = \frac{\rho AL}{6} \begin{bmatrix} 2 & 0 & 1 & 0 \\ 0 & 2 & 0 & 1 \\ 1 & 0 & 2 & 0 \\ 0 & 1 & 0 & 2 \end{bmatrix}$$

Consistent mass matrix of a **Space** *Truss element,* inclined to X-Y plane) – Same elements of 1-D mass matrix are repeated in three dimensions (along X, Y and Z directions) without sharing mass between them.

$$[M] = \frac{\rho AL}{6} \begin{bmatrix} 2 & 0 & 0 & 1 & 0 & 0 \\ 0 & 2 & 0 & 0 & 1 & 0 \\ 0 & 0 & 2 & 0 & 0 & 1 \\ 1 & 0 & 0 & 2 & 0 & 0 \\ 0 & 1 & 0 & 0 & 2 & 0 \\ 0 & 0 & 1 & 0 & 0 & 2 \end{bmatrix}$$

Consistent mass matrix of a Beam element

$$[M] = \rho A \left(\frac{L}{2} \right) \int \{H\}^T \{H\} d\xi \quad \text{with Hermite shape functions } \{H\} \text{ as used in a}$$

beam element.

$$= \frac{\rho AL}{128} \int \begin{bmatrix} 2(2 - 3\xi + \xi^3) \\ L(1 - \xi + \xi^2 + \xi^3) \\ 2(2 + 3\xi - \xi^3) \\ L(-1 - \xi + \xi^2 + \xi^3) \end{bmatrix} \times$$

$$\left[2(2 - 3\xi + \xi^3) \quad L(1 - \xi - \xi^2 + \xi^3) \quad 2(2 + 3\xi - \xi^3) \quad L(-1 - \xi + \xi^2 + \xi^3) \right] d\xi$$

$$= \frac{\rho AL}{420} \begin{bmatrix} 156 & 22L & 54 & -13L \\ 22L & 4L^2 & 13L & -3L^2 \\ 54 & 13L & 156 & -22L \\ -13L & -3L^2 & -22L & 4L^2 \end{bmatrix}$$

Consistent mass matrix of a CST element in a 2-D plane

$$[N]^T = \begin{bmatrix} N_1 & 0 & N_2 & 0 & N_3 & 0 \\ 0 & N_1 & 0 & N_2 & 0 & N_3 \end{bmatrix}$$

$$[M] = \int [N]\rho[N]^T\, dV = t \int [N]\rho[N]^T\, dA$$

$$= \frac{\rho t A}{12} \begin{bmatrix} 2 & 0 & 1 & 0 & 1 & 0 \\ & 2 & 0 & 1 & 0 & 1 \\ & & 2 & 0 & 1 & 0 \\ & & & 2 & 0 & 1 \\ & \text{Sym} & & & 2 & 0 \\ & & & & & 2 \end{bmatrix}$$

Note : Natural frequencies obtained using lumped mass matrix are LOWER than exact values.

Example 8.1 : Find the natural frequencies of longitudinal vibrations of the unconstrained stepped shaft of areas A and 2A and of equal lengths (L), as shown below.

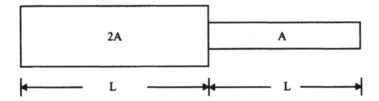

Solution : Let the finite element model of the shaft be represented by 3 nodes and 2 truss elements (as only longitudinal vibrations are being considered) as shown below.

$$[K]_1 = (2A)\left(\frac{E}{L}\right)\begin{bmatrix} 1 & -1 \\ -1 & 1 \end{bmatrix} = \left(\frac{AE}{L}\right)\begin{bmatrix} 2 & -2 \\ -2 & 2 \end{bmatrix};$$

$$[K]_2 = \left(\frac{AE}{L}\right)\begin{bmatrix} 1 & -1 \\ -1 & 1 \end{bmatrix}$$

Using consistent mass matrix approach

$$[M]_1 = \frac{\rho(2A)L}{6}\begin{bmatrix} 2 & 1 \\ 1 & 2 \end{bmatrix} = \frac{\rho AL}{6}\begin{bmatrix} 4 & 2 \\ 2 & 4 \end{bmatrix};$$

$$[M]_2 = \frac{\rho AL}{6}\begin{bmatrix} 2 & 1 \\ 1 & 2 \end{bmatrix}$$

Assembling the element stiffness and mass matrices,

$$[K] = \frac{AE}{L}\begin{bmatrix} 2 & -2 & 0 \\ -2 & 3 & -1 \\ 0 & -1 & 1 \end{bmatrix};$$

$$[M] = \frac{\rho AL}{6}\begin{bmatrix} 4 & 2 & 0 \\ 2 & 6 & 1 \\ 0 & 1 & 2 \end{bmatrix}$$

Eigenvalues of the equation $([K] - \omega^2 [M]) \{u\} = \{0\}$ are the roots of the characteristic equation represented by

$$\begin{vmatrix} 2AE/L - \omega^2 4\rho\rho AL/ & -2AE/L - \omega^2 2\rho\rho AL/ & 0 \\ 2AE/L - \omega^2 2\rho\rho AL/ & 3AE/L - \omega^2 6\rho\rho AL/ & -1AE/L - \omega^2 \rho AL/6 \\ 0 & -AE/L - \omega^2 \rho AL/6 & AE/L - \omega^2 2\rho\rho AL/ \end{vmatrix} = 0$$

Multiplying all the terms by (L/AE) and substituting $\beta = \dfrac{\rho L^2 \omega^2}{6E}$

$$\begin{vmatrix} 2(1-2\beta) & -2(1+\beta) & 0 \\ -2(1+\beta) & 3(1-2\beta) & -(1+\beta) \\ 0 & -(1+\beta) & (1-2\beta) \end{vmatrix} = 0$$

or $18\,\beta\,(\beta - 2)\,(1 - 2\beta) = 0$

The roots of this equation are $\beta = 0, 2$ or $\dfrac{1}{2}$ or $\omega^2 = 0,\ \dfrac{12E}{\rho L^2}$ or $\dfrac{3E}{\rho L^2}$

Corresponding eigenvectors are obtained from $([K] - \omega^2 [M]) \{u\} = \{0\}$ for different values of ω^2 as $\begin{bmatrix} 1 & 1 & 1 \end{bmatrix}^T$ for $\beta = 0$, $\begin{bmatrix} 1 & 0 & -2 \end{bmatrix}^T$ for $\beta = \dfrac{1}{2}$ and $\begin{bmatrix} 1 & -1 & 1 \end{bmatrix}^T$ for $\beta = 2$.

The first eigenvector implies rigid body motion of the shaft. One component (u_1 in this example) is equated to '1' and other displacement components

(u_2 and u_3 in this example) are obtained as ratios w.r.t. that component, following one method of normalisation. Alternatively, they may also be expressed in other normalised forms.

Note : Static solution for such an unconstrained bar, with rigid body motion, involves a singular [K] matrix and can not be solved for {u}, while dynamic analysis is mathematically possible.

Example 8.2

Find the natural frequencies of longitudinal vibrations of the same stepped shaft of areas A and 2A and of equal lengths (L), when it is constrained at one end, as shown below.

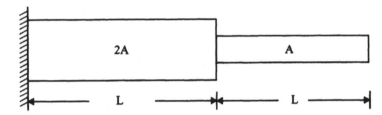

Solution

Let the finite element model of the shaft be represented by 3 nodes and 2 truss elements (as only longitudinal vibrations are being considered) as shown below.

$$[K]_1 = \left(\frac{2AE}{L}\right)\begin{bmatrix} 1 & -1 \\ -1 & 1 \end{bmatrix} = \left(\frac{AE}{L}\right)\begin{bmatrix} 2 & -2 \\ -2 & 2 \end{bmatrix}$$

$$[K]_2 = \left(\frac{AE}{L}\right)\begin{bmatrix} 1 & -1 \\ -1 & 1 \end{bmatrix}$$

Using consistent mass matrix approach

$$[M]_1 = \frac{\rho(2A)L}{6}\begin{bmatrix} 2 & 1 \\ 1 & 2 \end{bmatrix} = \frac{\rho AL}{6}\begin{bmatrix} 4 & 2 \\ 2 & 4 \end{bmatrix} ; \quad [M]_2 = \frac{\rho AL}{6}\begin{bmatrix} 2 & 1 \\ 1 & 2 \end{bmatrix}$$

Assembling the element stiffness and mass matrices,

$$[K] = \frac{AE}{L}\begin{bmatrix} 2 & -2 & 0 \\ -2 & 3 & -1 \\ 0 & -1 & 1 \end{bmatrix} ; \quad [M] = \frac{\rho AL}{6}\begin{bmatrix} 4 & 2 & 0 \\ 2 & 6 & 1 \\ 0 & 1 & 2 \end{bmatrix}$$

After applying boundary condition at node 1, 1st row and 1st column of the assembled matrix can be deleted. Eigenvalues of the equation ($[K] - \omega^2 [M]$) $\{u\} = \{0\}$ are the roots of the characteristic equation represented by

$$\begin{vmatrix} 3AE/L - \omega^2 6\rho AL/6 & -AE/L - \omega^2 \rho AL/6 \\ -AE/L - \omega^2 \rho AL/6 & AE/L - \omega^2 2\rho AL/6 \end{vmatrix} = 0$$

Multiplying all the terms by (L/AE) and substituting $\beta = \dfrac{\rho L^2 \omega^2}{6E}$

$$\begin{vmatrix} 3(1 - 2\beta) & -(1+\beta) \\ -(1+\beta) & (1 - 2\beta) \end{vmatrix} = 0$$

or $\qquad 11\,\beta^2 - 14\beta + 2 = 0 \qquad \Rightarrow \beta = \left(7 \pm \sqrt{27}\right)\big/11$

The roots of this equation are $\beta = 0.164,\ 1.109$ or $\omega^2 = \dfrac{0.984\,E}{\rho L^2}$ or $\dfrac{6.654\,E}{\rho L^2}$

Corresponding eigenvectors are obtained from $([K] - \omega^2 [M])\ \{u\} = \{0\}$ for different values of ω^2 as $[\ 0 \quad 1 \quad 1.732\ ]^{T}$.

Example 8.3

Find the natural frequencies of longitudinal vibrations of the constrained stepped shaft of areas A and 2A and of equal lengths (L), as shown below. Compare the results obtained using lumped mass matrix approach and consistent mass matrix approach.

Solution

Let the finite element model of the shaft be represented by 3 nodes and 2 truss elements (as only longitudinal vibrations are being considered) as shown below.

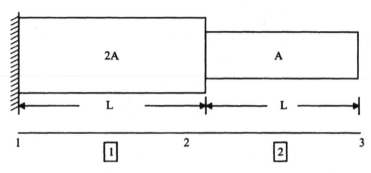

$$[K]_1 = \left(\frac{2AE}{L}\right)\begin{bmatrix} 1 & -1 \\ -1 & 1 \end{bmatrix} = \left(\frac{AE}{L}\right)\begin{bmatrix} 2 & -2 \\ -2 & 2 \end{bmatrix}$$

$$[K]_2 = \left(\frac{AE}{L}\right)\begin{bmatrix} 1 & -1 \\ -1 & 1 \end{bmatrix}$$

(a)   *Using lumped mass matrix approach*

$$[M]_1 = \frac{\rho(2A)L}{2}\begin{bmatrix} 1 & 0 \\ 0 & 1 \end{bmatrix} = \frac{\rho AL}{2}\begin{bmatrix} 2 & 0 \\ 0 & 2 \end{bmatrix}; \quad [M]_2 = \frac{\rho AL}{2}\begin{bmatrix} 1 & 0 \\ 0 & 1 \end{bmatrix}$$

Assembling the element stiffness and mass matrices,

$$[K] = \frac{AE}{L}\begin{bmatrix} 2 & -2 & 0 \\ -2 & 3 & -1 \\ 0 & -1 & 1 \end{bmatrix}; \quad [M] = \frac{\rho AL}{2}\begin{bmatrix} 2 & 0 & 0 \\ 0 & 3 & 0 \\ 0 & 0 & 1 \end{bmatrix}$$

Application of boundary condition (node 1 constrained) eliminates row 1 and column 1, thus reducing the size of stiffness and mass matrices to 2 × 2. Eigenvalues of the equation $([K] - \omega^2 [M]) \{u\} = \{0\}$ are the roots of the characteristic equation represented by

$$\begin{vmatrix} 3AE/L - \omega^2 3\rho AL/2 & -AE/L \\ -AE/L & AE/L - \omega^2 \rho AL/2 \end{vmatrix} = 0$$

Multiplying all the terms by (L/AE) and substituting $\beta = \dfrac{\rho L^2 \omega^2}{2E}$

$$\begin{vmatrix} 3(1-\beta) & -1 \\ -1 & (1-\beta) \end{vmatrix} = 0$$

or $3\beta^2 - 6\beta + 2 = 0$

The roots of this equation are $\beta = \dfrac{(3 \pm \sqrt{3})}{3}$ or 0.423, 1.577

Corresponding eigenvectors are $[0 \;\; -0.57734 \;\; 1]^T$ for $\beta = 1.577$

and $[0 \;\; 0.57734 \;\; 1]^T$ for $\beta = 0.423$

(b)   *Using consistent mass matrix approach*

$$[M]_1 = \frac{\rho(2A)L}{6}\begin{bmatrix} 2 & 1 \\ 1 & 2 \end{bmatrix} = \frac{\rho AL}{6}\begin{bmatrix} 4 & 2 \\ 2 & 4 \end{bmatrix}$$

$$[M]_2 = \frac{\rho AL}{6} \begin{bmatrix} 2 & 1 \\ 1 & 2 \end{bmatrix}$$

Assembling the element stiffness and mass matrices,

$$[K] = \frac{AE}{L} \begin{bmatrix} 2 & -2 & 0 \\ -2 & 3 & -1 \\ 0 & -1 & 1 \end{bmatrix}; \quad [M] = \frac{\rho AL}{6} \begin{bmatrix} 4 & 2 & 0 \\ 2 & 6 & 1 \\ 0 & 1 & 2 \end{bmatrix}$$

Application of boundary condition (node 1 constrained) eliminates row 1 and column 1, thus reducing the size of stiffness and mass matrices to 2 × 2. Eigenvalues of the equation ($[K] - \omega^2 [M]$) $\{u\} = \{0\}$ are the roots of the characteristic equation represented by

$$\begin{vmatrix} 3AE/L - \omega^2 6\rho AL/6 & -AE/L - \omega^2 \rho AL/6 \\ -AE/L - \omega^2 \rho AL/6 & AE/L - \omega^2 2\rho AL/6 \end{vmatrix} = 0$$

Multiplying all the terms by (L/AE) and substituting $\beta = \dfrac{\rho L^2 \omega^2}{6E}$

$$\begin{vmatrix} 3(1-2\beta) & -(1+\beta) \\ -(1+\beta) & (1-2\beta) \end{vmatrix} = 0$$

or $11 \beta^2 - 14 \beta + 2 = 0$

The roots of this equation are $\beta = \dfrac{(7 \pm 3\sqrt{3})}{11}$ or 1.10874, 0.16399

Corresponding eigenvectors are $[0 \quad -0.57734 \quad 1]^T$ for $\beta = 1.577$

and $[0 \quad 0.57734 \quad 1]^T$ for $\beta = 0.423$

Note : *Natural frequencies obtained with lumped mass matrices are LOWER* than those obtained with consistent mass matrices, while the mode shapes are practically same.

Example 8.4

Find the natural frequencies of vibrations of a simple cantilever beam.

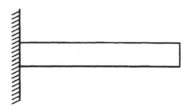

Solution

Let the finite element model of the beam be represented by 2 nodes and 1 beam element to facilitate manual calculation. After applying boundary conditions,

$$[K] = \left(\frac{EI}{L^3}\right)\begin{bmatrix} 12 & -6L \\ -6L & 4L^2 \end{bmatrix}; \quad [M] = \left(\frac{\rho AL}{420}\right)\begin{bmatrix} 156 & -22L \\ -22L & 4L^2 \end{bmatrix}$$

Eigenvalues of the equation $([K] - \omega^2 [M])\,\{u\} = \{0\}$ are the roots of

$$\begin{vmatrix} 12\alpha - 78\beta & -6L\alpha + 11L\beta \\ -6L\alpha + 11L\beta & 4L^2\alpha - 2L^2\beta \end{vmatrix} = 0$$

where $\alpha = \dfrac{EI}{L^3}$ and $\beta = \dfrac{\rho AL\omega^2}{210}$

or $35\,\beta^2 - 204\,\alpha\,\beta + 12\,\alpha^2 = 0$

The roots of this equation are $\beta = \dfrac{2\,\alpha}{35}$ or $\dfrac{202\,\alpha}{35}$

or $\omega^2 = \dfrac{12EI}{\rho AL^4}$ or $\dfrac{1212EI}{\rho AL^4}$

Corresponding eigenvectors are $[\,0.983 \quad 1.36/L\,]^T$ for $\beta = \dfrac{2\,\alpha}{35}$

and $[\,1.006 \quad 7.716/L\,]^T$ for $\beta = \dfrac{202\,\alpha}{35}$

8.4   SUMMARY

- A distributed mass system will have as many natural frequencies and mode shapes as the number of DOF, 'n'.

- Free undamped vibrations involve a set of n homogeneous equations. Such equations will not give a unique solution. A mode shape consists of relative displacement values at (n-1) DOF, obtained w.r.t. the chosen displacement value at one DOF. The mode shapes (eigen vectors) are usually normalised.

- The n natural frequencies may be real or complex (in pairs). Some of them may be zero (indicating rigid body mode) or repeated.

- Only first few frequencies (lower values) are significant and are usually calculated by iterative methods. Hence, a coarse mesh is adequate for dynamic analysis.

- They can be calculated using lumped mass matrix or consistent mass matrix, based on shape functions used for assumed displacement field. Each element of mass matrix of an element represents scalar mass, associated with a particular DOF and no transformation is involved between element (local) coordinate system and structure (global) coordinate system unlike stiffness matrix of vector elements.

- Lumped mass matrix is diagonal; has no components in the rotational DOF; mass of element is equally distributed at all the nodes of the element, irrespective of shape of the element; same mass at a node is taken along all translational DOF at the node and not shared in fractions.

- Consistent mass matrix is square, symmetric and banded, just like stiffness matrix.

OBJECTIVE QUESTIONS

1. An unconstrained 3-D frame with 4 nodes has ___ number of zero frequencies

    (a) 1   (b) 2   (c) 3   (d) 6

2. A frequency of value ___ indicates rigid body motion along one dof

    (a) zero   (b) 1   (c) infinity   (d) less than zero

3. Principal modes of vibration of a multi-dof system are

    (a) parallel   (b) orthogonal

    (c) integer multiples   (d) fractional multiples

4. With lumped mass matrix, the differential equation of vibration refers to

    (a) elastic coupling   (b) inertia coupling

    (c) mode superposition   (d) both inertia and elastic coupling

5. With consistent mass matrix, the differential equation of vibration refers to

    (a) elastic coupling   (b) inertia coupling

    (c) mode superposition   (d) both inertia and elastic coupling

6. Normalising eigenvector w.r.t. mass matrix is useful in

    (a) mode superposition   (b) evaluating natural frequencies

    (c) frequency response   (d) damped vibrations

7. An unconstrained 2-D frame with 4 nodes has ___ number of zero frequencies

   (a) 1   (b) 2   (c) 3   (d) 6

8. A 4-noded cantilever gives ___ number of frequencies

   (a) 3   (b) 4   (c) 6   (d) 9

9. A 3-noded simply supported beam gives ___ number of frequencies

   (a) 3   (b) 4   (c) 5   (d) 7

10. A natural mode of vibration represents ___ at each node

    (a) absolute displacements

    (b) relative displacements

    (c) proportional displacements

    (d) absolute strain.

CHAPTER 9

STEADY STATE HEAT CONDUCTION

Application of FEM is not limited to structural analysis. Availability of faster computers with large memory have facilitated in generalising mathematical concepts involved in finite element analysis and applying them to many different engineering fields. It is now possible to use the same finite element model of the component for steady state as well as transient thermal analysis, structural analysis due to static loads as well as dynamic loads etc.

The major difference between structural analysis and thermal analysis by FEM is in the number of unknowns. While in a structural analysis, the primary unknowns are *vector displacement components* ranging from 1 to 6 at any node in the model depending on the type of component and loads, thermal analysis deals with a single unknown, *scalar temperature*, at every node in the model.

In many practical situations, thermal load as well as mechanical loads will be simultaneously acting on a component. It is to be understood that *thermal expansion of a component induces stresses only when the expansion is partially or completely constrained.*

9.1 GOVERNING EQUATION

In Cartesian coordinates

Consider a small element (a cube of dimensions dx, dy and dz) in a solid body. The energy balance during time 'dt' can be stated by,

Heat inflow + Heat generated = Heat outflow + Change in internal energy

$$(q_X + q_Y + q_Z) \, dt + q \, (dx \, dy \, dz) \, dt$$

$$= (q_{X+dX} + q_{Y+dY} + q_{Z+dZ}) \, dt + \rho C_p \, dT \, (dx \, dy \, dz)$$

or $\quad [(q_X - q_{X+dX}) + (q_Y - q_{Y+dY}) + (q_Z - q_{Z+dZ})] \, dt + q \, (dx \, dy \, dz) \, dt$

$$= \rho C_p \, dT \, (dx \, dy \, dz)$$

where,   k = Thermal conductivity

              C_p = Specific heat at constant pressure

and   ρ = Mass density

Considering heat flow through the body along X-direction and using ***Fourier's law of heat conduction*** $q = -k.A\left(\dfrac{dT}{dx}\right)$ along each direction,

$$q_{x+dx} = q_x + \left(\frac{\partial q_x}{\partial x}\right) dx$$

$$= q_x - \left(\frac{\partial}{\partial x}\right)\left[k_X\, A_X \left(\frac{\partial T}{\partial x}\right)\right] dx$$

$$= q_x - \left(\frac{\partial}{\partial x}\right)\left[k_X \left(\frac{\partial T}{\partial x}\right)\right] dx\, dy\, dz$$

or $(q_x - q_{x+dx})\, dt = (\partial/\partial x) [\, k_X\, (\partial T/\partial x)\,]\, dx\, dy\, dz\, dt$

Considering similar expressions for q_{Y+dY} and q_{Z+dZ} and dividing all terms by dx dy dz dt

$$\frac{\partial}{\partial x}\left[k_X \left(\frac{\partial T}{\partial x}\right)\right] + \frac{\partial}{\partial y}\left[k_Y \left(\frac{\partial T}{\partial y}\right)\right] + \frac{\partial}{\partial z}\left[k_Z \left(\frac{\partial T}{\partial z}\right)\right] + q = \rho C_p \left(\frac{\partial T}{\partial t}\right)$$

For a homogeneous material, $k_x = k_y = k_z = k$ and hence

$$\frac{\partial^2 T}{\partial x^2} + \frac{\partial^2 T}{\partial y^2} + \frac{\partial^2 T}{\partial z^2} + \left(\frac{q}{k}\right) = \left(\frac{\rho C_p}{k}\right)\left(\frac{\partial T}{\partial t}\right) = \left(\frac{1}{\alpha}\right)\left(\frac{\partial T}{\partial t}\right) \qquad \text{.....(9.1)}$$

where, $\alpha = \dfrac{k}{\rho C_p}$ is called thermal diffusivity

For the steady state condition, the time differential on the right hand side becomes zero.

Then,

$$\frac{\partial^2 T}{\partial x^2} + \frac{\partial^2 T}{\partial y^2} + \frac{\partial^2 T}{\partial z^2} + \left(\frac{q}{k}\right) = 0 \qquad \text{.....(9.2)}$$

i.e., $k \nabla^2 T + q = \nabla \cdot (-k \nabla T) + q = 0$ ....(9.3)

where, $\nabla^2 T = \dfrac{\partial^2 T}{\partial x^2} + \dfrac{\partial^2 T}{\partial y^2} + \dfrac{\partial^2 T}{\partial z^2}$ for 3-D heat conduction

$$= \frac{\partial^2 T}{\partial x^2} + \frac{\partial^2 T}{\partial y^2} \qquad \text{for 2-D heat conduction}$$

$$= \frac{d^2T}{dx^2} \qquad \text{for 1-D heat conduction}$$

Laplace equation $\nabla^2 T = 0$ is a particular case of steady state problem when $q = 0$

Helmholtz equation

In general, a static field variable problem in terms of the *unknown scalar function* θ can be represented by Helmholtz equation, given by

$$\frac{\partial}{\partial x}\left(k_x \frac{\partial \theta}{\partial x}\right) + \frac{\partial}{\partial y}\left(k_y \frac{\partial \theta}{\partial y}\right) + \frac{\partial}{\partial z}\left(k_z \frac{\partial \theta}{\partial z}\right) + \lambda \theta + q = 0$$

This equation represents steady state heat conduction problem when θ represents nodal temperature; k_x, k_y and k_z are thermal conductivities of the material along x, y and z directions; $\lambda = 0$ and q is the heat source or sink. The equation can then be written as

$$\frac{\partial}{\partial x}\left(k_x \frac{\partial T}{\partial x}\right) + \frac{\partial}{\partial y}\left(k_y \frac{\partial T}{\partial z}\right) + \frac{\partial}{\partial z}\left(k_z \frac{\partial T}{\partial z}\right) + q = 0$$

Boundary conditions associated with a thermal analysis are:

- Specified temperature $\qquad\qquad\qquad$ $T = T_0$ $\qquad$ (at $x = x_i$)
- Specified heat flux (insulated boundary) $q = 0$ $\qquad$ (at $x = x_j$)
- Convection heat transfer $\qquad\qquad$ $q = h\,(T_L - T_\infty)$ $\quad$ (at $x = x_k$)
  (on fluid solid interface)

In Cylindrical coordinate system

In the case of solids of revolution, eq. **(9.1)** can be used more conveniently in cylindrical coordinates (r, θ, z coordinates) as given below

$$\frac{\partial^2 T}{\partial r^2} + \left(\frac{1}{r}\right)\frac{\partial T}{\partial r} + \left(\frac{1}{r^2}\right)\frac{\partial^2 T}{\partial \theta^2} + \frac{\partial^2 T}{\partial z^2} + \frac{q}{k} = \left(\frac{1}{\alpha}\right)\left(\frac{\partial T}{\partial t}\right) \qquad \text{....(9.4)}$$

9.2 1-D HEAT CONDUCTION

In this case, temperature is considered along the length of a rod or thickness of wall representing the direction of heat flow through conduction. Therefore, temperature is a function of only one linear dimension, x. Heat conduction in the other two directions is neglected. Steady state equation with no heat source then reduces to

$$\frac{dq}{dx} = \frac{d^2T}{dx^2} = 0$$

where, heat flux $q = -k\left(\dfrac{dT}{dx}\right)$ is the **Fourier's law** with −ve sign indicating reduction of temperature with increasing x.

The geometric model consists of 2-noded elements with heat conduction along the element. Such problems are broadly categorised into three types, depending on the possibility of convection heat transfer along the length of the element.

9.2.1 HEAT CONDUCTION THROUGH A WALL

In this case, conduction across the wall thickness through unit area of cross section is considered and the wall is assumed to have very large dimensions in the other two directions. On the two surfaces of the wall, specified temperature or specified convective heat transfer from the ambient fluid medium form the boundary conditions.

Using the iso-parametric method of derivation of element stiffness matrix, we can now obtain thermal conductivity matrix, designated as $[K_T]$, for an element of length L between nodes 1 and 2. A comparison of the method of deriving element stiffness matrix and element conductivity matrix for a 1-D element is given below for better appreciation.

Thermal conductivity matrix, $[K_T]$	Stiffness matrix, $[K]$
$Q/A = q = -k$ (dT/dx)	$P/A = \sigma = E$ (du/dx)
$T(\xi) = N_1\,T_1 + N_2\,T_2 = [N]^T\,\{T\}$	$u(\xi) = N_1\,u_1 + N_2\,u_2 = [N]^T\,\{q\}$
$dT/dx = [B_T]^T\,\{T\}$	$du/dx = [B]^T\,\{q\}$
$dx = (L/2)\,d\xi$	$dx = (L/2)\,d\xi$
$N_1 = (1 - \xi)/2;\quad N_2 = (1 + \xi)/2$	$N_1 = (1 - \xi)/2;\qquad N_2 = (1 + \xi)/2$
$[K_T] = \int [B_T]^T\,k\,[B_T]\,(L/2)d\xi$ $= \dfrac{k}{L}\begin{bmatrix} 1 & -1 \\ -1 & 1 \end{bmatrix}$	$[K] = \int [B]^T\,E\,[B]\,A\,(L/2)\,d\xi$ $= \dfrac{AE}{L}\begin{bmatrix} 1 & -1 \\ -1 & 1 \end{bmatrix}$

For a 2-element model (Ref. Figure 9.1),

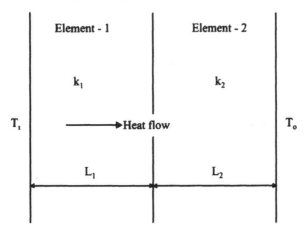

Element - 1   Element - 2

k_1 k_2

T_1 $\longrightarrow$Heat flow T_o

L_1 L_2

FIGURE 9.1 1-D heat conduction through a wall

Temperature T at any point in the element is defined by $T = [N]\ \{T_e\}$

where   [N] are the shape functions, as used for displacement function

and $\{T_e\}$ is the nodal temperature vector

$$\frac{dT}{dx} = \left(\frac{dT}{d\xi}\right)\left(\frac{d\xi}{dx}\right)$$

$$= \left(\frac{d[N]^T}{d\xi}\right) . \{T_e\} . \frac{2}{(x_2 - x_1)}$$

$$= \left(\frac{2}{L}\right) . [-1 \quad 1] . \{T_e\}$$

$$= [B_T]\ \{T_e\}$$

where $[B_T] = \left(\frac{2}{L}\right) . [-1 \quad 1]$

Element conductivity matrix,

$$[K_T]_e = \int_0^L [B_T]^T\ k\ [B_T]\,dx = \int_{-1}^{+1} [B_T]^T\ k[B_T]\left(\frac{dx}{d\xi}\right) d\xi$$

Since $x = N_1.x_1 + N_2.x_2 = (1 - \xi)x_1/2 + (1 + \xi)\ x_2/2$

and
$$\frac{dx}{d\xi} = -\frac{x_1}{2} + \frac{x_2}{2} = \frac{L}{2}$$

$$[K_T]_e = \frac{k}{L}\begin{bmatrix} 1 & -1 \\ -1 & 1 \end{bmatrix}$$

Element heat rate vector, $\{R_T\}_e = \dfrac{QL}{2}\displaystyle\int_{-1}^{+1}[N]^T\,d\xi = \dfrac{QL}{2}\begin{Bmatrix} 1 \\ 1 \end{Bmatrix}$

Assembling conductivity matrices and heat vectors of all the elements of a structure,

$$[K_T]\{T\} = \{R\}$$

Similar to the assembled stiffness matrix, assembled conductivity matrix is also symmetric, banded and singular. Number of these equations is reduced by applying boundary conditions, as detailed below, and the equations are solved for the unknown nodal temperatures.

(i) *Specified temperature T^* (at node m)*

- *Penalty approach:* A large value $C = \max{(K_{IJ})} \times 10^4$ is added to the mth diagonal element (in the mth row and m^{th} column) of the conductivity matrix $[K_T]$. In addition, $C \times T^*$ is added to the element in m^{th} row of $\{R\}$

- *Elimination method:* $K_{i\,m} \times T^*$ is added to the ith element of $\{R\}$, where 'i' ranges from 1 to total number of rows. In addition, m^{th} row and m^{th} column of the conductivity matrix $[K_T]$ and heat vector $\{R\}$ are deleted.

(ii) *Specified convection heat transfer from ambient fluid*

$q = h\,(T_m - T_\infty)$ from node 'm' to the ambient medium at T_∞

The film coefficient value 'h' is added to the element in the mth row and m^{th} column of the conductivity matrix and the value $h\,T_\infty$ is added to the mth element of $\{R\}$.

Example 9.1

Consider a brick wall of thickness 0.3 m, k = 0.7 W/m °K. The inner surface is at 28 °C and the outer surface is exposed to cold air at –15 °C. The heat transfer coefficient associated with the outside surface is 40 W/m² °K. Determine the steady state temperature distribution within the wall and also the heat flux through the wall. Use two elements and obtain the solution.

Solution

Considering a two-element model across half the thickness as shown, heat conduction matrix of the two elements is obtained as

$$[K_T]_1 = [K_T]_2 = \begin{bmatrix} k_1 & -k_1 \\ -k_1 & k_1 \end{bmatrix}$$

where $\qquad k_1 = \dfrac{k}{L} = \dfrac{0.7}{0.15} = \dfrac{14}{3} \ W/°C$

Assembled conductivity matrix is obtained by adding corresponding terms as,

$$[K_T] = \frac{14}{3} \begin{bmatrix} 1 & -1 & 0 \\ -1 & 1+1 & -1 \\ 0 & -1 & 1 \end{bmatrix}$$

We have 2 boundary conditions of constant temperature $T_1 = 28 \ °C$ and convection at free end (node 3) with heat flow q given by

$$q = h \, (T_3 - T_\infty). = 40 \, [T_3 - (-15)] = 40 \, T_3 + 600$$

To include these effects, $40 \, T_3$ is added on the left side while $- 600$ and contribution of T_1 are added on the right side.

Using these conditions, we get modified relations for the 2 unknown temperatures as

$$\left(\frac{14}{3}\right) \begin{bmatrix} 2 & -1 \\ -1 & 1+40 \times 3/14 \end{bmatrix} \begin{Bmatrix} T_2 \\ T_3 \end{Bmatrix} = \begin{Bmatrix} (14/3) \times T_1 \\ -600 \end{Bmatrix}$$

Solving them, we get $\quad T_2 = 7.68 \ °C \quad$ and $\quad T_3 = -12.63 \ °C$

Heat flow Q can be calculated from the first equation as

$$Q = \left(\frac{k}{L}\right)(T_1 - T_2) = \left(\frac{14}{3}\right)(28 - 7.68) = 94.83 \ W/m^2$$

Check : $\qquad Q = \left(\frac{k}{L}\right)(T_2 - T_3) = \left(\frac{14}{3}\right)[7.68 - (-12.63)] = 94.78 \ W/m^2$

$$Q = h(T_3 - T_0) = 40[(-12.63) - (-15)] = 94.8 \ W/m^2$$

Example 9.2

A composite slab consists of three materials with thermal conductivities of 20 W/m °K, 30 W/m °K, 50 W/m °K and thicknesses 0.3 m, 0.15 m and 0.15 m respectively. The outer surface is at 20 °C and the inner surface is exposed to the convective heat transfer coefficient of 25 W/m² °K and a medium at 800 °C. Determine the temperature distribution within the wall.

Solution

Since the plate can be considered infinite, heat transfer can be assumed to be one-dimensional across the thickness. Heat conduction matrices of the three elements covering the three materials are obtained as

$$
\underset{1}{\underline{\qquad \boxed{1} \qquad}} \underset{2}{\mathbf{x}} \underset{2}{\underline{\qquad \boxed{2} \qquad}} \underset{3}{\mathbf{x}} \underset{3}{\underline{\qquad \boxed{3} \qquad}} \underset{4}{\;} \begin{matrix} \text{Free} \\ \text{convection} \end{matrix}
$$

$$
[K_T]_1 = \begin{bmatrix} k_1 & -k_1 \\ -k_1 & k_1 \end{bmatrix} ; \quad [K_T]_2 = \begin{bmatrix} k_2 & -k_2 \\ -k_2 & k_2 \end{bmatrix} ; \quad [K_T]_3 = \begin{bmatrix} k_3 & -k_3 \\ -k_3 & k_3 \end{bmatrix}
$$

where $\quad k_1 = \dfrac{k}{L} = \dfrac{20}{0.3} = 66.7 \text{ W} / °C$

$\quad k_2 = \dfrac{k}{L} = \dfrac{30}{0.15} = 200 = 66.7 \times 3 \ \text{ W} / °C$

$\quad k_3 = \dfrac{k}{L} = \dfrac{50}{0.15} = 333 = 66.7 \times 5 \ \text{ W} / °C$

The equations after assembling conductivity matrix and heat vector are obtained by adding corresponding terms as,

$$[K_T] \{T\} = \{R\}$$

$$
\text{Or} \qquad 66.7 \begin{bmatrix} 1 & -1 & 0 & 0 \\ -1 & 1+3 & -3 & 0 \\ 0 & -3 & 3+5 & -5 \\ 0 & 0 & -5 & 5 \end{bmatrix} \begin{Bmatrix} T_1 \\ T_2 \\ T_3 \\ T_4 \end{Bmatrix} = \begin{Bmatrix} 0 \\ 0 \\ 0 \\ 0 \end{Bmatrix}
$$

We have 2 boundary conditions of constant temperature $T_4 = 20$ °C and convection at free end (node 1) with heat flow q given by

$$q = h \, (T_1 - T_\infty). = 25 \, (T_1 - 800) = 25 \, T_1 - 20000$$

$$= 66.7 \times 0.375 \, T_1 - 20000$$

Penalty method of applying boundary conditions

$$C = \max (K_{ij}) \times 10^4 = 66.7 \times (3 + 5) \times 10^4 = 66.7 \times 80000$$

Then

$$66.7 \begin{bmatrix} 1+0.375 & -1 & 0 & 0 \\ -1 & 1+3 & -3 & 0 \\ 0 & -3 & 3+5 & -5 \\ 0 & 0 & -5 & 5+80000 \end{bmatrix} \begin{Bmatrix} T_1 \\ T_2 \\ T_3 \\ T_4 \end{Bmatrix} = \begin{Bmatrix} 20000 \\ 0 \\ 0 \\ 66.7 \times 80000 \times 20 \end{Bmatrix}$$

Solving them, we get $T_1 = 304.6\,°C$; $T_2 = 119\,°C$; $T_3 = 57.1\,°C$; $T_4 = 20\,°C$.

Elimination method of applying boundary conditions

To include the effect of boundary conditions, $25\,T_1$ is added on the left side while 20000 and contribution of T_4 are added on the right side.

Using these conditions, we get modified relations for the 3 unknown temperatures as

$$66.7 \begin{bmatrix} 1+0.375 & -1 & 0 \\ -1 & 4 & -3 \\ 0 & -3 & 8 \end{bmatrix} \begin{Bmatrix} T_1 \\ T_2 \\ T_3 \end{Bmatrix} = \begin{Bmatrix} 20000 \\ 0 \\ 5\,T_4 \times 66.7 \end{Bmatrix}$$

Simplifying, we get

$$\begin{bmatrix} 1.375 & -1 & 0 \\ -1 & 4 & -3 \\ 0 & -3 & 8 \end{bmatrix} \begin{Bmatrix} T_1 \\ T_2 \\ T_3 \end{Bmatrix} = \begin{Bmatrix} 300 \\ 0 \\ 100 \end{Bmatrix}$$

or $T_1 = 304.6\,°C$

   $T_2 = 119\,°C$

   $T_3 = 57.14\,°C$

Example 9.3

Heat is generated in a large plate ($K = 0.4$ W/m °C) at the rate of 5000 W/m³. The plate is 20 cm thick. Outside surface of the plate is exposed to ambient air at 30 °C with a convective heat transfer coefficient of 20 W/m² °C. Determine the temperature distribution in the wall.

Solution

Since heat transfer through convection takes place on both the sides of the plate, half the plate can be considered for analysis with the mid-plane as the plane of symmetry. Heat transfer can be assumed to be one-dimensional across the thickness while the other dimensions of the plate can be considered infinite. Considering a two-element model across half the thickness as shown, heat conduction matrix and heat generation vector of the two elements are obtained for unit area of cross section.

$$[K_T]_1 = [K_T]_2 = \begin{bmatrix} k_1 & -k_1 \\ -k_1 & k_1 \end{bmatrix}$$

where

$$k_1 = \frac{k}{L} = \frac{0.4}{(5\times10^{-2})} = 8 \ W/°C$$

Assembled conductivity matrix is obtained by adding corresponding terms as,

$$[K] = \begin{bmatrix} 8 & -8 & 0 \\ -8 & 8+8 & -8 \\ 0 & -8 & 8 \end{bmatrix}$$

The nodal load vector consists of heat generation and is given by

$$\{R_1\} = \{R_2\} = \left(\frac{QL}{2}\right)\begin{Bmatrix} 1 \\ 1 \end{Bmatrix} = \begin{Bmatrix} 125 \\ 125 \end{Bmatrix}$$

Since

$$\frac{QL}{2} = 5000 \times \frac{(5\times10^{-2})}{2} = 125 \ W$$

We have the boundary condition of convection at free end (node 3) with heat flow q given by

$$q = h\,(T_3 - T_\infty) = 20\,(T_3 - 30) = 20\,T_3 - 600$$

To include this effect, $20\,T_3$ is added on the left side while 600 is added on the right side.

Using these conditions, we get modified relations for the 3 unknown temperatures as

$$\begin{bmatrix} 8 & -8 & 0 \\ -8 & 8+8 & -8 \\ 0 & -8 & 8+20 \end{bmatrix}\begin{Bmatrix} T_1 \\ T_2 \\ T_3 \end{Bmatrix} = \begin{Bmatrix} 125 \\ 125+125 \\ 125+600 \end{Bmatrix}$$

Solving them, we get $T_1 = 117.5\,°C$; $T_2 = 101.9\,°C$ and $T_3 = 55\,°C$

Caution : Previous type of check on nodal temperatures, based on heat conducted through different elements, is not applicable here, since the quantity of heat flowing through node-1 is increased by generation of heat between nodes 1 and 2 and so heat entering 2nd element at node-2 will be more. Similar inequality exists between heat flowing through node-2 and node-3.

Example 9.4

A composite bar of 3 different materials, rigidly fixed at both the ends, is subjected to a uniform temperature rise of 80 °C. In addition, axial loads are applied at two points on the bar as shown. Determine the displacements, stresses and support reactions.

Solution

The finite element model of this problem consists of 3 axial loaded elements as shown.

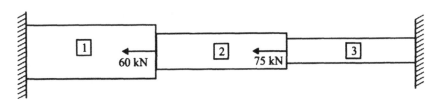

	Section-1	Section-2	Section-3
Material	Bronze	Aluminium	Steel
Area of cross section (mm²)	2400	1200	600
Length (mm)	800	600	400
Modulus of elasticity(GPa)	83	70	200
Coefficient of thermal expansion (/ °C)	18.9×10^{-6}	23×10^{-6}	11.7×10^{-6}

Stiffness matrices of elements 1, 2 and 3 (connected by nodes 1 & 2; 2 & 3 and 3 & 4 respectively) are given by,

$$[K]_1 = \begin{bmatrix} k_1 & -k_1 \\ -k_1 & k_1 \end{bmatrix}; \quad [K]_2 = \begin{bmatrix} k_2 & -k_2 \\ -k_2 & k_2 \end{bmatrix}; \quad [K]_3 = \begin{bmatrix} k_3 & -k_3 \\ -k_3 & k_3 \end{bmatrix}$$

where $k_1 = \dfrac{A_1 E_1}{L_1} = \dfrac{2400 \times 83 \times 10^3}{800} = 249 \times 10^3$

$$k_2 = \frac{A_2 E_2}{L_2} = \frac{1200 \times 70 \times 10^3}{600} = 140 \times 10^3$$

and $k_3 = \dfrac{A_3 E_3}{L_3} = \dfrac{600 \times 200 \times 10^3}{400} = 300 \times 10^3$

Assembled stiffness matrix is obtained by adding corresponding terms as,

$$[K] = \begin{bmatrix} k_1 & -k_1 & 0 & 0 \\ -k_1 & k_1 + k_2 & -k_2 & 0 \\ 0 & -k_2 & k_2 + k_3 & -k_3 \\ 0 & 0 & -k_3 & k_3 \end{bmatrix}$$

$$= 10^3 \begin{bmatrix} 249 & -249 & 0 & 0 \\ -249 & 249+140 & -140 & 0 \\ 0 & -140 & 140+300 & -300 \\ 0 & 0 & -300 & 300 \end{bmatrix}$$

The nodal load vector consists of loads applied at nodes 2 and 3 as well as loads due to constrained expansion. These loads are calculated for each element based on thermal expansion of that element.

Load in element 1

= $A_1 E_1 \alpha_1 \Delta T$ = 2400 × 83 × 10^3 × 18.9 × 10^{-6} × 80 = 301.2 kN

Load in element 2

= $A_2 E_2 \alpha_2 \Delta T$ = 1200 × 70 × 10^3 × 23 × 10^{-6} × 80   = 154.6 kN

Load in element 3

= $A_3 E_3 \alpha_3 \Delta T$ = 600 × 200 × 10^3 × 11.7 × 10^{-6} × 80 = 112.3 kN

Direction of load at the ends of each element should result in expansion of the element. At nodes 2 and 3, mechanical load also should be added to the thermal load from elements on either side.

Load-displacement relations can thus be written now as

$$10^3 \begin{bmatrix} 249 & -249 & 0 & 0 \\ -249 & 389 & -140 & 0 \\ 0 & -140 & 440 & -300 \\ 0 & 0 & -300 & 300 \end{bmatrix} \begin{Bmatrix} u_1 \\ u_2 \\ u_3 \\ u_4 \end{Bmatrix} = 10^3 \begin{Bmatrix} -301.2 + R_1 \\ 301.2 - 154.6 - 60 \\ 154.6 - 112.3 - 75 \\ 1123. + R_4 \end{Bmatrix}$$

Applying the condition that nodes 1 and 4 are fixed and hence displacements at these nodes are zero, load-displacement relations corresponding to the unknown displacements u_2 and u_3 can be written as

$$\begin{bmatrix} 389 & -140 \\ -140 & 440 \end{bmatrix} \begin{Bmatrix} u_2 \\ u_3 \end{Bmatrix} = \begin{Bmatrix} 301.2 - 154.6 - 60 \\ 154.6 - 112.3 - 75 \end{Bmatrix}$$

Solving these two simultaneous equations, we get

$u_2 = 0.222$ mm   and

$u_3 = -0.0012$ mm

Reactions can be calculated from the two deleted equations

$$\begin{bmatrix} 249 & -249 & 0 & 0 \\ 0 & 0 & -300 & 300 \end{bmatrix} \begin{Bmatrix} u_1 \\ u_2 \\ u_3 \\ u_4 \end{Bmatrix} = \begin{Bmatrix} -301.2 + R_1 \\ 112.3 + R_4 \end{Bmatrix}$$

Therefore. $R_1 = 245$ kN   and $R_4 = -111.9$ kN

Stresses in the elements,

$\sigma_1 = E_1 \varepsilon_1 = E_1 [B_1] \{q\}_{1\text{-}2}$

$$= 83 \times 10^3 \begin{bmatrix} -\dfrac{1}{800} & \dfrac{1}{800} \end{bmatrix} \begin{Bmatrix} 0 \\ 0.222 \end{Bmatrix}$$

$= 22.62$ N/mm^2

$\sigma_2 = E_2 \varepsilon_2 = E_2 [B_2] \{q\}_{2\text{-}3}$

$$= 70 \times 10^3 \begin{bmatrix} -\dfrac{1}{600} & \dfrac{1}{600} \end{bmatrix} \begin{Bmatrix} 0.222 \\ -0.0012 \end{Bmatrix}$$

$= -26.04$ N/mm^2

$\sigma_3 = E_3 \varepsilon_3 = E_3 [B_3] \{q\}_{3\text{-}4}$

$$= 200 \times 10^3 \begin{bmatrix} -\dfrac{1}{400} & \dfrac{1}{400} \end{bmatrix} \begin{Bmatrix} -0.0012 \\ 0 \end{Bmatrix}$$

$= 0.6$ N/mm^2

Check : $\sum F = 0$ or $R_1 + R_4 + P_2 + P_3 = 245 - 111.9 - 60 - 75 \approx 0$

9.2.2 HEAT TRANSFER THROUGH A FIN

A fin is of finite lateral dimensions (Fig. 9.2), unlike infinite lateral dimensions of a wall. A rod of small cross sectional area, an axi-symmetric plate around the periphery of an I.C. engine cylinder with heat flow in the radial direction and a flat plate with its ends across the width insulated can be analysed as 1-D fins. The heat transfer is essentially 1-D heat conduction along the length with heat loss through its periphery. Equations relating heat flow to nodal temperatures will therefore include additional matrix of convection heat transfer to represent heat flow Q through periphery of each element. This convection matrix [H] is a function of variable temperature over the element, which can be expressed in terms of nodal temperatures using shape functions, and is a square symmetric matrix of the same order as $[K_T]$. It forms the first (unknown) part 'h. T' of convective heat transfer $h\,(T-T_a)$ where T_a is the ambient temperature, while the known second part 'h T_a' is added to the heat flux vector on the right side of the equation $([K_T]+[H])\,\{T\} = \{R\}$.

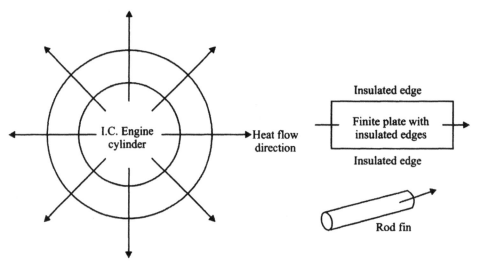

FIGURE 9.2 Different models of heat transfer by conduction and convection

$$\text{where,} \qquad [K_T] = \left(\frac{k\,A}{L}\right)\begin{bmatrix} 1 & -1 \\ -1 & 1 \end{bmatrix}$$

$$\{R\} = \left(\frac{PhT_\infty L}{2}\right)\begin{Bmatrix} 1 \\ 1 \end{Bmatrix}$$

$$[H] = \iint_S h [N]^T [N] dS = \int_0^L h \begin{Bmatrix} \dfrac{1-x}{L} \\ \dfrac{x}{L} \end{Bmatrix} \begin{bmatrix} \dfrac{1-x}{L} & \dfrac{x}{L} \end{bmatrix} P \, dx$$

$$= \frac{P h L}{6} \begin{bmatrix} 2 & 1 \\ 1 & 2 \end{bmatrix}$$

since $\displaystyle \int_0^L h \left(1 - \frac{x}{L}\right)^2 P \, dx = h P \int_0^L \left(1 - \frac{2x}{L} + \frac{x^2}{L^2}\right) dx$

$$= h P \left[\frac{x - x^2}{L} + \frac{x^3}{3L^2} \right]_0^L = \frac{h P L}{3}$$

and $\displaystyle \int_0^L h \left(1 - \frac{x}{L}\right)\left(\frac{x}{L}\right) P \, dx = h P \int_0^L \left(\frac{x}{L} - \frac{x^2}{L^2}\right) dx$

$$= h P \left[\frac{x^2}{2L} - \frac{x^3}{3L^2} \right]_0^L = \frac{h P L}{6}$$

where, P is the perimeter of the cross section of the element or fin

and A = Area of cross section of fin

The coefficients of $[K_T]$, [H] and {R} can all be divided by A. These coefficients are thus modified respectively to $\dfrac{k}{L}$, $\dfrac{PhL}{6A}$ and $\dfrac{PhT_\infty L}{2A}$ in some books.

For a fin of rectangular cross section of width 'w' and thickness 't', (Ref. Fig. 9.3)

$$P = 2(w + t) \approx 2w; \quad A = w.t$$

Then $\quad P/A \approx 2/t$

For a fin of circular section of radius 'r'

$$P = 2\pi r; \quad A = \pi r^2 \text{ and } P/A = 2/r$$

For a tapered fin of length 'L' and rectangular cross section,

$$A(x) = A_i + \frac{(A_j - A_i)x}{L}$$

and $\quad P(x) = P_i \, N_i(x) + P_j \, N_j(x)$

where P_i and P_j are the perimeters at nodes I and J

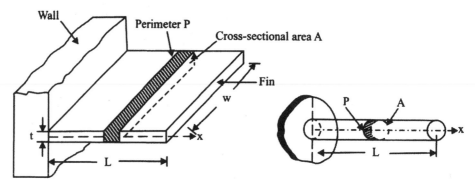

FIGURE 9.3 Rectangular and circular section fins

Following the same procedure as earlier explained and integrating the terms, we get

$$[K_T] = \left(\frac{k \hat{A}}{L}\right)\begin{bmatrix} 1 & -1 \\ -1 & 1 \end{bmatrix}; \quad [H] = \left(\frac{h L}{12}\right)\begin{bmatrix} 3P_i + P_j & P_i + P_j \\ P_i + P_j & P_i + 3P_j \end{bmatrix}$$

where, $\hat{A} = \dfrac{(A_i + A_j)}{2}$

and $\{R\} = \dfrac{h\, T_\infty L}{6}\begin{Bmatrix} 2P_i + P_j \\ P_i + 2P_j \end{Bmatrix}$

Example 9.5

A metallic fin, with thermal conductivity 360 W/m °K, 0.1 cm thick and 10 cm long extends from a plane wall whose temperature is 235 °C. Determine the temperature distribution along the fin if heat is transferred to ambient air at 20 °C with heat transfer coefficient of 9 W/m² °K. Take width of the fin as 1 m.

Solution

Neglecting temperature variation across thickness as well as along the width, conduction can be considered along the length of the fin only while heat loss through convection takes place around the periphery of the flat fin. Thus, using 1-D model, heat conduction and heat convection matrices of the three elements are obtained as

$$[K_T] = \frac{kA}{L}\begin{bmatrix} 1 & -1 \\ -1 & 1 \end{bmatrix}$$

$$[H] = \frac{PhL}{6A}\begin{bmatrix} 2 & 1 \\ 1 & 2 \end{bmatrix} = \frac{hL}{3t}\begin{bmatrix} 2 & 1 \\ 1 & 2 \end{bmatrix}$$

$$\{R\} = \frac{PhT_\infty L}{2A}\begin{Bmatrix} 1 \\ 1 \end{Bmatrix} = \frac{hT_\infty L}{t}\begin{Bmatrix} 1 \\ 1 \end{Bmatrix}$$

For each element, $L = \dfrac{0.1}{3}$ m

$$\frac{k}{L} = \frac{360}{(0.1/3)} = 10800$$

$$\frac{hL}{3t} = \frac{9(0.1/3)}{3 \times 0.001} = 100$$

$$\frac{hT_\infty L}{t} = 9 \times 20 \times \frac{(0.1/3)}{0.001} = 6000$$

Assembling the element matrices, we get $([K] + [H])\,\{T\} = \{R\}$

or

$$\left(10800 \begin{bmatrix} 1 & -1 & 0 & 0 \\ -1 & 1+1 & -1 & 0 \\ 0 & -1 & 1+1 & -1 \\ 0 & 0 & -1 & 1 \end{bmatrix} + 100 \begin{bmatrix} 2 & 1 & 0 & 0 \\ 1 & 2+2 & 1 & 0 \\ 0 & 1 & 2+2 & 1 \\ 0 & 0 & 1 & 2 \end{bmatrix} \right) \begin{Bmatrix} T_1 \\ T_2 \\ T_3 \\ T_4 \end{Bmatrix}$$

$$= 6000 \begin{Bmatrix} 1 \\ 1+1 \\ 1+1 \\ 1 \end{Bmatrix}$$

Substituting the boundary condition of constant temperature $T_1 = 235\ ^\circ C$ and dividing throughout by 100, we get

$$\begin{bmatrix} 220 & -107 & 0 \\ -107 & 220 & -107 \\ 0 & -107 & 110 \end{bmatrix} \begin{Bmatrix} T_2 \\ T_3 \\ T_4 \end{Bmatrix} = \begin{Bmatrix} 120 + 107\,T_1 \\ 120 \\ 60 \end{Bmatrix}$$

Solving these equations, we get

$$T_2 = 207.35\,°C; \quad T_3 = 190.2\,°C \quad \text{and} \quad T_4 = 185.6\,°C$$

Caution :   Previous type of check on nodal temperatures, based on heat conducted through different elements, is not applicable here, since part of the heat entering the fin through node-1 is lost through convection between nodes 1 and 2 and so heat entering 2nd element at node-2 will be less. Similar inequality exists between heat entering node-2 and node-3.

Example 9.6

A metallic fin, with thermal conductivity 70 W/m °K, 1 cm radius and 5 cm long extends from a plane wall whose temperature is 140 °C. Determine the temperature distribution along the fin if heat is transferred to ambient air at 20 °C with heat transfer coefficient of 5 W/m² °K. Take two elements along the fin.

Solution

Neglecting temperature variation across the cross section of the fin, conduction can be considered along the length of the fin only while heat loss through convection takes place around the periphery of the round fin. Thus, using 1-D model, heat conduction and heat convection matrices of the two elements are obtained as

$$[K_T] = \frac{k\,A}{L} \begin{bmatrix} 1 & -1 \\ -1 & 1 \end{bmatrix}$$

$$[H] = \frac{Ph\,L}{6} \begin{bmatrix} 2 & 1 \\ 1 & 2 \end{bmatrix}$$

$$\{R\} = \frac{PhT_\infty L}{6} \begin{Bmatrix} 1 \\ 1 \end{Bmatrix}$$

For each element, $L = \dfrac{0.05}{2} = 0.025\,\text{m}$

$$\frac{k}{L} = 70 \times 10^{-2} \times \pi \times \frac{(0.01)^2}{0.025} = 8.8 \times 10^{-3}$$

$$\frac{PhL}{6} = 2\pi \times (0.01) \times 5 \times 10^{-4} \times \frac{0.025}{6} = 13.1 \times 10^{-8}$$

$$\frac{PhT_\infty L}{2} = 2\pi \times (0.01) \times 5 \times 10^{-4} \times 20 \times \frac{0.025}{2} = 785.7 \times 10^{-8}$$

Assembling the element matrices, we get $([K_T] + [H]) \{T\} = \{R\}$

or

$$\left(8.8 \times 10^{-3} \begin{bmatrix} 1 & -1 & 0 \\ -1 & 1+1 & -1 \\ 0 & -1 & 1 \end{bmatrix} + 13.1 \times 10^{-8} \begin{bmatrix} 2 & 1 & 0 \\ 1 & 2+2 & 1 \\ 0 & 1 & 2 \end{bmatrix} \right) \begin{Bmatrix} T_1 \\ T_2 \\ T_3 \end{Bmatrix}$$

$$= 785 \times 10^{-8} \begin{Bmatrix} 1 \\ 1+1 \\ 1 \end{Bmatrix}$$

Substituting the given boundary condition of constant temperature $T_1 = 140\,°C$ and dividing throughout by 10^{-3}, we get

$$\begin{bmatrix} 17.6 & -8.8 \\ -8.8 & 17.6 \end{bmatrix} \begin{Bmatrix} T_2 \\ T_3 \end{Bmatrix} = \begin{Bmatrix} 1232 \\ 0.00786 \end{Bmatrix}$$

Solving these equations, we get

$$T_2 = 93.3\,°C \quad \text{and} \quad T_3 = 46.67\,°C.$$

9.3   2-D HEAT CONDUCTION IN A PLATE

Heat conduction through a finite plate needs a 2-D model for estimating temperatures at various points. Temperature variation across the thickness of a thin plate at any location is negligible and hence 2-D heat transfer is assumed. Similar to 2-D structural analysis, we use iso-parametric triangular elements for this analysis.

Temperature field within a triangular element is given by

$$T = N_1 T_1 + N_2 T_2 + N_3 T_3 = \xi T_1 + \eta T_2 + (1 - \xi - \eta) T_3$$

and

$$x = N_1 x_1 + N_2 x_2 + N_3 x_3 = \xi x_1 + \eta x_2 + (1 - \xi - \eta) x_3$$

$$y = N_1 y_1 + N_2 y_2 + N_3 y_3 = \xi y_1 + \eta y_2 + (1 - \xi - \eta) y_3$$

$$\begin{Bmatrix} \partial T / \partial \xi \\ \partial T / \partial \eta \end{Bmatrix} = \begin{bmatrix} x_{13} & y_{13} \\ x_{23} & y_{23} \end{bmatrix} \begin{Bmatrix} \partial T / \partial x \\ \partial T / \partial y \end{Bmatrix} = [J] \begin{Bmatrix} \partial T / \partial x \\ \partial T / \partial y \end{Bmatrix} \quad \text{where} \quad y_{ij} = y_i - y_j$$

$$\text{and} \quad x_{ij} = x_i - x_j$$

or $$\begin{Bmatrix} \partial T / \partial x \\ \partial T / \partial y \end{Bmatrix} = [J]^{-1} \begin{Bmatrix} \partial T / \partial \xi \\ \partial T / \partial \eta \end{Bmatrix} = \frac{1}{\text{Det } J} \begin{bmatrix} y_{23} & -y_{13} \\ -x_{23} & x_{13} \end{bmatrix} \begin{bmatrix} 1 & 0 & -1 \\ 0 & 1 & -1 \end{bmatrix} \begin{Bmatrix} T_1 \\ T_2 \\ T_3 \end{Bmatrix}$$

$$= \frac{1}{\text{Det J}} \begin{bmatrix} y_{23} & y_{31} & y_{12} \\ x_{32} & x_{13} & x_{21} \end{bmatrix} \begin{Bmatrix} T_1 \\ T_2 \\ T_3 \end{Bmatrix} = [B]\{T_e\}$$

$$\text{where,} \quad [B] = \frac{1}{\text{Det J}} \begin{bmatrix} y_{23} & y_{31} & y_{12} \\ x_{32} & x_{13} & x_{21} \end{bmatrix}$$

and $\qquad [K_T] = k \, A \, [B]^T \, [B]$

$$\{R\} = \frac{Q A}{3} \begin{Bmatrix} 1 \\ 1 \\ 1 \end{Bmatrix}$$

Convection along any edge I-J, $\quad [H] = \dfrac{h \, L_{I-J}}{6} \begin{bmatrix} 2 & 1 \\ 1 & 2 \end{bmatrix}$

Example 9.7

Two dimensional simplex elements have been used for modelling a heated flat plate. The (x, y) coordinates of nodes i, j and k of an interior element are given by (5,4), (8,6) and (4,8) respectively. If the nodal temperatures are found to be $T_i = 110\ °C$, $T_j = 70\ °C$ and $T_k = 130\ °C$, find

  (i) the temperature gradients inside the element and

  (ii) the temperature at point P located at $(x_P, y_P) = (6,5)$

Solution

A triangular element will have three natural or non-dimensional coordinates N_i, N_j and N_k such that $N_i + N_j + N_k = 1$ or $N_k = 1 - N_i - N_j$

  (a) Temperature gradient inside the element is given by

$$\begin{Bmatrix} \partial T/\partial x \\ \partial T/\partial y \end{Bmatrix} = \frac{1}{\text{Det J}} \begin{bmatrix} y_{jk} & y_{ki} & y_{ij} \\ x_{kj} & x_{ki} & x_{ij} \end{bmatrix} \begin{Bmatrix} T_i \\ T_j \\ T_k \end{Bmatrix}$$

$$= \frac{1}{\text{Det J}} \begin{bmatrix} 6-8 & 8-4 & 4-6 \\ 4-8 & 5-4 & 8-5 \end{bmatrix} \begin{Bmatrix} 110 \\ 70 \\ 130 \end{Bmatrix}$$

$$= \frac{1}{14} \begin{bmatrix} -2 & 4 & -2 \\ -4 & 1 & 3 \end{bmatrix} \begin{Bmatrix} 110 \\ 70 \\ 130 \end{Bmatrix}$$

since $\text{Det } J = x_{ik}\, y_{jk} - x_{jk}\, y_{ik} = (5-4)\,(6-8) - (8-4)\,(4-8) = 14$

$$\therefore \qquad \begin{Bmatrix} \partial T/\partial x \\ \partial T/\partial y \end{Bmatrix} = \left(\frac{1}{14} \right) \begin{Bmatrix} -200 \\ 20 \end{Bmatrix}$$

(b)   To find out temperature at point (x_P, y_P), the shape functions (N_i, N_j, N_k) of that point are calculated .as given below.

$$x_P = N_i\, x_i + N_j\, x_j + N_k\, x_k = N_i\, x_i + N_j\, x_j + (1 - N_i - N_j)\, x_k$$

$$6 = 5\, N_i + 8\, N_j + 4\,(1 - N_i - N_j)$$

$$= 4 + N_i + 4\, N_j \quad \text{or} \quad N_i + 4\, N_j = 2$$

Similarly, $y_P = N_i\, y_i + N_j\, y_j + N_k\, y_k = N_i\, y_i + N_j\, y_j + (1 - N_i - N_j)\, y_k$

$$5 = 4\, N_i + 6\, N_j + 8\,(1 - N_i - N_j)$$

$$= 8 - 4\, N_i - 2\, N_j \quad \text{or} \quad 4\, N_i + 2\, N_j = 3$$

which give $N_i = \dfrac{4}{7}; \;\; N_j = \dfrac{5}{14}; \;\; N_k = \dfrac{1}{14}$

Then temperature at (x_P, y_P) is given by

$$T_P = N_i\, T_i + N_j\, T_j + N_k\, T_k$$

$$= \left(\frac{4}{7} \right)(110) + \left(\frac{5}{14} \right)(70) + \left(\frac{1}{14} \right)(130)$$

$$= \frac{(880 + 350 + 130)}{14}$$

$$= 91.4\,^\circ\text{C}$$

9.4 SUMMARY

- Many engineering problems involve stresses due to change in temperature of the component or surrounding fluid : 1-D conduction problem through composite walls and fins; and 2-D conduction problems through plates.

- To solve for nodal temperature of a finite element model, element conductivity matrix (due to conduction within the component) is calculated for steady state problems, in the same way as element stiffness matrix.

The problem may also involve convection on the boundary or periphery with the ambient fluid.

- Thermal problems and stress analysis problems can be solved with the same finite element model, same distribution of nodes and elements in the component; Thermal problem involves only one nodal unknown temperature unlike 1 to 6 nodal DOF in structural analysis and hence size of the assembled matrix is smaller.

- Thermal problems deal with scalar temperature as the nodal unknowns and, hence, do not require transformation of element matrices from element (local) coordinate system to structure (global) coordinate system.

- Transient problems are solved in multiple time steps by iterative methods and accuracy depends on the selected time step duration.

OBJECTIVE QUESTIONS

1. Conductance matrix is the equivalent of stiffness matrix in
   (a)   thermal analysis   (b)   dynamic analysis
   (c)   fluid flow analysis   (d)   static structural analysis

2. _____ problem is solved through iterative method
   (a)   transient thermal   (b)   steady state thermal
   (c)   structure with thermal loads (d) static structural analysis

3. No. of DOF for a 4-noded quadrilateral thermal element is
   (a)   4   (b)   8   (c)   12   (d)   16

4. No. of DOF for a 3-noded triangular thermal element is
   (a)   3   (b)   6   (c)   9   (d)   12

5. No. of DOF for a 6-noded triangular thermal element is
   (a)   3   (b)   6   (c)   9   (d)   12

6. No. of DOF for a 4-noded tetrahedran thermal element is
   (a)   4   (b)   8   (c)   12   (d)   16

7. No. of DOF for a 8-noded quadrilateral thermal element is
   (a)   4   (b)   8   (c)   12   (d)   16

8. No. of DOF per node in a triangular thermal element is
   (a)   1   (b)   2   (c)   3   (d)   4

9. No. of DOF per node in a quadrilateral thermal element is
   (a)   1   (b)   2   (c)   3   (d)   4

CHAPTER 10

DESIGN VALIDATION AND OTHER TYPES OF ANALYSIS

10.1 COMPLIANCE WITH DESIGN CODES

Every component may not be analysed for displacements and stresses under the influence of external applied loads. Standard components like bearings, springs, bolts,.. are selected from design data books or manufacturer's catalogues. However, when components are to be manufactured to meet specific needs of customers, they need to be analysed. If the equipment in operation pose danger to the people working or living around those equipment, design of all components have to satisfy prescribed design (safety) codes. Different countries prescribe different **design codes** for equipment working in their countries. One of the most popular codes is Boiler & Pressure Vessel Code (Section VIII) for pressure vessels such as boiler drum, steam or gas turbine, valves, heat exchanger, condenser,... formulated by American Society of Mechanical Engineers (ASME). This code or its modified version is followed in many countries.

Many theories have been proposed to explain failure of components when subjected to external loads, based on maximum normal stress, maximum shear stress, maximum strain, maximum strain energy, maximum distortion energy etc.. ASME code is based on the most conservative theory i.e., maximum shear stress theory. According to this, a component fails when the maximum shear stress at any point exceeds prescribed limit for the particular material at the temperature of operation.

Limiting values of all materials are usually based on the values obtained from uni-axial tensile test. As explained in chapter-3, this specimen is subjected to tensile stress σ_x along its length by the application of a tensile load. Its other stress components σ_y and τ_{xy} are equal to zero. By recording the elongation corresponding to different values of load P, until the specimen breaks, yield stress (S_y), maximum tensile stress (S_t), maximum elastic strain, maximum plastic strain and modulus of elasticity (E) are calculated. These values are tabulated in the code for each material, at different temperatures. To obtain material property value at any other intermediate temperature, linear interpolation is carried out from the values at the nearest lower temperature and nearest higher temperature.

ASME code defines **stress intensity**, $S_m = \min (2S_y/3, S_t/3)$ for comparison with the maximum shear stress in the component, with built-in factor of safety. In the uni-axial tensile test, using Mohr's circle, we get maximum shear stress $\tau_{max} = S_m/2$. Maximum shear stress at any point in the component is equal to half of the difference between algebraically largest (σ_1) and algebraically smallest (σ_2) principal normal stresses.

$$(\sigma_1 - \sigma_2)/2 < S_m/2 \quad \text{or} \quad P_m < S_m$$

Thus, stress intensity (P) is defined as twice the maximum shear stress. Different components of stress intensity are calculated, as detailed below, from the three normal stresses and three shear stresses at each point of a component. Assuming that the magnitudes of stress components vary across the cross section of any component, mean stress is defined as the **membrane stress** while the varying part is defined as the **bending stress**. Stress intensity P calculated from the membrane part of the six stress components is called membrane stress intensity P_m while the maximum stress intensity calculated from the bending part (observed on the outermost layers) is called bending stress intensity P_b. These are checked against different limits as given below,

$$P_m < S_m \quad \text{and} \quad P_m \pm P_b < 1.5\, S_m$$

The reason behind specifying two different limits for these components can be easily understood from the following logic (Refer Fig.10.1). In the case of membrane stress, equal stress at every point in the cross-section will lead to simultaneous failure of the entire cross-section. In the case of bending stress, stress varies linearly throughout the cross-section. If the maximum bending stress exceeds the allowable limit, only the outer layer yields without failure of the entire cross section. As the load increases, more and more outer layers start yielding. Finally, at a particular load, the entire cross-section will reach the same stress value, in opposite directions in the top and bottom layers. Thus, the cross-section can withstand larger load in bending mode and hence the allowable limit can be more.

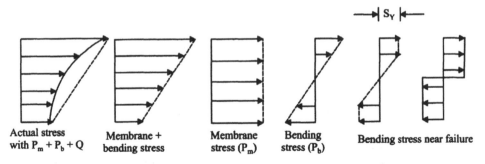

Actual stress with $P_m + P_b + Q$ Membrane + bending stress   Membrane stress (P_m)   Bending stress (P_b)   Bending stress near failure

FIGURE 10.1 Varying influence of membrane and bending stresses across thickness

The membrane stress is again classified as general membrane stress (P_m) applicable over a large area of the component and local membrane stress (P_L) applicable near the areas of structural discontinuity or stress concentration. Here again local yielding results in readjustment of stresses without causing failure of the entire component and thus a higher load can be sustained.

So, $P_L < 1.5\,S_m$

and $P_L \pm P_b < 1.5\,S_m$ .....(10.1)

In some components, stress variation across the cross section may not be linear. In such cases, total stress is considered as the sum of membrane stress, bending stress and **peak stress** (Q). Peak stress is not considered in design checks. It influences only fatigue and creep damages.

Then, $P_m \pm P_b + Q < 3\,S_m$

and $P_L \pm P_b + Q < 3\,S_m$ .....(10.2)

*All the above checks, in addition to checks related to fatigue failure or creep failure wherever applicable, have to be satisfied for validating design of any product. <u>Output of any FEM software includes stresses at various points of a product while code check is based on stresses across some typical cross sections of the product</u>. Hence, the six components of stresses (3 normal stresses and 3 shear stresses) obtained from Finite element analysis at every node point in a continuum structure need to be categorised considering their variation across each critical cross section into membrane(uniform), bending(linearly varying) and peak stress (non-linear part) components for validating the design by such codes. **Stress categorisation** procedure varies with relevant code. An example of typical **stress classification line** (A-B) across the thickness of a shell-nozzle junction and categorisation of total stress across the thickness into membrane, bending and peak stress components is shown in Fig.10.2*

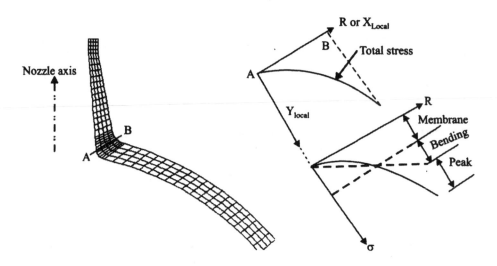

FIGURE 10.2 Categorisation of Stresses across thickness

The procedure is summarised below.

- Each of the six stress components calculated by the software in Cartesian coordinate system is categorised into membrane, bending and peak stress values.

- Principal normal stresses σ_1, σ_2 and σ_3 are calculated separately from the six stress components σ_x, σ_y, σ_z, τ_{xy}, τ_{yz} and τ_{zx} belonging to membrane, bending and peak categories in a 3-D stress analysis.

- Principal normal stresses σ_1, σ_2 and σ_3 are calculated from the four stress components σ_x, σ_y, σ_z (hoop) and τ_{xy} in an axi-symmetric analysis of 3-D component, with the hoop stress σ_z equal to the principal stress σ_3.

- Principal normal stresses σ_1 and σ_2 are calculated from the three stress components σ_x, σ_y and τ_{xy} in a 2-D stress analysis (plane stress or plane strain).

- Absolute maximum difference of any two principal stresses $|\sigma_1 - \sigma_2|$, $|\sigma_2 - \sigma_3|$ or $|\sigma_3 - \sigma_1|$ is calculated for the membrane, bending and peak categories and designated as P_m or P_L, P_b and Q respectively.

Principal normal stresses and stress intensities (P_m or P_L, P_b and Q) have to be calculated separately for membrane, bending and peak stress components. Depending on the type of analysis carried out using any general-purpose software, these values have to be compared with the allowable values of the material at the operating temperature.

Note : This categorisation of stresses is not relevant for discrete structures involving truss, beam, torsion or pipe elements, because stress is either constant across their cross section (truss or pipe elements) or varies linearly across every cross section and is maximum where bending moment or torque is maximum (beam, torsion or pipe elements).

10.2 TRANSIENT HEAT CONDUCTION

Time-dependent heat transfer problems are very common in engineering. A brief explanation is given about this topic. Detailed presentation is not included here, as this topic is outside the scope of syllabus in many universities. The solution follows iterative method in multiple time steps and the accuracy depends on the selected time step duration. The governing equation for transient heat conduction problem is

$$\frac{\partial^2 T}{\partial x^2} + \frac{\partial^2 T}{\partial y^2} + \frac{\partial^2 T}{\partial z^2} = \left(\frac{q}{k}\right) = \left(\frac{\rho C_p}{k}\right)\left(\frac{\partial T}{\partial t}\right) = \left(\frac{1}{\alpha}\right)\left(\frac{\partial T}{\partial t}\right) \quad .....(10.3)$$

where, $\alpha = k/\rho C_p$ is called thermal diffusivity

k = Thermal conductivity

C_p = Specific heat at constant pressure

and ρ = Mass density

In solving this problem, we get an additional matrix, related to the capacity of a material to absorb heat, called *capacitance matrix* [K_C]

$$[K_C] = \rho C_p \iiint [N]^T [N] dV$$

$$= \frac{\rho C_p AL}{6} \begin{bmatrix} 2 & 1 \\ 1 & 2 \end{bmatrix} \text{ for 1-D element of uniform section}$$

$$= \frac{\rho C_p \hat{A} L}{6} \begin{bmatrix} 1 & -1 \\ -1 & 1 \end{bmatrix} \text{ for 1-D element of varying section}$$

where, $\hat{A} = \frac{(A_i + A_j)}{2}$

and $[K_T]\{T\} + [K_C]\{T'\} = \{R\}$

where $\{T'\} = [dT_1/dt \quad dT_2/dt ......]^T$

10.3 BUCKLING OF COLUMNS

The truss element discussed in chapter 4 is assumed to be stable and extend or shorten, due to the tensile or compressive load applied along its axis. This is not always true. Slender columns, subjected to axial compressive load, are found to bend. This phenomenon is called *elastic instability* or *elastic buckling* (implying that the bending vanishes when load is removed) and occurs due to the difficulty of applying load exactly along the axis (without any eccentricity) or due to non-homogeneity of most practical materials (resulting in non-uniform stress distribution across the cross section).

A slender member AB of length 'L' and having hinged ends at A and B, subjected to axial load 'P', bends as shown in Fig.10.3. The bending behaviour depends on the end conditions of the member (free, hinged or fixed) as well as its dimensions.

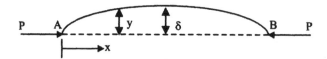

FIGURE 10.3 Buckling of a slender member due to axial compressive load

Governing equation for this deformation is $M / I = E / R$

$$\text{where, } M = -P.y \quad \text{and} \quad 1/R = d^2y/dx^2$$

or
$$\frac{d^2y}{dx^2} + \frac{P.y}{EI} = 0 \qquad \qquad .....(10.5)$$

The general solution is $y = C_1 \sin ax + C_2 \cos ax$ where $a = \sqrt{\dfrac{P}{EI}}$

The constants are evaluated from the end conditions.

$y = 0$ at $x = 0$ leads to $C_2 = 0$

while $y = 0$ at $x = L$ leads to $C_1 \sin aL = 0$

$C_1 = 0$ is a trivial solution; while $\sin aL = 0$ or $L\sqrt{\dfrac{P}{EI}} = n\,\pi$ where 'n' is any integer

For $n = 1$, $P_{cr} = \dfrac{\pi^2 EI}{L^2}$ is called *critical load* or *Euler load*.

This is an eigenvalue problem and the function $y = C_1 \sin ax$ is called an eigen function. Discrete values of buckling load $P = n^2\pi^2EI/L^2$ are called eigenvalues and corresponding mode (or displacement) shapes are eigen vectors, similar to natural modes of vibration, as shown in Fig. 10.4.

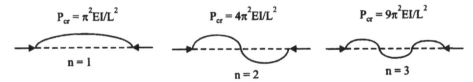

$P_{cr} = \pi^2 EI/L^2$ $P_{cr} = 4\pi^2 EI/L^2$ $P_{cr} = 9\pi^2 EI/L^2$

n = 1   n = 2   n = 3

FIGURE 10.4 Buckling mode shapes

Corresponding buckling stress,

$$\sigma_{cr} = \frac{P_{cr}}{A} = \frac{\pi^2 EI}{L^2 A} = \frac{\pi^2 E}{(L/\rho)^2} \qquad \qquad .....(10.6)$$

where $\rho = \sqrt{\left(\dfrac{I}{A}\right)}$ is the radius of gyration.

For loads smaller than P_{cr}, the deflection C_1 or δ is zero, implying that the column remains straight and shortens due to the applied compressive load.

For a given material of modulus of elasticity 'E', the critical stress σ_{cr} increases as the slenderness ratio (L/ρ) decreases i.e., as the column becomes shorter and thicker. Thus, below a particular value of 'L/ρ', for which $\sigma_{cr} > \sigma_y$, $\sigma_{cr} = \pi^2 E/(L/\rho)^2$ is not reached and the column does not buckle, before the column starts yielding. For mild steel, this limiting slenderness ratio is ≈ 100.

10.4 FATIGUE ANALYSIS

Fatigue life depends on fluctuating stress cycle, identified by stress amplitude over a mean stress at a point in the component, and the number of such cycles the product is designed for. All points of a component may not experience same stress amplitude or same mean stress. Also, a single point may experience stress cycles of different amplitude, for example due to cold start, warm start and hot start or due to load change in a steam turbine casing.

Some components are subjected to more than 10^6 stress cycles, for example due to flexural stress cycles of rotating components or stresses due to vibration in static components. They fall in the category of high cycle fatigue and are designed for infinite life by limiting stress amplitude below endurance limit (S_E). Components subjected to pressure and temperature fluctuations, such as due to different start-up procedures of power plant components, experience less than 10^6 stress cycles. Such components are designed for finite life, based on the stress amplitude ($>S_E$) at critical points of the component.

For example, a turbine casing or rotor experiences zero stress, when shut down and very high stress during start-up, when it comes in contact with steam at high temperature and pressure. Start-up procedure for a typical 110 MW steam turbine, from cold condition, is given in Fig. 10.5. Change of stress from stress-free cold condition to the high stress value during transient and back to zero stress after shut-down forms one stress cycle, with stress range (S_1) and the turbine is designed for a particular number of such cold starts (n_1) during its design life of about 40-50 years. Similarly other transients like warm start, hot start, load fluctuation due to varying demand between day and night etc.. all give different stress ranges (S_2, S_3,..) and each of them may occur for a different number of cycles (n_2, n_3, ..).

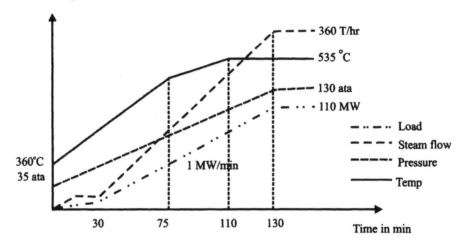

FIGURE 10.5 Typical cold start diagram of a 110MW thermal power plant

The stress range will vary at different points of the structure, for the same _transient_. Corresponding to each stress amplitude, number of cycles to cause fatigue failure (N_i) is read from Stress amplitude (S) Vs Number of cycles (N) curve of the material (Refer Fig.10.6, for a typical curve), assuming it acts alone. Due to large scatter in the experimental data of S-N values from uni-axial fluctuating load test, a factor of 2 on stress and 20 on number of cycles is used in some codes, while preparing S-N curve of a material. Then, fatigue usage fraction 'U' is obtained from the most widely used model (Palmgren-Miner hypothesis, popularly known as **_Miner's rule_**), for validating safe design of a component.

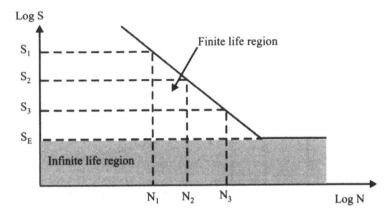

FIGURE 10.6 Typical S-N curve of a material

Fatigue usage fraction, $U_F = \sum_{i=1}^{m} \left(\dfrac{n_i}{N_i} \right) < 1$ .....(10.7)

The finite element analysis gives us 6 stress components (3 normal stresses and 3 shear stresses) at each node point of a 3-D continuum structure or 4 stress components (3 normal stresses along axial, radial and hoop directions and 1 shear stress) in an axisymmetric analysis, *while S-N curve of a material is given for one particular stress*. Usually vonMises stress or maximum principal stress difference is used with the S-N curve. This stress value is obtained at each point and at each time step of the transient. Stress amplitude (S_i), half of stress range, is calculated over the complete transient, consisting of many time steps. Corresponding to this stress amplitude, number of cycles to failure (N_i) is obtained from the S-N curve of the material and fatigue usage fraction calculated to validate the design.

10.5 CREEP ANALYSIS

Creep is a phenomenon in which a component, stressed well within its yield point by the applied loads, yields when the load is applied for a prolonged period. A constant stress state for a prolonged period is naturally feasible when the equipment is operating in steady state condition. The time in which the material starts yielding, at a particular stress level, is called *rupture time* (Refer Fig. 10.7). Creep is prominently observed in equipment operating at elevated temperatures.

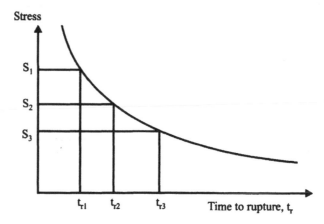

FIGURE 10.7 Typical Stress vs. rupture time curve of a material

If a structure, during its life time, works for periods of time (t_i) at different stress levels, life fraction rule (similar to Miner's rule) states that

$$\text{Creep usage fraction, } U_C = \sum_{i=1}^{m} \left(\frac{t_i}{t_{ri}} \right) < 1 \qquad .....(10.8)$$

The finite element analysis gives us 6 stress components (3 normal stresses and 3 shear stresses) at each node point of a 3-D continuum structure or 4 stress components (3 normal stresses along axial, radial and hoop directions and 1 shear stress) in an axisymmetric analysis, *while S-t, curve of a material is given for one particular stress*. Usually vonMises stress or maximum principal stress difference is used with the S-t, curve. This prolonged duration stress (S_i) is calculated at each point during steady state operation of the component. Corresponding to this stress value and operating temperature, rupture time (t_{ri}) is obtained for the particular material and the creep usage factor calculated to validate the design.

Cumulative damage due to fatigue and creep

There are components, as in a hydro turbine, which are subjected to fatigue only and creep at the operating temperature is insignificant. There are also components, as in a steam turbine, where *fatigue due to load transients* and *creep due to prolonged steady load operation at high operating temperature* are significant. It is well known that fatigue damage and creep damage are inter-related; but, the interaction effect is not well understood. Hence, some design codes suggest calculation of cumulative usage as the algebraic sum of damages due to fatigue and creep. Thus,

Cumulative usage fraction, $U = \sum_{i=1}^{m}\left(\dfrac{n_i}{n_i}\right) + \sum_{j=1}^{k}\left(\dfrac{t_i}{t_r}\right)$ .....**(10.9)**

and it is usually limited to ≈ 0.5 to account for the unknown interaction effects.

10.6 DAMPED FREE VIBRATION

Every structure, when excited with some force, vibrates with a particular frequency and amplitude. In most cases, amplitude of vibration reduces with passage of time and finally reaches zero value. This is called damping – inherent structural damping (usually considered as a function of mass and stiffness) or external damping with spring and/or viscous dashpot.

In a single degree of freedom system,

$$k\,u(t) + c\,\acute{u}(t) + m\,\ddot{u}(t) = 0 \qquad\qquad .....(10.10)$$

where $\acute{u} = \dfrac{du}{dt}$ is the velocity at that point

   c = damping coefficient, which is usually a function of velocity

and $\ddot{u} = \dfrac{d\acute{u}}{dt}$ or $\dfrac{d^2u}{dt^2}$ is the acceleration at that point

A solution of the form $u = ae^{\lambda t}$ will satisfy the equation.

Then, $m\,\lambda^2 + c\,\lambda + k = 0$ is the characteristic equation of the governing equation and its roots are

$$\lambda_{1,2} = \dfrac{-c \pm \sqrt{c^2 - 4\,km}}{2m} \quad \text{or} \quad -\left(\dfrac{c}{2m}\right) \pm \sqrt{\left(\dfrac{c}{2m}\right)^2 - \left(\dfrac{k}{m}\right)}$$

The type of motion depends on the nature of the roots or on the value of $(c/2m)^2 - (k/m)$

(i)   If $\left(\dfrac{c}{2m}\right)^2 > \left(\dfrac{k}{m}\right)$ or $c > 2\sqrt{km}$, the roots λ_1 and λ_2 are real and negative.

Then, $u(t)$ decays as a function of time and *no vibration occurs*

(ii) If $\left(\dfrac{c}{2m}\right)^2 < \left(\dfrac{k}{m}\right)$ or $c < 2\sqrt{km}$, the roots λ_1 and λ_2 are a pair of complex conjugate solutions. The real part represents exponential decay of the amplitude of vibration and imaginary part represents oscillatory part of motion with a *damped free vibration* frequency given by $\sqrt{(k/m)-(c/2m)^2}$.

(iii) $\left(\dfrac{c}{2m}\right)^2 = \left(\dfrac{k}{m}\right)$ or $c = 2\sqrt{km}$, is called *critical damping*

Any linear combination of these roots is also a solution. Thus, in general,

$$u = a_1\, e\lambda_1 t + a_2\, e\lambda_2 t \quad \text{is also a solution}$$

By assuming sinusoidal function for u(t) and expressing ú(t) and ü(t) in terms of u(t), the above equation can be rewritten as

$$(k + b\,c + b^2\,m)\, u(t) = 0$$

In a multi-degree of freedom system, the stiffness, damping and mass terms take the form of matrices and the governing equation can be expressed in the matrix form as

$$([K] + b\,[C] + b^2\,[M]\,)\,\{u(t)\} = \{0\} \qquad\qquad .....(10.11)$$

where, structural damping is assumed to be a function of mass and stiffness, given by

$$[C] = \alpha\,[M] + \beta\,[K] \quad \text{with } \alpha \text{ and } \beta \text{ being real constants.}$$

The elements C_{ij} of the damping matrix [C] have a physical significance analogous to elements k_{ij} of the stiffness matrix [K]. C_{ij} is equal to the external force required at node I in direction u_i to produce unit velocity at node J in direction u_j with velocities at all other masses zero. Analogous to stiffness matrix, $C_{ij} = C_{ji}$.

Damping force $\{F\}_D = -[C]\,\{\dot{u}\}$.

10.7 FORCED VIBRATION

When a structure is subjected to dynamic (time dependent) loads such as wind load, the displacements, strains and stresses will also vary with time. The solution can be obtained by marching in a series of time steps Δt and evaluating accelerations, velocities and displacements at each step. Modal matrix can be used in the mode superposition method to diagnose mass, damping and stiffness matrices and thus uncouple the equations of motion.

$$([K] + b\,[C] + b^2\,[M])\,\{u(t)\} = \{F(t)\} \qquad\qquad .....(10.12)$$

The calculation of nodal displacements Vs time is called *time history response*.

Sometimes, the dynamic load may not be known in the form of Force Vs time, but as random vibration (such as earthquake) in the form of frequency Vs acceleration. This frequency spectrum is treated as a linear combination of individual acceleration (amplitude) Vs time of specific frequencies. Each acceleration Vs time is equivalent to applying force $F_{ij}(t)$ at node I in J^{th} DOF given by F_{ij} = mass at node I x acceleration in J^{th} DOF. This analysis is called *response spectrum analysis*.

10.8   TORSION OF A NON-CIRCULAR BAR

A bar of circular cross section, subjected to pure torsion, is analysed using the assumption that plane section before applying torsion remains plane after applying the load. i.e. there is no warping of the section or all points on the cross section will have either zero or equal displacements along the axis of the bar. The same is not true with a bar of non-circular cross section. Displacements at different points of a bar of non-circular cross section subjected to torsion are defined by

$$u = -\theta\, zy; \quad v = \theta\, zx \quad \text{and} \quad w = \theta\, \psi(x,y)$$

where $\psi(x,y)$ is the movement of cross section in the axial direction (z-axis) per unit twist and is called *warping function*

and θ is the angle of twist per unit length

For pure torsion, $\varepsilon_x = \varepsilon_y = \varepsilon_z = \gamma_{xy} = 0$

$$\gamma_{xz} = \frac{\partial w}{\partial x} + \frac{\partial u}{\partial z} = \frac{\partial w}{\partial x} - \theta y$$

$$\gamma_{yz} = \frac{\partial w}{\partial y} + \frac{\partial v}{\partial z} = \frac{\partial w}{\partial y} + \theta x$$

Applying Hooke's law, the stress components become $\sigma_x = \sigma_y = \sigma_z = \tau_{xy} = 0$

$$\tau_{xz} = G\gamma_{xz} = G\left[\frac{\partial w}{dx} - \theta y\right]$$

$$\tau_{yz} = G\gamma_{yz} = G\left[\frac{\partial w}{dy} + \theta x\right]$$

The general equation of motion,

$$P_i \rho + \sigma_{ij,j} = \rho \left(\frac{du_i}{dt} \right)$$

simplifies for the case of body force $P_i = 0$ to

$$\frac{\partial \tau_{xz}}{\partial x} + \frac{\partial \tau_{yz}}{\partial y} = 0 \quad \text{since displacement } u_i \text{ is independent of time t}$$

Substituting for the shear stresses, from the above, yields Laplace's equation

$$\frac{\partial^2 w}{\partial x^2} + \frac{\partial^2 w}{\partial y^2} = 0$$

Using $\psi(x,y)$, we get $\dfrac{\partial^2 \psi}{\partial x^2} + \dfrac{\partial^2 \psi}{\partial y^2} = 0$

The stress-free condition of the periphery of the cross section requires that the tangential stress be zero on the lateral surface of the prism. Thus, the boundary condition is of Neumann type,

$$\tau_{xz} \lambda_x + \tau_{yz} \lambda_y = 0$$

where, $\lambda_x = \dfrac{dy}{ds}$ and $\lambda_y = \dfrac{-dx}{ds}$ are the direction cosines of the tangent vector

Substituting the above, we get

$$\left[\frac{\partial \psi}{\partial x} - y \right] \left(\frac{dy}{ds} \right) - \left[\frac{\partial \psi}{\partial y} + x \right] \left(\frac{dx}{ds} \right) = 0$$

Integral of the variational problem is given by

$$I = U = \left(\frac{1}{2} \right) \int_v \sigma_{ij} \, \varepsilon_{ij} \, dV$$

We use finite element idealisation over the cross section and interpolate the unknown solution function $\psi(x,y)$ by a polynomial over each element through nodal values ψ_i. From variational principle,

$$\delta I = 0$$

$$\text{or} \qquad \frac{\partial I}{\partial \psi_i} = 0 \qquad \forall \ i = 1, n$$

This condition leads to a system of linear algebraic equations to be solved for ψ_i in the usual manner.

Alternatively, a hybrid finite element model can be developed by introducing **stress function** $\varphi(x,y)$ interpolated over each element (satisfying equilibrium conditions) and warping function $\psi(x,y)$ specified along inter element boundary (satisfying compatibility conditions). This model gives more accurate stress solutions, particularly in the vicinity of external boundary. The stress components, in terms of the stress function, are given by

$$\tau_{xz} = \frac{\partial \varphi}{dy} \quad ; \quad \tau_{yz} = \frac{\partial \varphi}{dx}$$

Complete restraint of warping of a member, of non-circular cross section and subjected to torque, reduces rotation at the end but introduces axial normal stresses much larger than the torsional shear stress. Even in a simple portal frame, shown below, warping is at least partially restrained at nodes A, B, C and D. Such restraint influences response due to loads normal to plane of the frame. Full or partial restraint of warping is experienced at every joint of a frame and can be included with an additional DOF (rate of twist, $d\theta_x/dx$) at each node. Most commercial software do not include this 7th DOF at a node and warping restraint as well as associated axial stress are ignored.

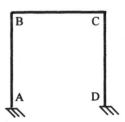

CHAPTER **11**

COMPUTATIONAL FLUID DYNAMICS

11.1 INTRODUCTION

The physics of almost every fluid flow and heat transfer phenomenon is governed by three fundamental principles – mass conservation, momentum conservation (or Newton's second law) and energy conservation – taken together with appropriate initial or boundary conditions. These three principles may be expressed mathematically in most cases through integral or partial differential equations (PDE) whose closed form solutions rarely exist. The ability to seek numerical solutions of these governing equations has led to the development of Computational Fluid Dynamics (CFD).

To obtain numerical solution to the physical variables of the fluid field, various techniques are employed:

- manipulating the defining equations

- dividing the fluid domain into a large number of small cells or control volumes (also called mesh or grid)

- transforming partial derivatives into discrete algebraic forms and

- solving the sets of linear algebraic equations at the grid points

Modern CFD can handle fluid flow associated with other phenomenon such as chemical reactions, multi-phase or free surface problems, phase change (melting, boiling, freezing), mass transfer (dissolution) and radiation heat transfer. A CFD code has three basic components – preprocessor, solver and postprocessor. The solver is the heart of a CFD code and is usually treated as a 'black box' while the other two components provide user/computer interface. Solver is based on one of the three major discrete methods – finite difference method (FDM), finite element method (FEM) or finite volume method (FVM).

Over 90% of CFD codes are based on FDM or FVM. FVM is now very well established and is used in most commercial CFD packages like FLUENT, FLOW-3D.

FEM was initially developed for structural analysis but has been extended to fluid flow problems as it offers the advantage of non-regular grid, capable of simulating complex boundary geometries. Also, the methodology used for describing flow conditions within each cell, though more complex than FDM, have a higher degree of accuracy. FVM draws together best attributes of FDM and FEM. It is capable of simulating complex boundary geometries while utilising relatively straight forward finite difference relationships to represent the governing differential equations. Complete presentation of FDM and FVM are not in the scope of this book. FDM is detailed below as some of these ideas are also used in FVM.

11.2 GOVERNING EQUATIONS

Partial differential equation is an equation involving one or more partial derivatives of an unknown function of two or more independent variables. *Order* of the highest derivative in the equation is the order of the equation. The partial differential equation is termed *linear* if the dependent variable (unknown function) and its partial derivatives are of first degree. The equation is termed *homogeneous* if each term of the equation contains either the dependent variable or one of its derivatives.

Some important linear partial differential equations in fluid dynamics are :

1. 1-D wave equation
$$\frac{\partial^2 u}{\partial t^2} = c^2 \frac{\partial^2 u}{\partial x^2}$$

2. 1-D heat conduction equation
$$\frac{\partial T}{\partial t} = \alpha \left(\frac{\partial^2 u}{\partial x^2} \right) \quad \text{where } \alpha = \frac{k}{(\rho C_p)}$$

3. 2-D Laplace equation
$$\frac{\partial^2 u}{\partial x^2} + \frac{\partial^2 u}{\partial y^2} = \nabla^2 u = 0$$

4. 2-D Poisson equation
$$\frac{\partial^2 u}{\partial x^2} + \frac{\partial^2 u}{\partial y^2} = \nabla^2 u = f(x,y)$$

5. 3-D Laplace equation
$$\frac{\partial^2 u}{\partial x^2} + \frac{\partial^2 u}{\partial y^2} + \frac{\partial^2 u}{\partial z^2} = \nabla^2 u = 0$$

A single partial differential equation can have more than one solution. Unique solution of a partial differential equation, corresponding to a given physical problem, depends on additional information like:

(a) *Boundary conditions:* values of the required solution on the boundary of some domain

(b) *Initial conditions:* values of the solution, when time t is an independent variable, at time = 0.

Superposition : For a homogeneous linear partial differential equation with known solutions u_1 and u_2 , any linear combination of these solutions is also a solution.

i.e. $u = c_1 u_1 + c_2 u_2$ is also a solution,

where c_1 and c_2 are constants.

Any PDE which is linear in the highest derivative is termed as *quasilinear.*

Ex : $A u_{xx} + B u_{xy} + C u_{yy} = F(x, y, u, u_x, u_y)$ .....(11.1)

where $u_x = \dfrac{\partial u}{\partial x}$; $u_y = \dfrac{\partial u}{\partial y}$; $u_{xy} = \dfrac{\partial^2 u}{\partial x\, \partial y}$

Using $du_x = \left(\dfrac{\partial u_x}{\partial x}\right).dx + \left(\dfrac{\partial u_x}{\partial y}\right).dy = \left(\dfrac{\partial^2 u}{\partial x^2}\right).dx + \left(\dfrac{\partial^2 u}{\partial x.\partial y}\right).dy$

$= u_{xx}\, dx + u_{xy}\, dy$

and $du_y = \left(\dfrac{\partial u_y}{\partial x}\right).dx + \left(\dfrac{\partial u_y}{\partial y}\right).dy = \left(\dfrac{\partial^2 u}{\partial x.\partial y}\right).dx + \left(\dfrac{\partial^2 u}{\partial y^2}\right).dy$

$= u_{xy}\, dx + u_{yy}\, dy$

eq. (11.1) can be rewritten as $\begin{bmatrix} A & B & C \\ dx & dy & 0 \\ 0 & dx & dy \end{bmatrix} \begin{Bmatrix} u_{xx} \\ u_{xy} \\ u_{yy} \end{Bmatrix} = \begin{Bmatrix} F \\ du_x \\ du_y \end{Bmatrix}$

For a non-trivial solution,

$A\,(dy)^2 - B\,(dx)\,(dy) + C\,(dx)^2 = 0$ or $A\left(\dfrac{dy}{dx}\right)^2 - B\left(\dfrac{dy}{dx}\right) + C = 0$

This can be expressed in matrix form as

$$\begin{vmatrix} A & B & C \\ dx & dy & 0 \\ 0 & dx & dy \end{vmatrix} = 0 \qquad\qquad .....(11.2)$$

which gives $\dfrac{dy}{dx} = \dfrac{\left[B \pm \sqrt{B^2 - 4AC} \right]}{2A}$

Characteristic curves, represented by this equation, can be real or imaginary depending on the value of $(B^2 - 4AC)$.

Partial differential equations (PDEs) (in particular, governing equations in fluid dynamics) are classified into three categories:

- Elliptic, when $B^2 - 4AC < 0$

  [Laplace equation, $\dfrac{\partial^2 u}{\partial x^2} + \dfrac{\partial^2 u}{\partial y^2} = 0$

  Poisson equation, $\dfrac{\delta^2 u}{\delta x^2} + \dfrac{\delta^2 u}{\delta y^2} = f(x, y)$]

- Parabolic, when $B^2 - 4AC = 0$

  [Heat equation, $\dfrac{\partial T}{\partial t} = \alpha \left(\dfrac{\partial^2 u}{\partial x^2} \right)$ and $\alpha > 0$]

- Hyperbolic, when $B^2 - 4AC > 0$

  [2nd order Wave equation, $\dfrac{\partial^2 u}{\partial t^2} = c^2 \dfrac{\partial^2 u}{\partial x^2}$]

11.3 FINITE DIFFERENCE METHOD (FDM)

Basic idea of FDM is that derivatives in differential equations are written in terms of discrete quantities of dependent and independent variables, resulting in simultaneous algebraic equations with all unknowns prescribed at discrete mesh points covering the entire domain. Appropriate types of differencing schemes and suitable methods of solution are chosen, depending on the particular physics of the flows which may include:

- inviscid or viscous flow
- incompressible or compressible flow
- irrotational or rotational flow

- laminar or turbulent flow
- subsonic or transonic or supersonic or hypersonic flow etc.

FDM utilises a time distance grid of nodes and a truncated Taylor series approach to determine the conditions at any particular node one time-step in the future based on the conditions at adjacent nodes at the current time.

There are two schemes of using Finite Difference type approximations to convert the governing Partial Differential Equation into an algebraic format – Explicit and Implicit. In the *explicit method,* required value of variable at one time-step in the future is calculated from known current values. It requires selection of a very small time-step based on the grid size and hence takes more computer time. *Implicit methods* allow arbitrarily large time-step thus reducing computer time required for solution. However, these methods do not give direct solution and require iterative solution which sometimes may lead to convergence problems (for control volumes with large aspect ratios). Also, these methods are not accurate for convective processes.

11.4 ELLIPTIC EQUATIONS (OR BOUNDARY VALUE PROBLEMS)

First boundary value (*Dirichlet*) problem –

   u is prescribed on the boundary curve C of region R

Second boundary value (*Neumann*) problem –

$$u_n = \frac{\partial u}{\partial n}$$ (normal derivative of u) is prescribed on the boundary curve C of

region R

   Third boundary value (*Mixed*) problem –

   u is prescribed on a part of the boundary curve C and u_n on the remaining part of C of region R

   C is usually a closed curve or sometimes consists of two or more such curves.

   To obtain numerical solution, partial derivatives are replaced by difference quotients. Using Taylor's expansion,

$$u(x + h, y) = u(x, y) + hu_x(x, y) + \left(\frac{h^2}{2}\right)u_{xx}(x, y) + \left(\frac{h^3}{6}\right)u_{xxx}(x, y) + ... \quad ..(11.3)$$

Similarly,

$$u(x - h, y) = u(x, y) - hu_x(x, y) + \left(\frac{h^2}{2}\right)u_{xx}(x, y) - \left(\frac{h^3}{6}\right)u_{xxx}(x, y) + ... \quad ..(11.4)$$

Subtracting eq. (4) from eq. (3) and neglecting higher order terms in h,

$$u_x(x,y) \approx \frac{[u(x+h,y) - u(x-h,y)]}{2h} \qquad .....(11.5)$$

Similarly, $u_y(x,y) \approx \dfrac{[u(x,y+k) - u(x,y-k)]}{2k}$ .....(11.6)

These are called ***central difference formulae*** for the first derivative.

On the same lines, two other forms can also be derived based on Taylor series expansion at some other points

$$u_x(x,y) \approx \frac{[u(x+y,h) - u(x,y)]}{h} \quad \text{or } \textbf{\textit{Forward difference formula}} \quad .....(11.7)$$

$$u_x(x,y) \approx \frac{[u(x,y) - u(x-h,y)]}{h} \quad \text{or } \textbf{\textit{Backward difference formula}} \quad .....(11.8)$$

Second derivative can also be expressed, in a similar way, as

$$u_{xx}(x,y) \approx \frac{[u_x(x+h,y) - u(x-h,y)]}{2h} \quad \text{or} \quad \frac{[u_x(x+h,y) - u_x(x,y)]}{h}$$

$$\approx \frac{[u(x+h,y) - 2u(x,y) + u(x-h,y)]}{h^2} \qquad .....(11.9)$$

$$u_{yy}(x,y) \approx \frac{[u_y(x,y+k) - u_y(x,y-k)]}{2k} \quad \text{or} \quad \frac{[u_y(x,y+k) - u_y(x,y)]}{k}$$

$$\approx \frac{[u(x,y+k) - 2u(x,y) + u(x,y+k)]}{k^2} \qquad .....(11.10)$$

These are called central difference formulae for the second derivative.

Substituting eq. 11.9 and eq.11.10 in Laplace equation and using h = k,

$$\nabla^2 u = \frac{\partial^2 u}{\partial x^2} + \frac{\partial^2 u}{\partial y^2} = u(x+h, y) + u(x-h, y)$$

$$+ u(x, y+k) + u(x, y-k) - 4u(x, y) = 0 \quad .....(11.11)$$

or u at (x, y) equals the mean of values of u at the four neighbouring mesh points.

This is called ***5-point regular operator***

Laplace equation can also be expressed using some other difference formulae as

$$\nabla^2 u = u(x + h, y + k) + u(x - h, y + k) + u(x + h, y - k) +$$

$$u(x - h, y - k) - 4u(x, y) = 0 \qquad .....(11.12)$$

This is called **5-point shift operator**

Combining eq.11.11 & eq.11.12, we get **9-point formula**, given by

$$u(x + h, y) + u(x - h, y) + u(x, y + k) + u(x, y + k) +$$

$$u(x + h, y + k) + u(x - h, y + k) + u(x + h, y - k) +$$

$$u(x - h, y - k) - 8u(x, y) = 0 \qquad .....(11.13)$$

The coverage of 5-point operators and 9-point operator are shown below on the model,

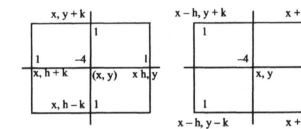 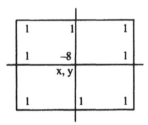

5-point regular operator   5-point shift operator   9-point formula

Similarly, Poisson equation can be rewritten in central difference form as

$$u(x + h, y) + u(x - h, y) + u(x, y + k) + u(x, y - k)$$

$$- 4u(x, y) = h^2 f(x, y) \qquad .....(11.14)$$

Use of polar coordinates (r-θ) may be more convenient with some geometries. For those situations, **Laplace equation in polar coordinates** can be obtained by substituting

$$\cos \theta = \frac{x}{r}; \quad \sin \theta = \frac{y}{r}; \quad \tan \theta = \frac{y}{x} \quad \text{and} \quad r^2 = x^2 + y^2$$

in $\qquad \nabla^2 u = \dfrac{\partial^2 u}{\partial r^2} + \left(\dfrac{1}{r}\right)\dfrac{\partial u}{\partial r} + \left(\dfrac{1}{r^2}\right) \cdot \dfrac{\partial^2 u}{\partial \theta^2} = 0 \qquad .....(11.15)$

where
$$\left(\frac{\partial u}{\partial r}\right)_i = \frac{(u_{i+2} - u_{i+4})}{2\nabla r}$$

$$\left(\frac{\partial^2 u}{\partial r^2}\right)_i = \frac{(u_{i+2} - 2u_i + u_{i+4})}{(\nabla r)^2}$$

$$\left(\frac{\partial^2 u}{\partial \theta^2}\right)_i = \frac{(u_{i+1} - 2u_i + u_{i+3})}{(\nabla \theta)^2}$$

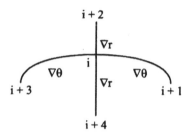

(a) Numerical solution of Dirichlet problem

A grid consisting of equidistant horizontal and vertical straight lines of distance h is introduced in the given region R. Then, the unknown values of u at each of the grid intersection points or mesh points are related to the neighbouring points. This yields a system of linear algebraic equations. The coefficients of the system form a sparse matrix. It can be changed to a band matrix by numbering mesh points in such a way that all non-zero elements are arranged around the principal diagonal. The system of linear algebraic equations can be solved by Gauss elimination method or Gauss-Siedel iteration method.

Alternating Direction Implicit (ADI) method uses Equation (11.9) and rewritten as

$$u(x + h, y) - 4u(x, y) + u(x - h, y) = - u(x, y + k) - u(x, y - k)$$

or $u_{i+1,j} - 4u_{i,j} + u_{i-1,j} = - u_{i, j+1} - u_{i, j-1}$ .....(11.16)

and $u(x, y + k) - 4u(x, y) + u(x, y - k) = - u(x + h, y) - u(x - h, y)$

$u_{i, j+1} - 4u_{i, j} + u_{i, j-1} = - u_{i+1, j} - u_{i-1, j}$ .....(11.17)

Solution is obtained by iteration method from an arbitrary starting value u_{ij} at all mesh points. Eq.(11.11) for a fixed row j, gives a system of n linear equations, corresponding to n columns, in n unknowns which can be solved for u_{ij} by Gauss elimination. This can be repeated, row by row, for all the rows. In the next step, Eq.(11.12) for a fixed column i, gives a system of m linear equations, corresponding to m rows, in m unknowns which can be solved for u_{ij} by Gauss elimination. This can be repeated, column by column, for all the columns.

Convergence can be improved by modifying eqs. 11.11 and 11.12 as

$$u_{i+1,j} - (2 + p)u_{i,j} + u_{i-1,j} = -u_{i,j+1} + (2 - p)u_{i,j} - u_{i,j-1} \quad .....(11.18)$$

$$u_{i,j+1} - (2+p)u_{i,j} + u_{i,j-1} = -u_{i+1,j} + (2-p)u_{i,j} - u_{i-1,j} \quad .....(11.19)$$

where p is a positive number

Example 11.1

A rectangular plate is subjected to temperatures on the boundary as shown. Find the temperature distribution inside the plate.

Solution

Let us divide the plate into a 3 × 3 mesh, for ease of calculation, as shown

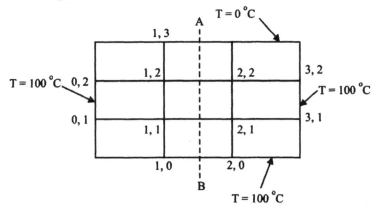

(a)   *Gauss elimination method*

Applying eq.(11.9) at (1,1) $-4 T_{1,1} + T_{2,1} + T_{1,2} = -(T_{1,0} + T_{0,1}) = -200$

Applying eq.(11.9) at (2,1) $T_{1,1} - 4 T_{2,1} + T_{2,2} = -(T_{2,0} + T_{3,1}) = -200$

Applying eq.(11.9) at (1,2) $T_{1,1} - 4T_{1,2} + T_{2,2} = -(T_{1,3} + T_{0,2}) = -100$

Applying eq.(11.9) at (2,2) $T_{2,1} + T_{1,2} - 4T_{2,2} = -(T_{3,2} + T_{2,3}) = -100$

Solving these 4 simultaneous equations, we get

$$T_{1,1} = T_{2,1} = 87.5 \, ^\circ C; \quad T_{1,2} = T_{2,2} = 62.5 \, ^\circ C$$

Note that the results also satisfy symmetry about the center line AB

(b) *Gauss-Siedel iteration method or Liebmann's method*

Starting with an initial approximation of $T_{1,1} = T_{2,1} = T_{1,2} = T_{2,2} = 100$, and substituting the most recent values on the right side of the modified equations, we get improved values for the unknowns on the left side, after each iteration.

$-4\,T_{1,1} + T_{2,1} + T_{1,2} + 4_{T2,2} = -200$ or $T_{1,1} = (T_{2,1} + T_{1,2} + 200)\,/\,4$

$T_{1,1} - 4\,T_{2,1} \qquad + T_{2,2} = -200$ or $T_{2,1} = (T_{1,1} + T_{2,2} + 200)\,/\,4$

$T_{1,1} \qquad - 4T_{1,2} + T_{2,2} = -100$ or $T_{1,2} = (T_{1,1} + T_{2,2} + 100)\,/\,4$

$\qquad T_{2,1} + T_{1,2} - 4T_{2,2} = -100$ or $T_{2,2} = (T_{2,1} + T_{1,2} + 100)\,/\,4$

Iteration	$T_{1,1}$	$T_{2,1}$	$T_{1,2}$	$T_{2,2}$
1	100	100	75	68.75
2	93.75	90.62	65.62	64.06
3	89.06	88.28	63.28	62.89
4	87.89	87.94	62.94	62.72
5	87.72	87.61	62.61	62.55
6	87.55	87.52	62.52	62.51
7	87.51	87.50	62.50	62.50
8	87.50	87.50	62.50	62.50

(c) *Alternating Direction Implicit (ADI) method*

Let us start with initial approximate solution of

$$T_{1,1} = T_{2,1} = T_{1,2} = T_{2,2} = 100$$

Iteration number is not indicated in the following equations for the terms whose values are specified (on the boundary and hence, constant).

Step 1 : Using $T_{i+1,j} - 4T_{i,j} + T_{i-1,j} = -\,T_{i,j+1} - T_{i,j-1}$ with different rows and columns,

For $j = 1$ and for $i = 1$, $T_{2,1}^{(1)} - 4T_{1,1}^{(1)} + T_{0,1} = -\,T_{1,0} - T_{1,2}^{(0)}$

for $i = 2$, $T_{3,1} - 4T_{2,1}^{(1)} + T_{1,1}^{(1)} = -\,T_{2,0} - T_{2,2}^{(0)}$

Solution of these two simultaneous equations in two unknowns give,

$$T_{1,1} = T_{2,1} = 100$$

For $j = 2$ and for $i = 1$, $T_{2,2}^{(1)} - 4T_{1,2}^{(1)} + T_{0,2} = -\,T_{1,1}^{(0)} - T_{1,3}$

for $i = 2$, $T_{3,2} - 4T_{2,2}^{(1)} + T_{1,2}^{(1)} = -\,T_{2,1}^{(0)} - T_{2,2}$

Solution of these two simultaneous equations in two unknowns give,

$$T_{1,2} = T_{2,2} = 66.67$$

Step 2 : Using $T_{i,j+1} - 4T_{i,j} + T_{i,j-1} = -T_{i+1,j} - T_{i-1,j}$ with different rows and columns,

For $i = 1$ and for $j = 1$, $T_{1,2}^{(2)} - 4T_{1,1}^{(2)} + T_{1,0} = -T_{0,1} - T_{2,1}^{(1)}$

for $j = 2$, $T_{1,3} - 4T_{1,2}^{(2)} + T_{1,1}^{(2)} = -T_{0,2} - T_{2,2}^{(1)}$

Solution of these two simultaneous equations give,

$$T_{1,1} = 91.11; \qquad T_{1,2} = 64.44$$

For $i = 2$ and for $j = 1$, $T_{2,2}^{(2)} - 4T_{2,1}^{(2)} + T_{2,0} = -T_{1,1}^{(1)} - T_{3,1}$

for $j = 2$, $T_{2,3} - 4T_{2,2}^{(2)} + T_{2,1}^{(2)} = -T_{1,2}^{(1)} - T_{3,2}$

Solution of these two simultaneous equations give,

$$T_{2,1} = 91.11; \qquad T_{2,2} = 64.44$$

Steps 1 & 2 are repeated until reasonably accurate solution for the unknowns is obtained

(b)   Numerical solution of Neumann problem and Mixed problem

In order to take into effect the values of normal derivatives specified on the entire (or part of the) boundary, the region is extended to some imaginary mesh points outside the given region and central difference formula for the derivative on the boundary is used to express the unknown value at the imaginary mesh point in terms of the values on the actual geometry. The simultaneous equations so obtained are solved by one of the standard numerical method for computing the required solution.

Example 11.2

Solve the equation $\dfrac{\partial^2 u}{\partial x^2} + \dfrac{\partial^2 u}{\partial y^2} = \nabla^2 u = f(x, y) = 12\,xy$ for a rectangular plate of 1cm × 1.5 cm with boundary conditions $u = 0$ on $x = 0$; $u = 0$ on $y = 0$; $u = 3y^3$ on $x = 1.5$ and $u_n = 6x$ on $y = 1$.

Solution :

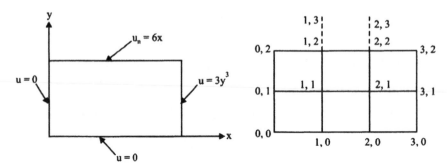

Plate of 1 cm × 1.5 cm is divided into a mesh of 2 × 3. So each cell is of 0.5 cm × 0.5 cm.

or $h = 0.5$ and $h^2 f(x, y) = \dfrac{12\,xy}{4} = 3\,xy$

From the given boundary condition $u = 3y^3$, $u_{3,0} = 0$; $u_{3,1} = \dfrac{3}{8} = 0.375$; $u_{3,2} = 3$

From the given boundary condition $u_n = 6x$,

at $u_{1,2} = 6 \times 0.5 = 3$

and at $u_{2,2} = 6 \times 1 = 6$

Using central difference formula, $\dfrac{\partial u_{1,2}}{\partial y} = \dfrac{(u_{1,3} - u_{1,1})}{2h} = 3$

or $u_{1,3} = u_{1,1} + 3$ .....(11.20)

Similarly, $\dfrac{\partial u_{2,2}}{\partial y} = \dfrac{(u_{2,3} - u_{2,1})}{2h} = 6$

or $u_{2,3} = u_{2,1} + 6$ .....(11.21)

Using 5-point regular operator and retaining only unknowns on the left side,

at   (1,1), $u_{2,1} + u_{1,2} - 4\,u_{1,1} = 3\,xy - u_{1,0} - u_{0,1}$

$\qquad\qquad\qquad\qquad = 3 \times 0.5 \times 0.5 - 0 - 0 = 0.75$ .....(11.22)

and at (2,1), $u_{1,1} + u_{2,2} - 4\,u_{2,1} = 3\,xy - u_{2,0} - u_{3,1} = 3 \times 1 \times 0.5 - 0 - 0.375$

$\qquad\qquad\qquad\qquad = 1.125$ .....(11.23)

To use normal derivative as the boundary condition, 5-point operator with imaginary points in the extended region is applied at (1,2) and (2,2) as follows:

At $(1, 2)$, $u_{1,1} + u_{2,2} + u_{1,3} - 4\,u_{1,2} = 3\,xy - u_{0,2} = 3 \times 0.5 \times 1 - 0 = 1.5$

or substituting $u_{1,3} = u_{1,1} + 3$ from eq.11.22,

$$2\,u_{1,1} + u_{2,2} - 4\,u_{1,2} \quad = 1.5 - 3 = -1.5 \qquad .....(11.24)$$

At $(2,2)$, $u_{1,2} + u_{2,1} + u_{2,3} - 4\,u_{2,2} = 3\,xy - u_{3,2} = 3 \times 1 \times 1 - 3 = 0$

or substituting $u_{2,3} = u_{2,1} + 6$ from eq.11.23,

$$2\,u_{2,1} + u_{1,2} - 4\,u_{2,2} \quad = 0 - 6 = -6 \qquad .....(11.25)$$

Eqns 11.22, 11.23, 11.24 and 11.25 can also be written in matrix form as

$$\begin{bmatrix} -4 & 1 & 1 & 0 \\ 1 & -4 & 0 & 1 \\ 2 & 0 & -4 & 1 \\ 0 & 2 & 1 & -4 \end{bmatrix} \begin{Bmatrix} u_{1,1} \\ u_{2,1} \\ u_{1,2} \\ u_{2,2} \end{Bmatrix} = \begin{Bmatrix} 0.75 \\ 1.125 \\ -1.5 \\ -6 \end{Bmatrix}$$

By solving them, we get $u_{1,1} = 0.077$; $u_{1,2} = 0.866$; $u_{2,1} = 0.191$ and $u_{2,2} = 1.812$

11.5 FINITE VOLUME METHOD (FVM)

This method utilises control volumes and control surfaces. Finite Volume formulations can be obtained either by a finite difference basis or by a finite element basis. The control volume, in a 1-D problem covering nodes $i - 1$, i and $i + 1$, for node i covers $\dfrac{\Delta x}{2}$ to the right of node i and $\dfrac{\Delta x}{2}$ to the left of node i with the control surface CS_1 and CS_2 being located at $i - \dfrac{1}{2}$ and $i + \dfrac{1}{2}$.

(a)   FVM via FDM

Consider the earlier example $\dfrac{d^2u}{dx^2} - 2 = 0$ for $0 < x < 1$

with Dirichlet Boundary conditions $u = 0$ at $x = 0$ & at $x = 1$

$$\int_0^1 \left[\frac{d^2u}{dx^2} - 2 \right] dx = 0 \quad 0 < x < 1$$

Integrating, $\left(\dfrac{du}{dx} \right) \Big|_0^1 - \int_0^1 2\, dx = 0$

$$\sum_{CS_1, CS_2} \left(\frac{\Delta u}{\Delta x} \right) - \sum_{CV} 2\, \Delta x = 0 \quad \text{at } i = 2$$

Diffusion flux $\left(\dfrac{du}{dx} \right)$ is conserved between i–1 and i through control

surface at $i - \dfrac{1}{2}$ or CS_1 and between i and i + 1 through control surface

at $i + \dfrac{1}{2}$ or CS_2

This can also be written, in terms of finite differences, as

$$\left(\frac{u_{i+1} - u_i}{\Delta x} \right) - \left(\frac{u_i - u_{i-1}}{\Delta x} \right) = 2\Delta x$$

or $u_{i+1} - 2\, u_i + u_{i-1} = 2\, \Delta x^2$

This is identical to the equation obtained by FDM.

(b)   FVM via FEM

In the same example, let $u = N_1\, u_1 + N_2\, u_2 \quad \left[1 - \dfrac{x}{h} \right] u_1 + \left[\dfrac{x}{h} \right] u_2$

$$\left(\frac{du}{dx} \right)_{CS1} = -\left(\frac{u_2 - u_1}{h} \right); \quad \left(\frac{du}{dx} \right)_{CS2} = -\left(\frac{u_3 - u_2}{h} \right)$$

$$\frac{d^2u}{dx^2} - 2 = \frac{\left(\dfrac{du}{dx} \right)_{CS1} - \left(\dfrac{du}{dx} \right)_{CS2}}{h} - 2 = \frac{(u_3 - 2u_2 + u_1)}{h^2} - 2 = 0$$

or $u_3 - 2u_2 + u_1 = 2\, h^2$

This is also identical to the equation obtained by FDM.

11.6 FDM vs. FEM

S.No.	Finite Difference Method	Finite Element Method
1.	Finite difference approximation from Taylor series expansion is used	Interpolation functions and polynomial expansion are used
2.	Equations are written for structured grids and less complicated. Hence, need lesser computer time	Equations are written for grids, not necessarily structured, with nodes irregularly connected around the entire domain resulting in a large sparse matrix system for solution
3.	Treatment of governing equation and different boundary conditions is not uniform. The solution method is tailor-made for each situation	Variational formulation employed not only for the governing equation but for all constraint conditions – useful for solution stability and accuracy

CHAPTER **12**

PRACTICAL ANALYSIS USING A SOFTWARE

12.1 USING A GENERAL PURPOSE SOFTWARE

Many general purpose software are readily available for analysis of mechanical, civil and aircraft structures based on FEM. Even though actual commands may vary from one software to another, the general features can be broadly classified under the following three categories.

(i) Pre-processor Phase

In this phase, data is input by the user regarding

(a) Idealised 1-D, 2-D or 3-D geometric model consisting of :

- element type (discrete structure with truss, 2-D beam, 3-D beam or pipe elements; continuum with 2-D plane stress, plane strain, thin shell, 3-D solid or thick shell elements)
- appropriate nodal coordinates
- element attributes and element connectivity.

In some large components, it is also possible to create a large 2-D or 3-D model using key points and Boolean operations on areas or volumes. These areas or volumes can be meshed by the software into many elements of equal or different sizes depending on user's choice as per the expected stress distribution. The software makes sure that the generated elements satisfy aspect ratio norms.

(b) Properties of materials such as

- modulus of elasticity, Poisson's ratio, mass density, coefficient of linear thermal expansion etc. for structural analysis

- thermal conductivity, specific heat etc. for a thermal analysis; with options of :
  - isotropic or orthotropic
  - constant or temperature-dependent material data.

Some softwares have standard material properties in their database. User need to specify the material type only. *Care must be taken to see that the units in the material database correspond to the units of the other data input by the user.*

(c) **Section properties** like:

- area for a truss element
- moment of inertia and section depth in a beam element
- thickness in a 2-D plate etc.

Some software include the facility of choosing a standard section shape such as C, I, L, H, from its database with their dimensions specified by the user and then calculate values like area, moment of inertia etc.

Material and section properties may be identified with material numbers and section numbers so that each element can be associated with a particular material number and a particular section number (called *element attributes*). In this way, material data and section data, common to many elements, need not be input repeatedly, saving considerable time and effort in data preparation.

(d) **Load particulars** such as:

- distributed loads due to self weight, wind load
- concentrated or point loads
- steady-state or transient temperature distribution over the entire model, for a structural analysis
- free-stream temperature, constant temperature on some part of boundary for a thermal analysis etc.

It is also possible to analyse the same structure for different sets of loads (defined in some software as *load steps or load cases)*

(e) **Boundary conditions or restraints** for translation or rotation DOF at various nodes (including restraints on rigid body motion), indication of symmetry for a structural analysis or insulated   wall for a thermal analysis etc.

Most software use consistent dimensions without any conversion of units inside the program. It is, therefore, user's responsibility to input data and interpret the output in appropriate units. For example, if coordinates are input in 'mm' and loads in 'Newton', then area of cross section should be in 'sq. mm', moment of inertia should be in 'mm^4' and modulus of elasticity should be in 'N/sq. mm', Correspondingly, the displacements calculated at nodes will be in 'mm' and stresses will be in 'N/sq.mm'.

Most programs also check for validity as well as sufficiency of the given data either in the pre-processor phase or in the solution phase before proceeding with the actual solution to save computer time as well as user's time. The pre-processor phase generates a data file in the sequence required by the solution phase

(ii) Solution Phase

In this phase, the program uses the data file generated by the pre-processor stage and carries out desired analysis.

Different options usually available are:

(a) static structural analysis, which calculates nodal displacements

(b) dynamic structural analysis, which calculates natural frequencies and mode shapes or time history response (corresponding to load Vs time data) or response to earthquake (corresponding to frequency Vs amplitude data)

(c) thermal analysis, which calculates nodal temperatures due to thermal conduction in a solid body with specified temperature and/ or convection boundary conditions.

(iii) Post-processor phase

The output of solution phase is a large set of nodal displacement or temperature values. The post processor phase reads these values as well as geometry data of pre-processor phase and presents in a more easily readable form such as iso-stress or iso-temperature contours, plots of deformed shape etc. Some software also has the facility of presenting output for a specific combination of different loads (or load steps).

Many general purpose software, such as ANSYS, ADINA, NASTRAN, PAFEC, NISA, PAFEC, STRUDL, etc., are commercially available in the market. The specific format and sequence of input data may vary between them. Also, modelling options as well as loads and boundary conditions that each of the software can handle may also vary. But, data to be input generally remains the same.

12.2 SOME EXAMPLES WITH ANSYS

An attempt is made, through simple exercises, to make the students understand various features of a general purpose finite element software. ANSYS software has been chosen for this purpose. The input commands or their format may change in different versions of this software or between different software. So, this should only be taken as a model. Till the students understand proper method of giving necessary data and using appropriate commands, they are advised to *cross-check the results obtained using any general purpose software with those calculated by conventional strength of materials approach*. This will ensure that the data input by them is interpreted by the software in the way they desired.

ANSYS is a general purpose software developed by Swanson Analysis Systems Inc, USA for analysis of many different engineering problems. Input sequence for solving the following problems are given here. Theoretical results from conventional methods are also given here for verifying the output of ANSYS and thus ensure that the features of this software are properly utilised. These problems are not exhaustive of the features of the software but only meant to link the theory covered so far with some typical problems.

ANSYS software has been used here for explaining some examples, with their explicit permission.

Example 12.1

Simple truss with concentrated loads

DATA : $A = 25 \text{ cm}^2$ $L_{1-2} = L_{2-3} = 100 \text{ cm}$ $L_{2-4} = 60 \text{ cm}$

$E = 2 \times 10^7 \text{ N/cm}^2$ $P = 10000 \text{ N}$ $\delta_{1X} = \delta_{1Y} = \delta_{3X} = 0$

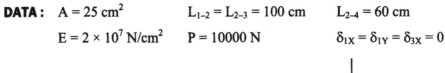

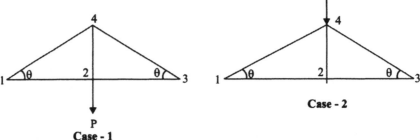

ANSYS model

The truss is assumed to be in X-Y plane with X in the horizontal direction, Y in the vertical direction and origin at node 1. The truss is analysed for two steps (or load cases), with a concentrated load P acting along –Y direction at node 2 (case-1) and same load acting at node 4 (case-2). In both the cases, node 1 is fixed in all DOF while node 3 is fixed in Y direction only (roller support). To explain different input commands, this problem is solved 2 times with one load acting at a time while problem 3 is solved in a single run with loads acting separately in 2 different load steps.

The method of solving as multiple load cases gives results for the these load cases separately, while saving computer time in the calculation of element stiffness matrices in element local coordinate system, transformation of these matrices to the common global coordinate system, assembling these element matrices, and applying boundary conditions on the assembled stiffness matrix.

Input Data In ANSYS

Preferences – Structural

Preprocessor –

Element type – Add – Structural link – 2D spar 1

Real constants – Add – Set No. 1; Area 25

Material props – Constant Isotropic – Material No 1; EX 2e7

Modeling create – Nodes – on Working plane –
(0,0),(100,0),(200,0),(100,60)

Elements – Thru Nodes – (1,2),(2,3),(1,4),(2,4),(3,4)

Loads – Loads Apply – Structural Displacement – on Nodes – 1 FX, FY
– 3 FY

** **Case-1** ** Structural Force/Moment – on Nodes - 2
FY   Constant value   -10000

** **Case-2** ** Structural Force/Moment – on Nodes - 4
FY   Constant value   -10000

Solution – Analysis type – New Analysis – Static
Solve current LS – Solution is done – Close

General Postproc – Plot results – Deformed shape – Def + undeformed
List results – Nodal solution – DOF solution – All DOFs

Node	UX	UY
1,2,3,4,5	---	---

Element solution – Line Elem results – Structural ELEM

EL	MFORX	SAXL
1,2,3,4,5	---	---

Reaction solution – All items

Node	FX	FY
1,3	---	---

Results Obtained : Consistent with units of input data, the displacements are read in 'cm' and the member forces/reactions are read in 'N'.

<table>
<tr><td align="center">**Case 1**</td><td align="center">**Case 2**</td></tr>
<tr><td align="center">$\delta_{2Y} = -0.0083834$</td><td align="center">$\delta_{2Y} = -0.0083834$</td></tr>
<tr><td align="center">$\delta_4 = -0.0071834$</td><td align="center">$\delta_4 = -0.0071834$</td></tr>
<tr><td align="center">$R_{1Y} = R_{3Y} = 5000$</td><td align="center">$R_{1Y} = R_{3Y} = 5000$</td></tr>
<tr><td align="center">$F_{1-2} = F_{2-3} = 8333.3$</td><td align="center">$F_{1-2} = F_{2-3} = 8333.3$</td></tr>
<tr><td align="center">$F_{1-4} = F_{3-4} = -9718.3$</td><td align="center">$F_{1-4} = F_{3-4} = -9718.3$</td></tr>
<tr><td align="center">$F_{2-4} = 10000$</td><td align="center">$F_{2-4} = 0$</td></tr>
</table>

Check of Results : Solving by the method of joints and applying the conditions

$$\sum F_x = 0 \text{ and } \sum F_y = 0 \text{ at each joint or node.}$$

<table>
<tr><td></td><td align="center">**Case 1**</td><td align="center">**Case 2**</td></tr>
<tr><td>At node 2</td><td>$F_{2-4} = 10000 \text{ N}$</td><td>$F_{2-4} = 0 \text{ N}$</td></tr>
<tr><td>At node 4</td><td>$F_{1-4} = F_{3-4} = F_{2-4} / 2 \sin\theta$</td><td>$F_{1-4} = F_{3-4} = P / \sin\theta$</td></tr>
<tr><td></td><td>$= 9718.25 \text{ N}$</td><td>$= 9718.25 \text{ N}$</td></tr>
<tr><td>At node 3</td><td>$F_{2-3} = F_{3-4} \cos\theta = 8333.3 \text{ N}$</td><td>$F_{2-3} = F_{3-4} \cos\theta = 8333.3 \text{ N}$</td></tr>
<tr><td></td><td>$R_{3Y} = F_{3-4} \sin\theta$</td><td>$R_{3Y} = F_{3-4} \sin\theta$</td></tr>
<tr><td></td><td>$= 5000 \text{ N}$</td><td>$= 5000 \text{ N}$</td></tr>
<tr><td>At node 2</td><td>$F_{1-2} = F_{2-3} = 8333.3 \text{ N}$</td><td>$F_{1-2} = F_{2-3} = 8333.3 \text{ N}$</td></tr>
<tr><td>At node 1</td><td>$R_{1Y} = F_{1-4} \sin\theta = 5000 \text{ N}$</td><td>$R_{1Y} = F_{1-4} \sin\theta = 5000 \text{ N}$</td></tr>
<tr><td></td><td>$R_{1X} = F_{1-4} \cos\theta _- F_{1-2}$</td><td>$R_{1X} = F_{1-4} \cos\theta - F_{1-2}$</td></tr>
<tr><td></td><td>$= 0 \text{ N}$</td><td>$= 0 \text{ N}$</td></tr>
</table>

Example 12.2

Stepped shaft subjected to temperature change.

DATA :   Element 1 - $A = 24 \text{ cm}^2$; $\alpha = 20 \times 10^{-6} / °C$; $E = 1 \times 10^7 \text{ N/cm}^2$

Element 2 - $A = 18 \text{ cm}^2$; $\alpha = 12 \times 10^{0-6} / °C$; $E = 2 \times 10^7 \text{ N/cm}^2$

Element 3 - $A = 12 \text{ cm}^2$; $\alpha = 12 \times 10^{-6} / °C$; $E = 2 \times 10^7 \text{ N/cm}^2$

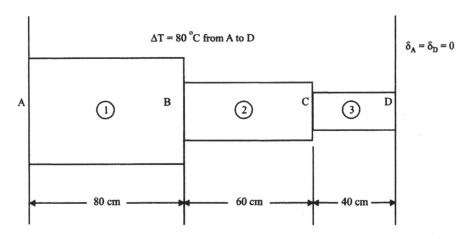

$\Delta T = 80\ ^{\circ}C$ from A to D $\qquad$ $\delta_A = \delta_D = 0$

ANSYS Model

R_1, M_1 R_2, M_2 R_3, M_2

1 ① 2 ② 3 ③ 4   R – Real constant set
x x x x M – Material properties set

Input Data in ANSYS

Preferences – Structural

Preprocessor –

 Element type – Add – Structural link – 2D spar 1

 Real constants – Add – Set No. 1 ; Area 24

          Set No. 2 ; Area 18

          Set No. 3 ; Area 12

 Material props – Constant Isotropic – Material No 1; EX 1e7 ; ALPX 20e-6

             Material No 2 ; EX 2e7 ; ALPX 12e-6

 Modeling create – Nodes – on Working plane – (0,0),(80,0),(140,0),(180,0)

       Elements –Elem attributes – Real const. Set no.1, Matl No.1

             Thru Nodes – (1,2)

       Elem attributes – Real const. Set no.2, Matl No.2

             Thru Nodes – (2,3)

       Elem attributes – Real const. Set no.1, Matl No.1

             Thru Nodes – (3,4)

  Loads – Loads Apply – Structural Displacement – on Nodes – 1,4 FX

    Structural Temperature – on Nodes – Pick ALL

           Temp   Constant value   80

 Solution – Analysis type – New Analysis – Static

Solve current LS – Solution is done – Close

General Postproc – Plot results – Deformed shape – Def + undeformed

List results – Nodal solution – DOF solution – Translation UX

Node	UX
1,2,3,4	---

Element solution – Line Elem results – Structural ELEM

EL	MFORX	SAXL
1,2,3	---	---

Reaction solution – Structural force FX

Node	FX
1,4	---

Results Obtained : $\delta_B = 0.016$ $\delta_C = 0.0176$

$R_A = -R_D = 336000$ $F_1 = F_2 = F_3 = -336000$

Example 12.3

Beam With Concentrated & Distributed Loads

DATA : $A = 20 \text{ cm}^2$ $L_{1-2} = L_{2-3} = 100 \text{ cm}$

$I = 50 \text{ cm}^4$ $h = 5 \text{ cm}$ $E = 2 \times 10^7 \text{ N/cm}^2$

$P = 10000 \text{ N (Case-1)} \quad p = 60 \text{ N/cm (Case-2)} \quad \delta_{1y} = \delta_{3y} = 0$

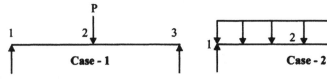

Case - 1   Case - 2

Input Data In ANSYS

Preferences – Structural

Preprocessor –

Element type – Add – Structural Beam – 2D Elastic 3

Real constants – Add – Set No. 1; Area (A) 25 ; Moment of Inertia(IX) 50
                    Height of section(h) 5

Material props – Constant Isotropic – Material No 1; EX 2e7

Modeling create – Nodes – on Working plane – (0,0), (100,0), (200,0)

Elements – Thru Nodes – (1,2),(2,3)

Loads – Loads Apply – Structural Displacement – on Nodes – 1,3 FY

Structural Force/Moment – on Nodes - 2

FY   Constant value   – 10000

Write LS file – LS file No. 1

Loads – Delete – Structural Force/Moment – on Nodes – 2 - All

Apply – Pressure – on Beams – 1,2 – Face No. 1 – value = – 60

Write LS file – LS file No. 2

Solution – Analysis type – New Analysis – Static

Solve from LS files – Start file No 1; End file No. 2; Increment 1

General Postproc – Read First set -

Plot results – Deformed shape – Def + undeformed

List results – Nodal solution – DOF solution – All DOFs

Node	UX	UY
---	---	---

Element solution – Line Elem results – Structural ELEM

EL	MFORX	SAXL
---	---	---

Reaction solution – All items

Node	FX	FY
---	---	---

Plot Ctrls – Animate – Deformed shape – Def + undeformed – Play

Read Next set -

Plot results – Deformed shape – Def + undeformed

List results – Nodal solution – DOF solution – All DOFs

Node	UX	UY
---	---	---

Element solution – Line Elem results – Structural ELEM

EL	MFORX	SAXL
---	---	---

Reaction solution – All items

Node	FX	FY
---	---	---

Plot Ctrls – Animate – Deformed shape – Def + undeformed – Play

Results Obtained :

Case 1	Case 2
$\delta_2 = -1.6667$	$\delta_2 = -1.25$
$\theta_1 = -\theta_3 = -0.025$	$\theta_1 = -\theta_3 = -0.02$
$R_{1Y} = R_{3Y} = 5000$	$R_{1Y} = R_{3Y} = 6000$

Check of Results :

$$\delta_{max} = \frac{P L^3}{48 E I}$$
$$= 1.6667 \text{ cm}$$

$$\delta_{max} = \frac{5 p L^4}{384 E I}$$
$$= 1.25 \text{ cm}$$

$$\theta_{max} = \frac{P L^2}{16 E I}$$
$$= 0.025$$

$$\theta_{max} = \frac{p L^3}{24 E I}$$
$$= 0.02$$

$$R_{1Y} = R_{3Y} = \frac{P}{2} = 5000 \text{ N} \qquad R_{1Y} = R_{3Y} = \frac{p L}{2} 60 \times \frac{200}{2} = 6000 \text{ N}$$

Deformed shape
as plotted

True shape of
deformed plot

Note : Deformed plot of the beam in both the cases will show straight lines between nodes 1 & 2 as well as between nodes 2 & 3, eventhough a cubic displacement polynomial is used while calculating stiffness matrix of each beam element. This is not reflective of the formulation used but a *limitation of the post processor to represent displacement distribution between any two points*. Mathematically, with the displacement data at two nodes, only a straight line can be fit.

Example 12.4

Natural frequencies and mode shapes of a cantilever

DATA : $A = 20 \text{ cm}^2$ $I = 50 \text{ cm}^4$ $h = 5 \text{ cm}$

$L_{1-2} = L_{2-3} = L_{3-4} = L_{4-5} = 25 \text{ cm}$

$E = 2 \times 10^7 \text{ N/cm}^2$ $\rho = 8 \times 10^{-3} \text{ kg/cm}^3$ $\delta_{1X} = \delta_{1Y} = \theta = 0$

Input Data In ANSYS

Preferences – Structural

Preprocessor –

Element type – Add – Structural Beam – 2D Elastic 3

Real constants – Add – Set No. 1; Area(A) 20 Moment of Inertia(Ix) 50
Height of section(h) 5

Material props – Constant Isotropic – Material No 1; EX 2e7; Density 8e-3

Modeling create–Nodes–on Working plane –
(0,0),(25,0),(50,0),(75,0),(100,0)

Elements – Thru Nodes – (1,2),(2,3),(3,4),(4,5)

Loads – Loads Apply – Structural Displacement – on Nodes – 1 ALL

Solution – Analysis type – New Analysis – Modal

Analysis options – Subspace ; No. of modes to extract – 4 ;
No. of modes to expand – 4

Expansion pass - on

Solve current LS – Solution is done – Close

General Postproc – Results summary– Freq 1 to 4

Read First set – Plot Ctrls – Animate – Mode shape – Play !
Repeat for all modes

Read Next set – Plot Ctrls – Animate – Mode shape – Play !

Results Obtained : Natural frequencies –   4.4215, 27.645, 77.476, 125.80

Check of Results : In this problem, number of frequencies calculated and the accuracy of results depend on the number of elements used in the model.

$$\text{Least frequency, } \omega_1 = \left(\frac{1}{2\pi}\right)\sqrt{(g/\delta)}$$

where δ = max deflection of the cantilever

$= \dfrac{p\,L^4}{8\,EI}$ and $p = \rho\,A\,g$ is the distributed load

Then, $\omega_1 = 3.557$ rad/sec

Example 12.5

1-D heat conduction through a composite wall

DATA : L_1=30 cm L_2=15 cm L_3=15 cm

$K_1 = 20$ W/m °C $K_2 = 30$ W/m °C $K_3 = 50$ W/m °C

$E = 2 \times 10^7$ N/cm² $h = 25$ W/m² °C $T_1 = 800$ °C ; $T_5 = 20$ °C

ANSYS Model

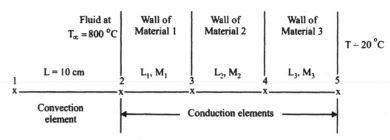

Real constants – Area of cross section, $A = 1$ cm^2 for all the 4 elements

M – Material properties set

Input Data in ANSYS

Preferences – Structural

Preprocessor – Element type – Add – Thermal link – Convection;

2D conduction

Real constants – Add – Set No. 1 ; Area 25

Material props – Constant Isotropic – Material No 1 ; HF 25

Material No 2 ; KX 20

Material No 3 ; KX 30

Material No 4 ; KX 50

Modeling create – Nodes – on Working plane –

(0,0),(0.1,0),(0.4,0),(0.55,0),(0.7,0)

Elements – Elem attributes – Elem type 1 ; Matl No. 1

Thru Nodes – (1,2)

Elem attributes – Elem type 2 ; Matl No. 2

Thru Nodes – (2,3)

Elem attributes – Elem type 2 ; Matl No. 3

Thru Nodes – (3,4)

Elem attributes – Elem type 2 ; Matl No. 4

Thru Nodes – (4,5)

Loads – Loads Apply – Temperature – on Nodes – 1   800; 5   20

Solution –   Analysis type – New Analysis – Steady state

Solve current LS – Solution is done – Close

General Postproc – Plot results – Nodal solution - DOF solution - Temperatures

List results – Nodal solution – DOF solution – Temperatures

Node	Temperature
---	---

Element solution – Line Elem results – Heat flow

EL	Heat flow
---	---

Reaction solution – All items

Node	Heat flow
---	---

Other option

Preprocessor - Element type – Add – Thermal link

Convection – Option – K3 – SFE command - 2D conduction

Loads – Loads Apply – Temperature – on Nodes – 1   800 ;   5   20

Convection – on elem – 1 - HF 25 ; TBulk 800

Results Obtained : $T_2 = 304.76\,^{\circ}\text{C}$; $T_3 = 119.05\,^{\circ}\text{C}$; $T_4 = 57.14\,^{\circ}\text{C}$

Heat flow = 12380.95 W

Check of Results

Overall thermal resistance, $U = \dfrac{1}{\left[\dfrac{1}{h} + \dfrac{L_1}{K_1} + \dfrac{L_2}{K_2} + \dfrac{L_3}{K_3}\right]} = 15.873$

Heat flow, $Q = U\,(T_1 - T_5) = 15.873\,(800 - 20) = 12380.95$

$$Q = h\,(T_1 - T_2) \qquad \Rightarrow \quad T_2 = 304.76\,^{\circ}\text{C}$$

$$= K_1 \frac{(T_2 - T_3)}{L_1} \qquad \Rightarrow \quad T_3 = 119.05\,^{\circ}\text{C}$$

$$= K_2 \frac{(T_3 - T_4)}{L_2} \qquad \Rightarrow \quad T_4 = 57.14\,^{\circ}\text{C}$$

$$= K_3 \frac{(T_4 - T_5)}{L_3} \qquad \Rightarrow \quad T_4 = 57.14\,^{\circ}\text{C}$$

Results with scaling correction factors of the program

$$T_2 = 798.31\ ^\circ C \qquad T_3 = 290.72\ ^\circ C \qquad T_4 = 121.52\ ^\circ C$$

When SFE command option is used for the convection element, effective film coefficient, $h_f^{eff} = T_B\ h_f$ (where, T_B is the Bulk temperature value input in SFE command and h_f is the film coefficient value input in SFE command) is used. This results in a higher temperature drop across wall thickness and consequently in higher thermal stresses. Design based on these temperatures will be conservative.

Example 12.6

Stress concentration factor in a plate with hole

DATA: L = 160 cm   H = 100 cm   Plate thickness, t = 0.8 cm

Hole dia = 20 cm   E = 2×10^7 N/cm^2 Poisson's ratio = 0.3

P = 10240 N

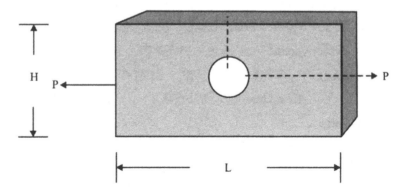

ANSYS Model: Since the geometry as well as loads are symmetric about the two major dimensions of the plate, a quarter plate can be modeled for analysis. To ensure uniform loading along the small side, the load P is applied as uniform pressure. $p = \dfrac{P}{H\,t}$

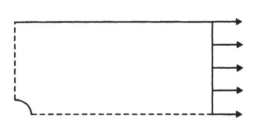

Note: In the case of continuum analysis, unlike in the case of discrete structures, accuracy of results obtained by Finite Element Method improves in general by the use of more number of elements as well as by the use of higher order elements such 8-noded quadrilateral or 6-noded triangle. Due to the limitations of number of DOF in the educational version of ANSYS, refinement of solution is not attempted here.

Input Data In ANSYS

Preferences – Structural

    Preprocessor –

    Element type – Add – Structural solid – Quad 4 node

– option – Plane stress w/thk

    Real constants – Add – Set No. 1 ;   Thickness 0.8

    Material props – Constant Isotropic – Material No 1; EX 2e7 ; NUXY 0.3

    Modeling create – Rectangle – By 2 corners - X,Y, L, H   0,0,80,50

                       Circle – Solid circle – X,Y, Radius   0, 0, 10

                       Operate – Boolean subtract – Areas

                          – Base area ;   Area to be subtracted

      Loads – Apply – Structural Displacement – Symmetry B.C. – on lines

                     Pressure – on line – constant value ;   80

    Meshing – Size cntrls – Global size – Element edge length   3

    Mesh – Areas Free

Solution – Solve current LS – Solution is done - close

General postproc - Plot results – Deformed shape – Def + Undeformed shape
                   Plot results – Nodal solution – DOF solution – Translation UX

                          Element solution – Stress – X-direction SX

       Sorted listing – Sort Nodes – Descending order – Stress X-direction

       List results - Element solution – Stress – X-direction SX

               Max value 238.63 N/cm^2

       Sorted listing – Sort Nodes – Descending order – Stress Y-direction

       List results - Element solution – Stress – Y-direction SY

               Max value 29.147 N/cm^2

Alternative method of creating model

Preprocessor – Modeling create – Lines – Arcs – By Cent & Radius – (0,0),
(10,0)

Arc length in degrees 90

Key points – On Working plane - (80,0), (80,50), (0,50)

Lines – Straight line - By key points

Area – Arbitrary - By lines

Results Obtained :

Normal stress along X-axis (SX) Max value = 238.63 N/cm^2

Max normal stress, in the absence of stress concentration

$$= \frac{P}{(H-d)t} = \frac{10240}{(100-20)} \times 0.8 = 160 \text{ N/cm}^2$$

Stress concentration factor $= \dfrac{238.63}{160} = 1.4914375$

Check of Results : For D/H = 0.2, Stress Conc. factor for a rectangular plate with circular hole = 2.51 (from Mechanical design Handbooks).

Example 12.7

Centrifugal stresses in an axisymmetric solid

Data :   E = 2 × 10^7 N/cm^2 Poisson's ratio = 0.3   Mass density = 8 gm/cm^3

Speed, N = 3000 rpm   All dimensions are in mm

ANSYS Model :   Since the geometry and loads are axi-symmetric, any one section in axis-radius plane can be modeled. Also, since the geometry as well as loads is symmetric about the mid plane along the axis, half the flywheel can be modeled for analysis with symmetry boundary conditions applied on the plane of symmetry. ANSYS program assumes X-axis to be along the radius while Y-axis represents the axis of symmetry. Also, the program requires that the model be input in the right handed coordinate system (1st quadrant of X-Y plane is more convenient). For this purpose, a quarter of the component on the bottom side of the right half is modeled as shown. This quarter section is rotated by 90° to represent this in the familiar way with horizontal x-axis and vertical y-axis. Angular velocity $\omega = \dfrac{2\pi N}{60} = 314 \text{ rad/sec}$ is input about the axis of revolution (Y-axis).

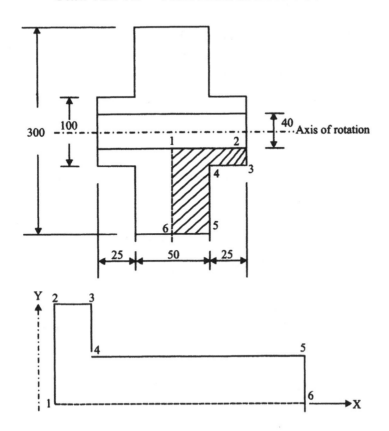

Input Data in ANSYS

Preferences – Structural

    Preprocessor – Element type – Add – Structural solid – Quad 4 node

                                         – option – Axisymmetric

      Material props – Constant Isotropic – Material No 1

                       EX 2e5 ; NUXY 0.3 ;   DENS 8e-3

    Modeling - create – Key points – On Working plane

                 (20,0),(150,0),(150,25),(50,25),(50,50),(20,50)

          Lines – Straight line - By key points –

               (1,2),(2,3),(3,4),(4,5),(5,6),(6,1)

          Area – Arbitrary - By lines – Pick lines 1,2,3,4,5,6

        Meshing – Size cntrl – Global size – Element edge length 5

           Mesh – Areas – Free – Pick area 1

    Loads – Apply – Structural Displacement – Symmetry B.C. – on lines - 1

        Others – Angular velocity – OMEGY about Y-axis 314

Solution – Solve current LS – Solution is done - close

General postproc – Plot results – Deformed shape – Def + Undeformed shape
Nodal solution – DOF solution – Translation UX

Translation UY

Element solution – Stress – X-direction SX

Y-direction SY

Z-direction SZ

Sorted listing – Sort Nodes – Descending order – Stress X-direction SX

Y-direction SY

Z-direction SZ

Results Obtained :

	Max	Min
UX (mm)	7.767	1.477
UY (mm)	0	−5.054
SX (N/mm^2)	0.337e3	−0.119e2
SY (N/mm^2)	0.121e3	−0.145e3
SZ (N/mm^2)	0.543e3	0.6005e2

OBJECTIVE QUESTIONS

1. ANSYS uses
   (a) frontal solution   (b) banded matrix solution
   (c) Cramer's rule   (d) Cholesky decomposition

2. A single analysis with 3 similar load steps takes ____ time compared to 3 analyses with single load case in each
   (a) 3 times more   (b) < 3 times less
   (c) same   (d) not related

3. An analysis with 1 load step takes ____ time compared to analysis with 3 similar load cases
   (a) 1/3 times less   (b) > 1/3 times less
   (c) same   (d) not related

4. Consistent loads are based on
   (a) stress equilibrium   (b) displacement continuity
   (c) energy equivalence   (d) force balance

5. Within elastic limit, results due to a combination of loads is same as linear superposition of results by each of those loads

   (a) always true   (b) always false

   (c) sometimes true   (d) needs repeated analysis

6. A truss element in ANSYS is identified as

   (a) line element   (b) spar element

   (c) truss element   (d) beam element

7. A uniformly distributed load on a beam is indicated in ANSYS as

   (a) udl   (b) uvl

   (c) pressure   (d) equivalent nodal loads

8. Uniform pressure along an edge of a plate element is specified in ANSYS as

   (a) pressure on element   (b) pressure along edge

   (c) pressure at each node   (d) same pressure at all nodes

9. Deformed shape of a simply supported beam with concentrated load at the mid-point appears in ANSYS as

   (a) circular arc

   (b) triangle with max displacement at mid-point

   (c) parabolic arc

   (d) straight line

10. Deformed shape in ANSYS is drawn with

   (a) actual nodal displacements

   (b) normalised nodal displacements

   (c) magnified nodal displacements

   (d) reduced nodal displacements

11. Loads command in ANSYS includes

   (a) loads & displacements   (b) loads & stresses

   (c) loads only   (d) loads or displacements

12. As a default option, mesh is refined in ANSYS using

   (a) g-method   (b) h-method   (c) p-method   (d) r-method

13. Real constants in ANSYS indicate

(a) material properties   (b) section properties

(c) thermal properties   (d) nodal loads

14. "Attributes" in ANSYS refer to

(a) section & material properties   (b) section properties

(c) material properties   (d) applied loads

15. Basic shapes of area / volume in ANSYS modelling are called

(a) Basics   (b) Primitives

(c) Primaries   (d) Areas and volumes

16. Most FEM software reduce computer memory requirement by storing

(a) half of symmetric stiffness matrix

(b) half of banded matrix

(c) stiffness matrix as a column vector

(d) complete stiffness matrix

17. Most FEM software use

(a) displacement method   (b) force method

(c) stress method   (d) hybrid method

18. Stresses in most FEM software are given in

(a) N/mm^2 (b) Pascal

(c) units based on input data (d) user specified units

19. Most FEM software analyse a structure using

(a) displacement method   (b) stress method

(c) force method   (d) mixed method

20. Displacements in most FEM software are given in

(a) mm   (b) m

(c) units based on input data (d) user specified units

21. Distributed load along an edge of a plane stress element is usually specified as

(a) pressure at nodes along the edge

(b) pressure along the edge

(c) equivalent nodal loads at the nodes on the edge

(d) different values of pressure applied at all nodes of the element

22. A tensile distributed load along an edge of a plane stress element is represented by _____ at the nodes

(a) +ve pressure   (b) –ve pressure

(c) +ve nodal loads   (d) –ve nodal loads

23. Generalised load means

(a) load   (b) displacement

(c) load or displacement (d) temperature

24. Attributes in ANSYS refer to ___ for the elements

(a) material property set number

(b) section property set number

(c) material & section property set numbers

(d) load set number

25. ANSYS accepts section properties set based on ___

(a) element type   (b) element size

(c) type of load   (d) type of material

ANSWERS

CHAPTER - 1

1.	(b)	2.	(c)	3.	(a)	4.	(a)	5.	(a)	6.	(c)
7.	(a)	8.	(b)	9.	(a)	10.	(a)	11.	(a)	12.	(c)

CHAPTER - 4

1.	(c)	2.	(d)	3.	(a)	4.	(b)	5.	(a)	6.	(d)
7.	(b)	8.	(c)	9.	(b)	10.	(d)	11.	(d)	12.	(b)

CHAPTER - 5

1.	(c)	2.	(d)	3.	(b)	4.	(b)	5.	(a)	6.	(b)
7.	(b)	8.	(b)	9.	(a)	10.	(c)	11.	(a)	12.	(b)
13.	(c)	14.	(c)	15.	(d)	16.	(b)	17.	(d)	18.	(b)
19.	(a)	20.	(b)	21.	(c)	22.	(b)	23.	(a)	24.	(c)
25.	(d)	26.	(d)	27.	(d)	28.	(c)	29.	(a)	30.	(b)
31.	(c)	32.	(c)	33.	(b)	34.	(a)	35.	(b)	36.	(a)
37.	(a)	38.	(d)	39.	(a)	40.	(a)	41.	(a)	42.	(b)
43.	(a)	44.	(b)	45.	(c)	46.	(b)	47.	(a)	48.	(b)
49.	(c)	50.	(c)	51.	(b)	52.	(c)	53.	(c)	54.	(b)

CHAPTER - 6

1.	(d)	2.	(a)	3.	(a)	4.	(b)	5.	(c)	6.	(d)
7.	(b)	8.	(c)	9.	(d)	10.	(b)				

CHAPTER - 7

1.	(d)	2.	(b)	3.	(b)	4.	(b)	5.	(d)	6.	(a)
7.	(a)	8.	(c)	9.	(a)	10.	(a)	11.	(d)	12.	(c)
13.	(d)	14.	(b)	15.	(d)	16.	(a)	17.	(c)	18.	(d)
19.	(c)	20.	(b)	21.	(c)	22.	(a)	23.	(b)	24.	(b)
25.	(c)	26.	(d)	27.	(d)	28.	(a)	29.	(c)		

CHAPTER - 8

1.	(d)	2.	(a)	3.	(b)	4.	(a)	5.	(b)	6.	(a)
7.	(c)	8.	(c)	9.	(b)	10.	(b)				

CHAPTER - 9

1.	(a)	2.	(a)	3.	(a)	4.	(a)	5.	(b)	6.	(a)
7.	(b)	8.	(a)	9.	(a)						

CHAPTER - 12

1.	(b)	2.	(b)	3.	(b)	4.	(c)	5.	(a)	6.	(b)
7.	(c)	8.	(c)	9.	(b)	10.	(c)	11.	(a)	12.	(b)
13.	(b)	14.	(a)	15.	(b)	16.	(b)	17.	(a)	18.	(c)
19.	(a)	20.	(c)	21.	(a)	22.	(b)	23.	(c)	24.	(c)
25.	(a)										

REFERENCES FOR ADDITIONAL READING

1. "Introduction to Finite Element Analysis – Theory and application" - Harold C Martin & Graham F Carey   - McGraw-Hill Book Company.

2. "Theory of Matrix structural analysis" - Przemieniecki J S - McGraw-Hill Book Company.

3. "The Finite element method By Zienkiewicz O C" - Tata McGraw-Hill Publishing Company Ltd.

4. "Numerical methods in Finite element analysis" - Bathe K J & Wilson K L - Prentice Hall of India.

5. "Introduction to the Finite element method" - Desai C S & Abel J F - Van Nostrand.

6. "The finite element method in engineering" - Singiresu S Rao - Butterworth Heinemann.

7. "Introduction to Finite elements in engineering" - Tirupathi R Chandrupatla & Ashok D Belegundu - Prentice Hall of India.

8. "Finite element analysis – Theory and programming" - C S Krishna Moorthy - Tata McGraw-Hill Publishing Company Limited.

9. "An Introduction to the Finite element method" - Reddy J N - McGraw-Hill Book Company.

10. "Concepts and applications of Finite Element Analaysis" - Robert Cook etal - John Wiley & Sons.

11. "Fundamentals of Finite Element Analysis" - David Hutton - Tata McGraw-Hill.

12. "A first course in the Finite Element Method" - Daryl L. Logan - Thomson, Brooks, cole.

13. "Theory of elasticity" - Timoshenko S & Goodier J N - McGraw-Hill Book Company.

14. "Dynamics of structures" - Walter C Hurty & Moshe F Rubinstein - Prentice Hall of India.

15. "Computational Fluid Dynamics" - T.J.Chung - Oxford University Press.

16. "Classification of Finite Element stresses according to ASME Section III stress categories" - W.C.Kroenke – Pressure vessels and piping; analysis and computers, ASME, June 1974.

17. "A Computational approach for the classification of FEM axisymmetric stresses as per ASME code" - G.Lakshmi Narasaiah et al – ASME Pressure Vessel & Piping Conference, 19-23 June 1988, Pittsburgh, USA.

18. "Design Check of Pressure Vessel Analysis by FEM" - G.Lakshmi Narasaiah – National Conference on State of the art technologies in Mechanical Engineering (NCSAME-2006) held at JNTU College of Engineering, Hyderabad.

INDEX

A

Approximate method 4

Area coordinates 175

Aspect ratio 138

Attributes 309

Assembling element matrices 94, 209
- banded matrix 209
- skyline method 212

B

Bandwidth
- half-bandwith 210
- minimization 211

Boundary conditions 96, 224
- cyclic symmetry 227
- elimination method 96, 263
- multi-point constraints 98
- penalty approach 97, 263
- sector symmetry 227
- symmetry 226

Buckling
- elastic buckling 282

Bulk modulus 73

C

Capacitance matrix 281

Characteristic equation 239

Collocation method 8
- point collocation 8
- sub-domain collocation 8

Compatibility 78, 134, 139

Conductivity matrix 258

Constitutive equations
- stress-strain relations 65

Continuum structures 27, 62

Convergence 27, 133

Coordinate system
- global / structure 90, 92
- local / element 90, 92
- natural /non-dimensional 169
- transformation matrix 92

Critical load 282
- elastic buckling 282

D

D'Alembert's principle 237

Damped free vibration 287

Damping 287

Deformed plots 231

Degree of fixity 2, 27

Degrees of freedom (DOF) 61, 85
- uncoupled 101

Design codes 277

Discrete structures 3, 26, 62

Displacement field 12, 90

Displacement method 31, 32

Distortion energy 72

Distributed loads 201
- consistent 204
- equivalent 201

Dynamic coupling 241

E

Eigen values & Eigen vectors
- matrices 35

Elastic Buckling 282
- critical load 282
- mode shapes 283

Elastic instability 282

Elements 27
- complex 128
- constant strain triangle (CST) 129
- linear strain triangle (LST) 167
- multiplex 128
- quadratic strain triangle (QST) 167
- simplex 128
- transition 228

Elements - shapes
- hexahedron 143, 168, 181
- quadrilateral 178, 180, 185, 187
- tetrahedron 143, 168, 187
- triangular element 129, 175, 177, 184, 273

Elements - types
- 3-D solid 143
- axisymmetric 144
- beam 87, 98, 173
- general beam 100
- pipe 103
- plate bending 141
- thick shell 143
- thin shell 143
- torsion 86
- truss 86, 90, 159, 170, 172

Equilibrium equations 79

Equivalent stress
- vonMises stress 71

Errors 30
- discritisation 31
- modelling 31
- numerical 31

Exact method 4

Extensive property 89

F

Fatigue 283

Finite difference method 293

Finite element method (FEM) 4, 15
- displacement method 31
- force method 32
- hybrid method 32
- mixed method 32

Finite volume method 293

Flexibility 64

Forced vibration 288
- time-history response 289
- response spectrum analysis 289

Fourier's law 256, 258

Frequencies
- natural frequencies 311

G

Galerkin method 5

Gaussian points 216

Geometric Isotropy 134, 136

H

Helmholtz equation 257

Higher order elements 167

Hybrid method 32

Hyper cube 135

I

Interpolation functions
- hermite 169
- lagrange 169

Isoparametric elements 169, 170

J

Jacobian 131, 182

L

Lagrange elements 137

Laplace equation 257, 290

Least squares method 10

Linear analysis 63

Lower bounds 27

M

Mass matrix 238, 240
- consistent 243
- lumped 241

Mathematical model 1, 2, 4

Matrices-operations
- characteristic equation 55, 59
- cholesky method 51
- cramer's rule 45
- direct method 44
- eigen values 55, 238
- eigen vectors 55, 238
- normalisation 239

- orthogonal 240
- gauss elimination method 47
- gausss Jordan method 43, 46, 53
- iterative method 53, 58
- LU factorization 48
- method of cofactors 42
- solution of simultaneous equations 41

Matrices – properties 36
- determinant 39
- inversion 41
- multiplication 37
- positive definite 55
- quadratic form 38
- transpose 36, 37

Matrices-types 35

Matrix method 26

Mesh generation 214
- h-method 215
- optimum mesh 215
- p-method 215
- r-method 215

Miner's rule 284

Mixed method 32

Modelling 220, 240

Modulus of elasticity 64

Mohr's circle 68

N

Natural coordinates 170

Natural frequencies 237
- modal matrix 239
- mode shapes 237, 239
- spectral matrix 239

Nodes 27

Non-dimensional coordinates 169

Non-linear analysis 63
- geometric non-linearity 64
- material non-linearity 64

Normalisation 239

Numerical integration 216

O

Optimum design 1, 33

Orthotropic material 77

P

Parametric elements 169

Pascal tetrahedron 135

Pascal triangle 134

Plane strain 72

Plane stress 72

Poisson's ratio 72

Postprocessor 311

Potential energy 12, 14

Preprocessor 309

Principal stress 68
- stress at a point 79

R

Rayleigh-Ritz method 12

Response spectrum analysis 289
- forced vibration 288

Rigid body motion 62

Rigidity modulus 63

Rupture time 285

S

S-N curve 284
- fatigue 284

S-t_r curve 286
- creep 285

Serendipity elements 136

Solution phase 311

Static condensation 230

Stiffness 14, 64
- infinite/zero 2
- relative 3

Strain at a point 67
- hoop 145
- normal 67
- shear 67

Strain-displacement relations 73, 184

Strain energy 22

Strength of materials approach 2, 3

Stress at a point 65
- complementary shear 70
- max shear 71
- normal 66
- principal 68

- pure shear 70
- shear 66
- vonMises 71

Stress categorisation 279
- bending 278
- classification line 279
- membrane 278
- peak 279

Stress function 291

Stress intensity 278

Stress-strain relations 74
- plane strain 76
- plane stress 75, 78

Structural analysis 32

Substructuring 230

Super elements 230

T

Thermal analysis 32

Thermal stress 73

Time history response 289, 311
- forced vibration 288

Transformation matrix 92

Transition element 228

U

Upper bounds 27

Usage fraction 285
- creep 286
- cumulative 286
- fatigue 284
- miner's rule 284

- cumulative 286
- fatigue 284
- miner's rule 284

V

Variational method 4, 12

W

Weighted residual methods 4

Warping function 289

Y

Young's modulus
- modulus of elasticity 63, 64

T - #0629 - 101024 - C0 - 244/181/19 - PB - 9781138118096 - Gloss Lamination